Innovative Applications of Educational Technology Tools in Teaching and Learning

Blessing Foluso Adeoye, Ph.D.

Edited by:

Blessing F. Adeoye, PhD
Department of Science and Technology Education
Faculty of Education, University of Lagos, Lagos, Nigeria
docadeoye@gmail.com

Prof. Joel B. Babalola
Department of Educational Management,
University of Ibadan, Ibadan, Nigeria
&
Prof. Michele T. Cole
School of Business
Robert Morris University, Pittsburgh, USA

Order this book online at www.trafford.com
or email orders@trafford.com

Most Trafford titles are also available at major online book retailers.

Print information available on the last page.

ISBN: 978-1-4907-6574-7 (sc)
ISBN: 978-1-4907-6575-4 (hc)
ISBN: 978-1-4907-6581-5 (e)

Library of Congress Control Number: 2015916519

Trafford rev. 10/06/2015

 www.trafford.com

North America & international
toll-free: 1 888 232 4444 (USA & Canada)
fax: 812 355 4082

Contents

SECTION 1
The Concept of Educational Technology, Need, and Significance

SECTION 2
Utilisation of Educational Technology
Tools for Teaching and Learning

SECTION 3
Adoption and Integration of Educational
Technology Tools in the Classroom

SECTION 7
Blended Learning

List of Reviewers

Dr. Blessing F. Adeoye, University of Lagos, Lagos, Nigeria

Pai Obanya, Emeritus Professor, University of Ibadan, Nigeria

Prof. Joel B. Babalola, University of Ibadan, Nigeria

Prof. Michele T. Cole, Robert Morris University, USA

Dr. Georgy O. Obiechina, University of Port Harcourt, Nigeria

Dr. Golda Ekenedo, University of Port Harcourt, Nigeria

Dr. Gbenga Peter Sanusi, University of Ibadan, Nigeria

Dr. Bamidele Ogunlade. Ekiti State University Ado-Ekiti, Nigeria

Dr. Florence Folami-Adeoye, Millikin University, Decatur, Illinois, USA

Dr. Abubakar, Abubakar Sadiq, Abubakar Tafawa Balewa University, Nigeria

Matthew David Fazio, Robert Morris University, Pittsburg, Pennsylvania, USA

Robert Loyal Siedenburg, University of Illinois. Urbana-Champaign, Illinois, USA

Adeneye O. A. Awofala, PhD, University of Lagos, Lagos. Nigeria

Michelle Adebukola Adeoye, University of San Francisco, California, USA

Foreword

Prof. Pai Obanya

In today's technology-driven world, education (the development of human potentials for integration into a creatively productive life of in an ever-changing world) cannot escape the influence of both the methods and products of technology. This is the truism that lies at the heart of this book.

The book has assembled the sound knowledge and practical experience-induced insight of a galaxy of specialists in the discipline to present in perfect logical sequencing and easy-to-follow language the essentials of educational technology and the creative applications of the methods and tools of technology to rethinking and redoing education.

The twenty-one-chapter book in seven distinct but closely interrelated sections is pedagogically structured. Section 1 deals with the basics and contains papers dealing with conceptual issues. This is followed by the section on the utilisation of educational technology tools for promoting teaching and learning. The third section brings the application of the tools down to where the action really is—the classroom level, with a strong emphasis on adaptation by classroom teachers to a variety of situations.

Cultural issues that should not be ignored in the application of technology to education are discussed in section 4. Section 5 highlights emerging trends in the discipline. This is intended to draw attention to the fact that educational technology is a fast-growing field, in a rapidly changing world environment. Users of this book should therefore always be on their intellectual toes, ever willing and ready to update their knowledge of the subject at top-speed frequency. Section 6 is concerned with teacher education in which the methods and tools of technology must today be an integral element, while the very last section delves into blended learning—the growing trend in a knowledge-driven world that must enrich conventional education delivery methods with the rich fruits of technology to universalise educational opportunities.

The book is rich in innovative ideas backed by rigorous research and reports by knowledgeable and reflective practitioners. Readers are tasked to reflect on the materials presented and to learn from the experience of the authors. Most importantly, readers are also expected to plough back the innovative ideas that run through the book into their own professional activities—teaching, research, and education-related social engagement activities.

Preface

We are living and working in a world where people frequently talk about knowledge, innovation, and technology. It is generally believed that dynamics of work and life are aided and perhaps driven by the acquisition and application of knowledge, innovation and technology. *Innovative Applications of Educational Technology Tools in Teaching and Learning* is a creative combination of these three concepts in a book written by carefully selected academics so as to drive in the need to apply creativity, care, and caution in adopting, adapting, and integrating the modern technology tools such as computers, modems, telephones, and the Internet into the process of teaching and learning at a time when these globalizing and unifying concepts of knowledge, innovation and technology have widened the existing divides (in the form of the knowledge divide, innovation divide, technological divide, and digital divide) and deepened the existing gulfs between the developed and the developing nations, between the rich and the poor as well as between urban and rural institutions of learning.

Educational technology as an effective use of technological tools in learning is not a new concept. What is new is the speed at which new technology tools (such as media for text, audio, video, satellite TV, CD-ROM, computer-based learning, machines; make-contact or networking hardware and methods of e-learning comprising digitisation approaches) are being rolled out for the use of teachers and learners who are in most cases have to be on the fast lane of capacity building of knowledge or theoretical perspectives and that of innovation for their effective application. This capacity issue aside, teachers, learners, and school administrators in developing countries like Nigeria fight some constant battles to overcome internal weaknesses such as widespread poverty and external threats such as lack of electricity and access to the Internet that have accompanied the integration and application of these education technology tools. Although the book does not contain the application of theories of human behaviour such as the instructional theory, learning theory, educational psychology, media psychology, and human performance technology to education technology. The preliminary sections of the book have not only been devoted to building of capacity on relevant concepts but have earmarked a substantial space to identification of and suggestions on how to deal with practical obstacles hindering the effective assimilation and application of educational technology tools in this part of the world.

The book, being an opening of a matrix of publishing for the author, has not been designed to cover all the settings such as preschool, primary, secondary, tertiary education technology tools. It has systematically and logically followed a thematic approach or issue-specific approach with the aim of providing answers to pertinent questions such as What makes up the process of teaching and learning? What are the old and the modern educational technology tools? Why are technology tools becoming increasingly necessary in the process of teaching and learning? What are the general and specific (innovative) uses of educational technology tools? Considering that educational technology is as old as pedagogy itself, why are we now talking about integration of technology into the process of education especially teaching and learning? What are the challenges for integrating foreign technology into and creating and applying local technology in the process of education (learning and teaching) in Africa in general and particularly in Nigeria? Why do we need to integrate and use technology tools innovatively in the process of teaching and learning? How do we integrate and use technology tools innovatively in the process of teaching and learning? Consequently, the initial section has dealt with basic concepts involved in educational technology, integration of technologies in education; availability of the tools; effectiveness in their application; indigenous design of these tools through creative science and technology; absorption of foreign tools using adaptive science and technology; and challenges faced by teachers, learners, and educational administrators in creating indigenous education technology tools and in capturing and adapting the foreign ones as well as in effective utilisation of the available tools in a technology-endangered sociocultural environment.

This author has made a unique contribution by providing a solid platform through which future contributors to the debates of knowledge, innovation, and technology can navigate into a deeper sea where we can find the most needed theoretical perspectives required to transform the education technology as a discipline and as a practice from the current perspective concentrating on absorption, adaptation, and application of foreign tools to an enviable higher stage of designing locally made but internationally competitive tools for the use of our learners and teachers at all levels of schooling in Nigeria. The book has the capacity to inform a useful technical, professional, and policy dialogue. I hope the insights this book has provided will stimulate further production of other debates and documents on the same theme, deepen the understanding of learners and teachers not only in Nigeria, but in all other developing countries of Africa, and further stimulate adaptive research on innovative ways of creating local and adapting foreign education technology tools to serve the vulnerable in the society and our rural areas in a better form than hitherto.

Joel B. Babalola

Professor of Educational Management

Former Dean of Education, Faculty of Education, University of Ibadan, Ibadan, Nigeria

Founding President of Higher Education Research and Policy Network (HERPNET)

Fellow of Nigeria Association of Educational Administration and Planning (NAEAP)

Acknowledgements

I would like to express my appreciation to the three giants in education, technology, and business (Profs. Pai Obanya, Joel Babalola, and Michele Cole), all the reviewers, and colleagues who assisted me with this book, those who provided support, read, wrote, offered comments, and assisted in the editing, proofreading, and design. Thanks to all of you.

I would like to thank my wife, children, and the rest of my family, who supported and encouraged me in spite of all the time it took me away from them. It was a long and difficult journey for them.

Last and not least: I ask for forgiveness of all those who have been with me over the course of years and whose names I have failed to mention. I say thanks to all.

Dr. Blessing Foluso Adeoye

Introduction

In this digital age, technology has become a very vital factor of development in all disciplines. Every day new software or device is being developed to improve lives in one way or another (Mojidra, 2013). Technology in the broadest terms could include the collection of tools, machineries, devices, modifications, arrangements, and procedures used by humans. However, in the context of educational technology as presented in this book, it is understood as technologies that have arrived with the "Information Revolution" i.e., those associated with computers and information communication technology. Examples of such technologies are electronics devices, computer, video, collaborative writing tools, social networking, and the Internet.

Educational technology is the application and organisation of people, methods, techniques, devices, equipment and materials systematically and scientifically in order to solve teaching and learning problems (Adeoye, 2015, Delhin, 2012). Educational technology also provides students with the opportunity to learn about the processes and knowledge needed to use technology as a tool to enhance the teaching and learning processes (Umeh and Nsofor, 2014).

Historically, educational technology has embraced both the traditional and digital tools and methods in education. The concepts of ICT, meaning mainly communication technology, and digital data networks have appeared as the latest phase of development. Due to the trend of merging different technologies, educational technology embraces all the latest technologies used for communication, data processing, and data storage. All these media and technologies have become tools in the hands of educationist and students; they must be applied innovatively for effective teaching and learning.

Innovative applications of technology in the classroom mean more than teaching basic computer skills and software programs in the class. It must happen across the disciplines and curriculum in ways that teaching and learning processes can be enhanced. It must also support active engagement, group participation, local and global collaboration, and interaction.

The technology used in the classroom is very beneficial for both students and teachers. For instance, since there are a number of students who are visual learners, visual materials could be used in classrooms to let the students see their notes as opposed to a simply teacher-centred method of teaching. For students that cannot see, audio materials will be much better while audiovisual materials will be helpful for students who can see and hear simultaneously. Also, there is a number of good educational software that can be used to supplement the class curriculum. These can make available to students quizzes, tests, activities, and study questions that could help the students continue with the learning process when they are out of the classroom.

With the continuing advances in technology, students are getting improved access to such educational opportunities. New technology emerges daily, and the price of the existing one decreases, making it much more accessible in the educational setting. Technology has greatly grown to the point that it is available to assist at all educational levels. There are a number of educational technology tools, including gaming media for small children that assist them in getting ready for school and in a number of situations also give them a head start on their education.

With all the benefits above, most schools lag far behind when it comes to integrating technology into the classroom learning. Many are just beginning to explore the true potential technology offers for teaching and learning (Mojidra, 2013). If technology is integrated properly, it will help students acquire the skills they need to survive in this digital age.

This book explores the innovative applications of educational technology tools in educational management; sociology, psychology, and philosophy of education; curriculum studies; science, technical and vocational education; adult education; technology education; human kinetics and health education; guidance and counseling; and other areas of education. The book will equally publish papers that examine utilisation, methods and techniques, and integration of educational technology in the classrooms.

It is possible that not all readers will be equally interested in all chapters covered in this book. However, we have tried to weave many threads into a single fabric to meet the needs of many readers. However, it intends to help readers chart their own path through various contributions from many researchers that make up this book.

SECTION 1: The Concept of Educational Technology, Need, and Significance

SECTION 2: Utilisation of Educational Technology Tools for Teaching and Learning

SECTION 3: Adoption and Integration of Educational Technology Tools in the Classroom

SECTION 4: Cultural Factors Affecting Integration of Educational Technology Tools in Education

SECTION 5: Emerging Trends in Educational Technology Tools in Education

SECTION 6: Integration of Educational Technology in Health Education

SECTION 7: Blended Learning

References

Adeoye, B. F. (2015). Technology guide for teaching and learning. Ibadan, His Lineage Publishing House.

Delhin, P. C. (2012). Computers instructional technology in primary and secondary schools, Ibadan, International Publishers.

Mogidra, R. (2013). Effective educational technology and its application in 21 . . . (n.d.). Retrieved from http://raijmr.com/wp-content/uploads/2013/03/22_130-137-Ravindra-Mogidra.pdf.

Project SMaRT » Blog Archive » Why integrate technology . . . (n.d.). Retrieved from http://weblogs.pbspaces.com/projectsmart/2009/09/30/why-integrate-technology-into-the-curriculum.

Umeh, A. E., and Nsofor, C. C. (2014). Modern trends in the use of educational technology in the classroom. Retrieved from http://www.eajournals.org/wp-content/uploads/Modern-Trends-In-The-Use-Of-Educational-Technology-In-The-Classroom.pdf.

SECTION 1

The Concept of Educational Technology, Need, and Significance

Chapter 1

The Impact of Educational Technology: Students' Experiences and Perceptions of Educational Technology in Higher Education

Neba Nfonsang and Gary Schnellert

Introduction

The world of today relies on technology to operate more efficiently and effectively as technology is required in every context and walk of life to perform almost every task, to work faster and more accurately in ways previously not possible. Technology has been integrated into the fabric of modern society due to its vital role and purpose in the world. Kim (2009) asserted that the purpose of learning technology is to challenge the status quo of education, to provide learning environments that support multiple intelligence and different learning styles, to enhance the sharing of educational material with learning communities, and to enable students to be creators of knowledge and not just consumers of knowledge. To emphasise the need of technology in education, Daggett (2010) stated, "If the American education system is to prepare students to meet the demands of an increasingly technological world, indeed if it is to be effective at all, it must integrate technology into the academic curriculum" (p. 1).

It is essential to explore students' technology skill and preference levels as well as students' experiences and perceptions of educational technology for the following reasons. First, technological proficiency should be a characteristic of this digital age and students have indicated that they needed more technology training and skills instead of new or more technology (Dahlstrom, 2012). It is crucial for educational institutions to continuously assess the technology skill levels of students to understand the types and levels of technology training that should be offered to students to enhance their academic success. Second, since educational technology has historical roots and has evolved from the printing press to today's digital tools (Saettler, 2004) and will continue to rapidly evolve (Arora, 2013), it is imperative to understand students' preferences for educational technology. Third, though findings of studies carried out at an elementary school system at Southwestern Ontario, with a sample population of 106 students, showed that students had positive attitudes towards modern technology (Hurley and Vosburg, 1997), there is very limited research on the attitude of students towards technology especially in higher education. Fourth, the perceptions of usefulness of educational technology help faculties and institutions to make decision concerning integration of technology into the learning process (Moseley, 2010). Also, "Institutions of higher education must be aware of how students employ technology, and must consider student perspectives regarding how technology can best be integrated for instruction and communication" (Surry et al., 201, p. 31).

Purpose of the Study

Schools across the United States have spent large amounts of money for educational and computer-based technologies (Pearson and Young, 2002). In the year 2009, 63 billion dollars was spent on technology across all levels of education in the United States (Daggett, 2010). It is not enough to spend millions of dollars on educational technology. It is more important to find out whether the available institutional technologies are relevant and how students can benefit most from these technologies. Therefore, the purpose of this was to explore the relationships among students' technology skill levels, preference levels of educational technology, attitudes towards educational technology, and perceptions of the impact of educational technology on learning. The following research questions guided this study:

1. What skill levels, preference levels, attitudes, and perceptions do students have in using educational technology?

2. What are the relationships among students' technology skill levels, preference levels of educational technology, attitudes towards educational technology, and perceptions of the impact of educational technology on learning?

3. How do students' technology skill levels, preference levels of educational technology, and perceptions of the impact of educational technology on learning predict students' attitudes towards educational technology?

Significance of the Study

In this era, technology integration into teaching and learning is mandatory as educational institutions believe that technology has the potential to improve learning outcomes. The significance of this study lies in the investigation of higher education students' experiences of technology in an effort to understand the valuable technologies and technology skills required for academic success at postsecondary institutions. Kelly (2010) stated that though most high school students in the United States have grown up immersed in technology and are often called digital natives, many of them lack the skills to apply the latest technology to its maximum potential in an educational setting. Therefore, it is important for high school students to acquire skills for using educational technology before entering college as well and be prepared with the technology skills necessary to survive the higher education learning environment.

Research shows that only 45% of high school students who enter college graduate with a bachelor's degree; this high dropout rate of freshmen is attributed to lack of required skills to survive the college environment (United States Department of Education, 2011). The United States Department of Education (2013) explained that the application of digital technology to teaching and learning will increase the quality of education by improving high school graduation rate and preparing students to be college ready. Therefore, this present study would unveil the relevant technologies and technology skill levels necessary for students to enhance their learning experience.

Since technology is useful for college education, the reauthorized Elementary and Secondary Education Act requires students to be technologically literate before they enter college. The new goal for the educational system in the United States expects every student to be college and career ready after graduation from high school, as proposed by President Obama (United States Department of Education, 2011). The new goal was proposed because the standards required under the No Child Left Behind Act do not reflect the skills needed for college success. For high school students to be college ready, they need to use the types of technologies similar to those used in higher education institutions.

Moreover, this study is very necessary as it adds to existing literature about students' experiences and perceptions of the impact of educational technology. Currently, there is little research evidence and data to demonstrate the impact of technology on learning. Therefore, "there is a need to investigate whether education technology impacts on the teaching and learning experience in a positive way" (Joseph, 2012, p. 3) compared to learning that involves no educational technology.

Finally, this study should provide higher education institutions with actionable recommendations on how to address students' issues in relation to students' technology experiences. Generally, the results of this study should provide information on how technology services in high schools, colleges, and universities could be improved to enhance student's technology experiences. Specifically, this study should inform pre-K-12 schools in the United States about the technology skills and educational technologies that should be integrated into pre-K-12 curriculum to prepare high school students for college. Therefore the results of this study will benefit both high schools and colleges assuming that higher education leaders will establish relationships with pre-K-12 education leaders to facilitate the achievement of the college readiness. Pre-K-12 students will benefit most from educational technology if higher education leaders communicate the skills (including technology skills) required to prepare high school students for college success and contribute in developing the curriculum and standard requirements for pre-K-12 schools ("Postsecondary Readiness," 2013).

Methodology

This research used a nonexperimental, quantitative, cross-sectional, and survey study to analyze the relationships among students' technology skill levels, preference levels of educational technology, attitudes towards educational technology, and perceptions of the impact of educational technology on learning.

Participants

The participants of this study were students enrolled in the University of North Dakota during the 2013–2014 academic years. Two hundred student participants completed the survey. The participants consisted of 35% male, 65% female. The most representative age group consisted of students between thirty and thirty-nine years (43%), followed by students between forty and forty-nine (27%) years. A majority of respondents were graduate students (73%) and ninety-eight students indicated their disciplines representing about forty-three majors (mode = 16 educational leadership). Almost all the students (99%) completed their high school education in the United States, and most participants (40.59%) graduated from high school between 1991 and 2000 while 23.76% graduated between 1981 and 1990. Half of the students did not own any digital device during their high school education; however, 45% owned desktop computers. As university students, most respondents indicated that they owned laptop computers (92.46%), smartphones (75.40%), tablet devices (62.96%), and desktop computers (63.78%).

Procedures

The researcher started the study by meeting the requirements of the Institutional Review Board (IRB) at the University of North Dakota. An online survey was created through the Qualtrics Survey Software, and a list of one thousand randomly selected student e-mail addresses was collected from the Office of Institutional Research at the University of

North Dakota. The e-mail addresses were selected using the Statistical Analysis System (SAS), a software system for data analysis. On November 24, 2013, a survey invitation e-mail containing the link to the survey was sent to the selected sample. On November 28, 2013, a follow-up e-mail message was sent to the selected students again to encourage those who did not respond to the questionnaire to complete the online survey. Completion of the survey took an average of five minutes. Survey data submitted by respondents was collected through the Qualtrics software and the last date for data collection was December 1, 2013.

Instrument

A survey instrument created by the Office of Information Technology, University of Minnesota, was modified and used to collect the data for this study. Through personal communication, the researcher obtained permission from the research associate of Academic Support Services at the University of Minnesota to use their technology surveys to guide the development of the instrument for this study. The technology surveys were obtained from the University of Minnesota Web site (http://z.umn.edu/techsurveys). The instrument for this study was used to measure the demographic variables of respondents, such as gender, age group, academic level, and program of study. More important, the instrument used Likert-type items to measure the main variables of this study, including students' technology skill levels (eight items), preference levels of educational technology (nine items), attitudes towards educational technology (nine items), and perceptions of the impact of educational technology on learning (seven items). The Likert scale was preferred because of its suitability for studies that involve the combination of Likert-type items into single composite variables (Boone and Boone, 2012).

Measures

Likert-type and Likert scales were used to measure the research variables: students' technology skill levels (1 = *Never used*, 5 = *Very comfortable*), preference levels of educational technology (1 = *No preference*, 5 = *Very strongly prefer*), attitudes towards educational technology (1 = *Very useless*, 7 = *Very useful*), and perceptions of the impact of educational technology on learning (1= *Strongly disagree, 5 = Strongly agree*). For the various scales, participants were asked the following:

- to rate their comfort level in using technology such as blackboard (students' technology skill levels scale);

- to select which communication methods—for instance, social media—they would prefer their instructors to use (students' preference levels of educational technology scale);

- to rate how useful educational technologies such as digital videos have been in their coursework (students' attitudes towards educational technology scale); and

- to rate statements concerning the effect of technology in their learning—for example, "educational technology has enabled me to access course material from anywhere at any time" (students' perceptions of the impact of educational technology scale).

To test the quality of each scale, an exploratory factor analysis with oblimin rotation was conducted for the items of each scale. A visual inspection of the scree plots (fig. 1, fig. 2, fig.3, and fig. 4) for each scale indicated that the data for each scale supported a one-factor solution. A final factor analysis of each scale with number of factors constrained to 1 showed that all the items of the students' technology skill levels, attitudes towards educational technology, and perceptions of the impact of educational technology on learning scales were good for further analysis while two items of the preference levels of educational technology scale were removed as these items (preference 1 and preference 3) had a factor loadings below .30. Internal reliability was found to be sufficient for students' technology skill levels scale ($\alpha = .76$), preference levels of educational technology scale ($\alpha = .73$), attitudes towards educational technology ($\alpha = .83$), and perceptions of the impact of educational technology on learning ($\alpha = 94$). The scale distributions all approached normality based on the visual inspection of histograms for each scale. Items with factor loadings above .30 were then summed up into their respective composite variables.

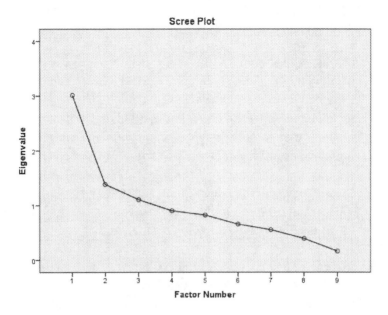

Figure 1. An illustration of scree plot for students' technology preference levels items

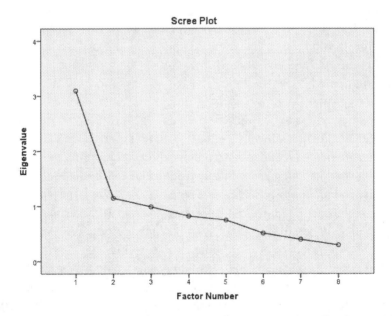

Figure 2. An illustration of scree plot for students' technology skill levels items

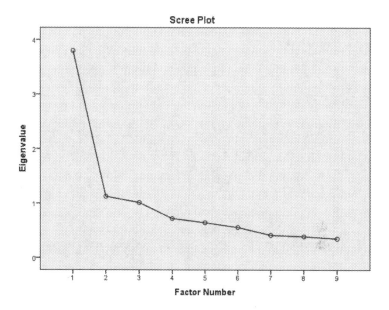

Figure 3. An illustration of scree plot for students' attitude towards educational technology items

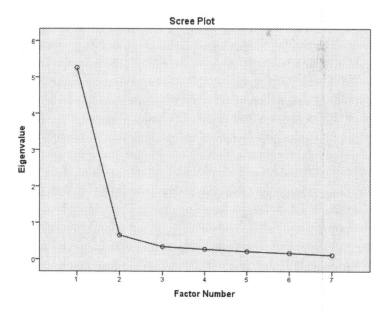

Figure 4. An illustration of scree plot for students' perception of the impact of educational technology items

Results

Research question 1

What skill levels, preference levels, attitudes, and perceptions do students have in using educational technology?

Students' technology skill levels. To measure technology skill levels, students were asked about their comfort levels in using specific educational technologies. Table 1 summarizes the number of students with different comfort levels in using different educational technologies.

Table 1. Student's comfort level in using educational technology

#	Activity	Very comfortable	Comfortable	Uncomfortable	Very uncomfortable	Never used/Don't know	Total responses
1	Accessing course material through blackboard	152	41	4	2	2	201
2	Completing and submitting quizzes/assignments online	144	50	3	1	3	201
3	Creating a PowerPoint slide	137	48	13	0	3	201
4	Contributing to wikis or blogs	68	78	21	5	26	198
5	Editing a digital video with multimedia programs such as movie maker	24	30	55	17	72	198
6	Creating animations in Photoshop or flash	9	24	56	19	90	198
7	Using e-books or library Web sites for studies	75	84	18	8	16	201
8	Using online discussion tools	70	100	16	5	8	199

The highest number of students (96.52%) reported that they were comfortable or very comfortable using online assignment tools to complete and submit assignments. Also, most of the students rated their comfort levels in accessing course material through blackboard (92.02%), creating a PowerPoint slide (92.04%), and using e-books or library Web sites for studies (79.10%) as very comfortable or comfortable.

Student's preference levels of educational technology. Students were asked which communication methods they would prefer their instructors to use. Findings of the study showed that most students very strongly preferred or strongly preferred their instructors to use e-mails (76.65%) and blackboard technology (54%). Besides, 68.75% of the students very strongly preferred or strongly preferred face-to-face interaction compared to using only educational technology for instruction. Very few students indicated that they would very strongly or strongly prefer their instructors to use social media and Facebook for communication. Table 2 shows a summary of these responses for students' preference levels of communication methods.

Table 2. Communication method used by instructor

#	Communication method used by instructor	Very strongly prefer	Strongly prefer	Prefer	Slightly prefer	No preference	Total responses (n)
1	Face-to-face interaction	97	35	29	13	18	192
2	Blackboard technology	44	61	45	21	22	193
3	E-mail	82	69	31	11	4	197
4	Text messaging	14	16	24	52	80	186
5	Online charting	12	15	26	38	90	181
6	Social media	3	11	17	41	113	185
7	Facebook	2	7	7	36	129	181
8	Phonelike communication over the Internet	9	13	19	45	100	186
9	Phone conversation	20	28	39	32	67	186

Additionally, in comparison with other students, participants were asked to rate their overall levels of technology skills. The findings showed that 44% of the students reported that they were about the same skill level with other students, 23% were more skilled, 22% less skilled, 9% much more skilled, and 2% much less skilled.

Students' attitudes towards educational technology. To assess the students' attitudes towards educational technology, students were asked to rate how useful different educational technologies have been to their coursework. Table 3 represents a summary of these responses in terms of number of students. A majority of students evaluated blackboard (94.50%), e-mails (84.76%), PowerPoint presentations (73.23%), and Web-based quizzes/assignments (62.56%) as very useful or useful for educational purposes.

Table 3. Students' attitudes towards educational technology

#	Educational technology	Very useful	Useful	Somewhat useful	Neutral	Somewhat useless	Useless	Very useless	Total responses (N)
1	Blackboard	114	45	19	8	8	2	4	200
2	Web-based quizzes or assignments	65	57	33	24	10	4	2	195
3	PowerPoint presentations	83	62	29	14	4	5	1	198
4	Wikis or blogs	29	31	43	49	17	15	13	197
5	E-mails	109	58	25	3	1	0	1	197
6	Digital video	44	45	45	47	7	6	2	196
7	Animations or simulations	25	36	44	68	8	13	2	196
8	E-books or Library Web site	64	49	30	40	4	8	4	199
9	Online discussion tools	34	45	46	38	15	15	4	197

Students' perceptions of the impact of educational technology. The study evaluated students' perceptions of the effects of educational technology on their learning experiences. The results of the study indicated that 88.50% of students strongly agreed with the statement "Educational technology has enabled me to access course material from anywhere at any time," followed by the statement "Educational technology has helped me to interact with instructors and other students" (79.5%). Table 4 illustrates the number of students with different levels of agreement on statements about the impact of educational technology.

Table 4. Students' perceptions of the impact of educational technology on learning

#	The effects of educational technology	Strongly Disagree	Disagree	Neither agree nor Disagree	Agree	Strongly agreeagreeAgree	Total Responses (N)
1	Educational technology has helped me to interact with instructors and other students	13	7	21	88	71	200
2	Educational technology has helped me to complete assignments on time	14	10	23	80	73	200
3	Educational technology has enabled me to access course material from anywhere at anytime	11	3	9	62	115	200
4	Educational technology has help me to improve the organisation of my course work	13	16	45	72	54	200
5	In relation to my studies, educational technology has increased my participation, motivation, and satisfaction	14	20	52	75	39	200
6	Educational technology has enhanced my overall learning experience	12	8	25	98	56	199
7	I prefer taking courses that use educational technology	15	12	47	74	52	200

Research question 2

What are the relationships among students' technology skill levels, preference levels of educational technology, attitudes towards educational technology, and perceptions of the impact of educational technology on learning?

Pearson correlation. The results of relationships revealed significant correlations among students' preference levels of educational technology and students' attitudes towards educational technology (r = .42, p = .01), students' preference levels of educational technology and perceptions of the impact of educational technology on learning (r = .23, p = .01), students' technology skill levels and students' attitudes towards educational technology (r = .35, p = .01) as well as students' technology skill levels and perceptions of the impact of educational technology on learning (r =.32, p = .01). Table 5 summarizes the correlation results.

Table 5. Pearson correlations

		Preference levels	Attitudes	Skill levels	Perceptions
Preference levels	Pearson Correlation				
	Sig. (2-tailed)				
	N	174			
Attitudes	Pearson Correlation	.417**			
	Sig. (2-tailed)	.000			
	N	167	186		
Skill levels	Pearson Correlation	.071	.347**		
	Sig. (2-tailed)	.363	.000		
	N	167	180	191	
Perceptions	Pearson Correlation	.232**	.322**	.107	
	Sig. (2-tailed)	.002	.000	.141	
	N	172	185	190	199

**. Correlation is significant at the 0.01 level (2-tailed).

Research question 3

How do students' technology skill levels, preference levels of educational technology, and perceptions of the impact of educational technology on learning predict students' attitudes towards educational technology?

Multiple regression analysis. The final quantitative analysis examined how the students' technology skill levels, preference levels of educational technology, and perceptions of the impact of educational technology on learning scales predicted students' attitudes towards educational technology using multiple regression. The results of the regression indicated the three predictors explained 57.20% of the variance (R^2 = .33, F (3,159) = 25.25, $p < .01$). As shown on table 6, it was found that students' attitudes towards educational technology was significantly predicted by students' technology skill levels ($\beta = .27, p < .01$), preference levels of educational technology ($\beta = .36, p < .01$), and perceptions of the impact of educational technology on learning ($\beta = .25, p < .01$).

Table 6. Multiple regression results showing significant prediction of students' attitude towards educational technology

Model		Coefficients[a]				
		Unstandardized Coefficients		Standardized coefficients	t	Sig.
		B	std. error	beta		
1	(Constant)	19.193	3.912		4.906	.000
	Preference levels	.542	.103	.355	5.265	.000
	Skill levels	.417	.101	.272	4.129	.000
	Perceptions	.337	.089	.254	3.769	.000

a. Dependent Variable: Attitudes

Discussions

Based on the study findings, the following discussions and conclusions were reached. First, the findings suggest that a majority of students have gained technology skills in using relevant educational technologies such as blackboard assignment tool, PowerPoint slides, e-books and library Web sites. Most of the students also felt that blackboard and e-mails were very useful or useful for education. Using blackboard to enhance education is very crucial because it has several resources and interactive tools to support learning. Students also benefited from e-mail as this is the major means of communication among students and professors. This probably explains why the greatest proportion of students very strongly or strongly preferred their instructors to use e-mails for communication compared to the other means of communication. Equally important, technology's role in fostering access to course material anywhere at any time emerged as the most important perceived impact of educational technology. Students have benefited from technologies and devices such as mobile phones, tablets, and laptops as these tools have enabled them to access the learning environment at their convenience from a distance.

Second, the findings also revealed that students' preference levels of educational technology significantly and positively correlated with students' attitudes towards educational technology and students' perceptions of the impact of educational technology. Similarly, students' technology skill levels significantly and positively correlated with attitudes towards educational technology and students' perceptions of the impact of educational technology as shown in figure 5.

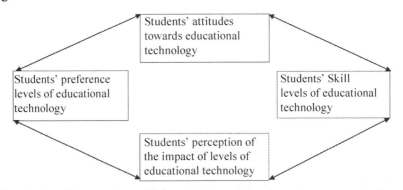

Fig. 5. An illustration of the relationships found among students' technology skill levels, preference levels of educational technology, attitudes towards educational technology, and perceptions of the impact of educational technology on learning

Third, these studies found a model that could predict students' attitudes towards technology. With this model, it is possible to estimate a student's attitude towards technology given the student's technology skill level, preference level of educational technology, and perception of the impact of educational technology. It will be important to investigate the relationship between students' attitudes towards educational technology and students' academic success in a further study so as to understand how the predictors of students' attitudes towards educational technology indirectly influence student success.

Implications and Limitations

The single largest implication of this study is its potential to inform education leaders about student's experiences and perceptions of the impact of educational technology. As a result, research-based educational policies and curricula could be developed, taking into consideration the relevant educational technologies and technology skills required by students to perform efficiently in the academic environment. These research results could be useful both in high schools and higher education institutions. Since high school students are required to be college ready before they graduate, it is important for them to immerse themselves in the educational technology environment required in college. Therefore, it is strongly recommended that educational technologies such as laptops, tablets, blackboards, e-mails, and PowerPoint technologies be introduced to high school students as these technologies have highly benefited college students and will possibly prepare high school students for the college academic environment.

Though this study suggests some potential solutions, it also has some limitations. The research was a cross-sectional survey and not longitudinal in nature because it was conducted at a single point in time. "Education Week" (2011, para 12) states that "the kinds of studies that produce meaningful data often take several years to complete—a timeline that lags far behind the fast pace of emerging and evolving technologies." Therefore, to improve the validity of the data, an ongoing or longitudinal survey should be considered. However, the rate at which educational technology is changing poses a challenge for an ongoing survey to be conducted. Also, the rapid changes in technology may make it difficult to implement the findings of this study in the future since educational technologies used today may be obsolete in the future. However, this research can be replicated in any era and educational context, taking into account the available technology in that era or context.

References

Arora, K. (2013). The evolution of educational technology. Retrieved from http://edtechreview.in/news/news/trends-insights/insights/171-edtech-evolution-infographic.

Boone, D., and Boone, H. (2012). Analyzing Likert data. *Journal of Extension*, 50(5), 1–5. Retrieved from http://www.joe.org.

Daggett, W. (2010). Preparing student for their technological future. (Doctoral dissertation). Retrieved from http://www.leadered.com/pdf/Preparing%20Students%20for%20Tech%20Future%20white%20paper.pdf.

Dahlstrom, E. (2012). ECAR study of undergraduate students and information technology, 2012 (Research report). Louisville, CO: *EDUCAUSE Center for Applied Research.* Retrieved from http://www.educause.edu/ecar.

Education Week (September 1, 2011). Technology in education. Retrieved from http://www.edweek.org/.

Hurley, N., and Vosburg, D. (1997). Modern technology: The relationship between student attitudes toward technology and their attitudes toward learning using modern technology in an everyday setting. Retrieved from http://www.u.arizona.edu/~ldonahue/student_attitude.pdf.

Joseph, J. (2012). The barriers of using education technology for optimizing the educational experience of learners. *Procedia—Social and Behavioral Sciences, 64*(0), 427–436. doi: http://dx.doi.org/10.1016/j.sbspro.2012.11.051. Postsecondary Readiness (2013).

Kelly, P (2010). Building Technology Skills in Secondary Schools Improves College Readiness. Retrieved from http://www.certiport.com/Portal/Common/DocumentLibrary/WNE Winter 2010 Newsletter Ray Kelly.pdf.

Kim, J. (2009). The purpose of learning technology. Retrieved from https://www.insidehighered.com.

Moseley, W. (2010). Student and faculty perceptions of technology's usefulness in community college general education courses. (Doctoral dissertation). Retrieved from http://digitalcommons.unl.edu/.

Pearson, G., and Young, A. (2002). Technically speaking: Why all Americans need to know more about technology. Washington, DC: National Academy Press.

Saettler, P. (2004). The Evolution of American Educational Technology Greenwich, CT: Information Age Publishing.

Surry, D., Gray, R., Stefurak, J., Kowch E., McPheron M., Jameson J., and Ensminger, D. (2011). Technology integration in higher education: Social and organizational aspects. Hershey, PA: Information Science Reference.

United States Department of Education (2011). College- and career-ready standards and assessment. Retrieved from http://www2.ed.gov/policy/elsec/leg/blueprint/faq/college-career.pdf.

United States Department of Education, Office of Educational Technology (2013). Expanding evidence approaches for learning in a digital world. Retrieved from http://www.ed.gov/edblogs/technology/files/2013/02/Expanding-Evidence-Approaches.pdf.

Chapter 2

Education, Technology, and Universalizing Quality Outcomes

Augustine Obeleagu Agu

Introduction

The fact that Africa, especially Nigeria, has an educational problem is no secret. Delivering quality education that is relevant is still a challenge. Many countries after independence in the '60s made education a priority sector for national development and have pushed for endless reforms and resources to increase access and boost student achievement. Countries have made progress; but not significant enough to make the continent globally competitive. The question has always been and still is how we can prepare the African children to compete very effectively with children from other continents in the twenty-first century. We argue in this chapter that Technology can be of enormous benefit in the quest for Africa to prepare her children for the twenty-first-century skills.

Some Facts We Need to Face in Africa

- Since the adoption of the Education for All (EFA) in 1990 and Millennium Development Goals (MDGs) in 2000, access to education has increased significantly throughout most the nations of Africa except the countries that have witnessed long periods of internal conflict such as Somalia and Republic of South Sudan. EFA is a global movement aiming to meet the learning needs of all children, youth and adults by 2015 (UNESCO, 1990). MDGs are eight international development goals established at the millennium summit. Two of the eight goals (Goal 2: Universal Primary Education, and Goal 3: Gender Equality and Empowerment are educational related). Many countries are on track to achieve Universal Primary Education and gender equality in education by 2015. However, progress is threatened by among other factors, complex demands placed on systems struggling to cope with increased enrollment. Amidst determined efforts by governments, as of 2007, 32 million children in sub-Saharan Africa remained un-enrolled, and in 2008, an estimated 29 million children across the continent were not enrolled (UNESCO, 2010). These figures suggest that much work remains to be done for all children in Africa to fulfill their rights to education.

- Despite the encouraging trend, some countries, including three of the five most populated of the continent (Nigeria first, Democratic Republic of Congo fourth, and South Africa fifth) have shown declining enrollment over the same period even though education is free and compulsory. Nigeria is the home to the largest number of out of school children in the world, 8.8 million in 2008 (UNESCO, 2010, p. 2). In North Africa, four countries, Sudan, Morocco, Mauritania and, to a lesser extent, Egypt, face critical challenges in terms of access, and are not on track to achieve Universal Primary Education.

- There is consensus that the most significant underlying causes for the current shortcomings and challenges to education in Africa are due to the lack of equity in the delivery of education services, which overwhelmingly affects poor and marginalized populations, and the low quality of education, especially in most of the African countries, which affects retention and learning outcomes, and contributes to grade repletion and dropping out (UNESCO/UNICEF/UNECA, 2011, p. 2).

- In order to sustainably achieve EFA and MDGs 2 and 3, a double focus on equity and education quality, including teaching, is required. Innovative strategies which promote enrollment and increased retention of students need to be scaled up in the most underserved communities. At the same time the quality of teaching and learning needs to improve and disparities in learning outcomes addressed to mitigate student dropouts so that primary school graduates acquire the necessary knowledge and life-skills, and values to become productive and active members of their society.

• Students in many African countries are graduating from both high school and university unprepared for the world of work. Most employees have complained about the work readiness of high school and higher education products. In fact, a lot of organisations in Nigeria tend to embark on intensive training of new employees to get them up to speed.

• Today, most African countries and Nigeria particularly, is at an inflection point. Over the next 20 years, the country will experience huge growth in the number of young people in its society. If these young people are healthy, educated and find productive environment, the country could emerge as a great nation. But if not, she could go down as a failed state given the turmoil that will befall her. Nigeria's median age is 18.2 years (2014 estimate), compared with a world average of 29.7 (2014 estimate) (United Nations Department of Economic and Social Affairs Population, 2014). The combination of a large population and a low median age means Nigeria has an opportunity to capitalize in the medium term on the economic and social potential. Failure to capitalize on such opportunities can contribute to state fragility where young people do not receive education and develop skills to contribute productively, both economically and socially. Investment in quality education that targets the poor and hard to reach groups will be very critical in influencing the direction the country will take.

• In order to achieve the rights of all children to quality education, it will require resources prioritized for the poorest of the poor and most marginalized communities, including orphans and vulnerable children, those in rural areas, pastoralists, children with disabilities, and those affected by natural disasters and conflict and also commonsense. It will need utilisation and application of technologies currently available to all and sundry.

• Increasingly in Africa today, there are two achievement gaps in our educational systems. The first is the gap between the quality of schooling that most children from the middle class get and the quality of schooling available for most poor and vulnerable children and the consequent disparity in outcomes. The second one is the global achievement gap. This is the gap between what even our best schools are teaching and testing versus what all students will need to succeed as learners, workers and citizens in today's global economy.

An Emerging Context for Schooling

The information above, suggest that our education system needs to be transformed to benefit all. In today's highly competitive global "knowledge economy," all students need good quality education for college, work and citizenship. The failure to give all students good quality education that equips them with life-skills leaves the African youth, and our various countries at a very competitive disadvantage. The world has changed; but our education system has not, and is increasingly becoming obsolete.

Throughout much of the twentieth century, the basic skills of reading, computation and writing were the focus of schooling. Three R's (reading, writing, arithmetic), became essential in the workplace. However, in the twenty-first century, mastery of the basic skills of reading, writing and arithmetic is increasingly becoming no longer enough. Effectiveness in work, learning, and citizenship in the twenty-first century demand that we all know how to think, reason, analyze, weigh evidence, solve problems, and communicate well (Wagner, 2008). These are essential life-skills for all of us.

With globalisation, African children will be competing for well-paying jobs with other children from all over the world. For Nigeria, the issue of 'Federal Character' will be a thing of the past in private company jobs that will be paying more than the public sector jobs. Furthermore, with technology, a lot of jobs will be automated with the obvious implication of drastic cut in labour. The decrease in oil and the attendant reduction in finance and the introduction of austerity measures will accelerate the process of mass retrenchment of people with old skills and the employment of people with new skills.

All the above represent key challenges and opportunities to our educational system. They compel a fundamental reconsideration of all of our assumptions about what children need to learn and the delivery of learning.

Conceptualisations of Quality Universal Education

Universal education means having a singular, common educational experience with the same broad purpose for all. The practice of universal education started in Europe in the sixteenth century in the wake of the emergence of the concept of the individual (Meyer and Jepperson, 2000). The idea of the individual created great social waves that undermined the corporate character of medieval society (Sachar, 1990, 4–5). The establishment of universal education is best understood as an extended historical process in which the education of the young moved out of the home and church and into the public sphere of differentiated schooling (Benavot and Resnik, 2006). Ramirez and Boli (1993) describe this process as the institutionalisation of Western models of socialisation with three stages of development: (a) universal education was a part of reformation movement to enhance individual faith among protestant families; (b) mass education was part of a movement to weaken family socialisation and home based instruction by establishing schools with fairly standardized curriculum; (c) universal education in which the nation state became the central initiator, and administrator of the of an interconnected system of schools. The European states found universal education to be a practical mechanism to form citizens as autonomous individuals who will obey laws and support the public good because of reason. Thus, schools to provide a universal education evolved into an integral part of the modern European nation state (Meyer, Ramirez and Soysal, 1992).

The European states for a great part of the 20th century institutionalized the concept of universal education and with the help of the Christian Missionaries promoted it in their colonies in Africa. Most African states at independence introduced UPE policies (Agu, 1986). In almost all countries, the concept of universal education has the following institutional commonalities (a) a common aim for all to become individuals who take on particular responsibilities (b) serving the interests of the states, the collective decision maker of the curriculum and (c) implemented through formal schooling. Transnational networks and international organisations played influential roles in the promotion of mass schooling in Africa, Asia, and Latin America in the last since the 1960's. The practical challenge is how to conceptualize a common core education programme that can be implemented for all and be of high quality for all.

The definition of quality education has made three evolutions. The first stage, it was defined by characteristics of inputs (such as qualified teachers, adequate school resources and good curriculum and materials). The second stage was the focus on outcomes such as student achievement, educational attainment, graduation rates etc. The third stage looks at quality from the perspective of relevance to the needs of the society. The issue for the third stage of quality is whether the existing system of school is preparing young people for the demands of life today. This raises the issue of defining universal education from the perspective of competence. Competence is defined as "the ability to successfully meet complex demands in a particular context through the mobilization of psychological prerequisites (including both cognitive and noncognitive aspects)" (Rychen and Salganik, 2003, p. 43). This definition does three things. One, it includes the demands placed on individuals in different areas of life, including in the family, the workplace, social and personal life. Two, it recognizes that a range of internal and external conditions interact to allow individuals to meet demands. Third, it recognizes the role of context/environment. These are all important considerations in charting the role of technology in education.

Should Africa adopt the policy of "Going beyond Education for All" To "Education Above All" as suggested by Pai Obanya. My answer is YES. Education above all according to Obanya in practical terms will mean placing Education above everything else and would involve concerted action on (a) promoting the enduring humane values of traditional Africa in wider society, and (b) investing more meaningfully on Education by channelling appropriate level of funding to inputs and processes that have positive multiplier effects on the system: infrastructure, teachers, instructional materials, etc. A policy of going beyond education for all to education above all should adopt tools made available by the surging technological innovations that are readily accessible to all and sundry. New technologies offer hope for more effective ways of teaching and learning. Technology has the power to free us from those limitations to make education far more portable, flexible and personal; thereby ensuring "education above all." As the modern world grows smaller, flatter, and more competitive, the achievement of African students must be measured not only against national and regional standards, but also against international ones. Work and jobs move around the world to whoever can perform them at the highest quality and the lowest price. This century should be African century for rapid development and transformation. The challenge is whether Africans will have the capacities to lead the continents development or will outsource the leadership of African development to others. The outcome will depend on how we educate our youths and children.

With educational technology the following are possible:

- Curricula can be customized to meet the learning styles and life situations of individual students giving them productive alternatives.

- Education can be freed from geographic constraints allowing students and teachers to be anywhere doing their work at any time.

- Students can have more interaction with their teachers and with one another, including students and teachers in other countries and continents.

- Data systems can put the spotlight on performance, make progress or lack of progress transparent to all concerned, and increase accountability.

- Schools can be operated at lower cost, relying more on technology and less on labour.

The Life Skills or Survival Skills for the Twenty-first Century

African educators should always keep this question at the back of their minds: what year are they preparing the students for? 1950? 1960? etc.? Can we honestly say that the education programs are preparing our students for 2020 or 2030? Are we even preparing them for today or yesterday or 50 years ago? Dewey could be right in asserting that "if we teach today's students as we taught yesterday's we rob them of tomorrow" (Dewey, 1944: 167). As educators, it is our responsibility to prepare the learners in our care for their world and their future. Many policy makers, politicians, scholars, and educationists have commented/discussed twenty-first-century skills and tools for our learners (Paul, 2013; Wagner; 2008; Moe and Chubb, 2009; and Jacobs, 2010). For Africans it is important to note that as of the time of writing this chapter almost fifteen years has already passed from this century. The key question here is: what does it mean to be educated in the twenty-first century?

Today's student, globally, lives in a technological era in which Cell phones, Internet and Google and text messaging are normal. Technologies have made it possible for almost everyone in particular children to know things very easily. There is no choice. We should embrace technology. Let's remember that it was technology—the development and adoption of a symbolic alphabet—that ended an era of "orality" and began an era of "literacy." It was technology— the development of movable type and printing presses—that ended an era of scholastic authority by a selected priesthood and created mass literacy in the vernacular of every culture (Wilmarth, 2010). Now technology is again redefining the landscape of education and what will eventually be the important future life skills.

What skills will be important for this century and beyond? As indicated above, it seems that there is a consensus, on the life or survival skills needed for success in the twenty-first-century world. The core life-skills will include:

1. *Creativity and Innovation.* This is using knowledge and understanding to create new ways of thinking in order to find solutions to new problems and to create new products and services. Creativity and innovation are key factors not only in solving problems but also in developing new products and services.

2. *Critical Thinking and Problem* Solving. This is applying higher-order thinking to new problems and issues, using appropriate reasoning in effectively analyzing the problem and making decisions about the most effective ways to solve the problem. Critical thinking and problem solving (CTPS) consists of (a) the ability to ask good questions, (b) the capacity to be effective in teams (c) the ability to look at problems systemically.

3. *Communication:* This is communicating effectively in a wide variety of forms and contexts for a wide range of purposes and using multiple media and technologies. The ability to express one's views clearly and to communicate effectively across cultures is increasingly becoming very important.

4. *Information Management.* This includes accessing, analyzing, synthesizing, creating, and sharing information. Employers in the twenty-first century have to manage huge amounts of information flowing into their lives on a daily basis.

5. *Effective Use of Technology.* This is the capacity to identify and use technology efficiently, effectively, and ethically as a tool to manage information.

6. *Lifelong Learning.* This is no longer an option but a skill that is in demand.

7. *Initiative and Entrepreneurialism.*

Most African countries have been trying for the past sixty years to close the first achievement gap in education (quality and quantity of schooling) with some success. But where the continent is very deficient is in the second global achievement gap represented by the core life-skills summarized above. It has become increasingly clear that in the best schools in Africa, students are not learning the skills that matter most for the twenty first century. The system of education in many African countries—curricula, teaching methods, and the tests and evaluation techniques—were created in a different century for the needs of another era. The global achievement gap for Africa remains hidden because it is fueled by globalisation and the technological changes that have taken place over the last three decades. But these changes are very deep and fundamental, and we need to understand them and use them to reshape what our children need to know in the twenty-first century. This new world, in which children must compete and succeed, has been transformed by rapidly evolving technologies (Wagner 2008). To deal with the rapidly evolving world and technologies, we need to ensure that students are adequately educated for the future. Technology offers great possibilities to deal with the challenges.

It is clear that new skills and approaches are needed if our students are to compete effectively with others. Jobs are changing, and most sectors are becoming globalized. Students are competing not just with fellow citizens but with job seekers from around the world. To prepare our students adequately, schools must use technology not only to deliver education programmes but also to deploy alternative approaches to instruction.

Understanding of Education and Technology

Digital technology is now an increasingly prominent feature of education provision and practice in many countries and contexts. Mobile telephones, Internet use, and other forms of computing are familiar, everyday tools for a sizable proportion of the global population. The pace and scale of the digital innovation has made many scholars and practitioners to describe digital technology as a key driver of societal development around the world (Selwny, 2013: 1). Manuel Castells, succinctly put this way, "we know that technology does not determine society: it is society." (Castells, 2006: 3). For many people, digital technologies have led to a greatly improved era of living—the so-called digital age. The ever-expanding connectivity of the digital technology is recasting social arrangements and relations in a more open, democratic, and ultimately empowering manner.

There is no question that we are living in a technologically reordered world—a world that is structured and arranged along significantly different lines than was the case few decades ago. We are now in the period called "information society," "information age," or "preindustrial" era. These ideas all point towards the growing importance of the production and consumption of information and knowledge as key sources of power and competitiveness in the global economy, also called the "knowledge economy." As pointed out, the "knowledge economy" "symbolizes a transition from the manual/ machine-assisted production line of material things to an abstract, placeless interaction between human and electronic brains for the production of services" (Chakravartty and Sarikakis, 2006: 22). Following this logic, the production and distribution of knowledge and information is now a core component of contemporary economic growth and, therefore, changes in employment with implications and applications to education.

The theory or idea of knowledge economy implies that individuals and organisations face major educational challenges in adjusting to these new dispensations. Education as has always been an integral part of the changing contemporary world. As correctly noted, "The knowledge economy is intrinsically related to education" (Dale, 2005: 118). Within the contexts of "knowledge economy," "information age," and "network society," education is seen as a continuous concern; that is to say "lifelong learning" that embraces not only compulsory phases of schooling, but also education and training throughout the life-course.

This brings us to the concern of this chapter the importance of the use of digital technologies in the transformation of education in Africa. Generally, there is a widespread acceptance that digital technologies must play an integral role in the provision of all aspects of education— from the integration of computers in school, to the virtual delivery of online courses and training.

How Technology Is Transforming Education

Technology is already transforming the educational process from the management of education to the instructional delivery and assessment with the adoption of various tools and strategies. At this juncture, it is important to explain what educational technology is. Educational technology is not a single entity, but a diverse array of technological devises and technology-based activities and practices (Selwyn, 2013: 5). Educational uses of digital technology encompass the use of Internet-connected computing devices such as laptop and tablet computers and "smartphones," as well as the institutional uses of these technologies in the form of virtual learning environments, electronic smart boards and so on. These technological devices could be used in the educational systems to support a diversity of forms of educational provision across the educational subsectors and the work-based training. Within the institutional contexts of school and university, much effort is put into the classroom instructional processes alongside the use of "blended" forms of online and offline provision of teaching as well as fully virtual provision.

The use of available technology in education will make education real. Sticking to the twentieth-century model of block-and-mortar schooling makes little sense when there are twenty-first-century technologies available that enable different instructional approaches and delivery systems. The key for educators is to figure out how to use technology to engage and instruct all children ensuring that learning takes place. Children love Facebook, Twitter, YouTube, texting, the Internet, and anything in the digital technology domain. The question is how can education become more personalized and adapted to individual needs?

Blogs, Wikis, and Social Media

The development of new tools such as blogs, wikis, and social media has altered the way people and organisations convey information to one another. People are increasingly not dependent on experts to share information. Anyone who knows how to use these tools can now access most information unhindered. Following the wind of change, education will gradually shift away from experts and capacity building to focus on networks, which is the future. New dissemination tools, such as education blogs, wikis, and social media, give educators the potential to liberate education and learning from the clutches of experts. The issue is how to utilize this potential.

Blogs and the liberation of learning

Weblogs (blogs) is a tool for people to share information and express their views. Many Internet sites offer features that allow individuals to write opinions, make comments on daily events, and provide news coverage of breaking developments (Kline and Burstein, 2005). The blogs when used in an instructional mode can strengthen the interactive quality of issues being discussed. Furthermore, it can also enable multidirectional communications among educators, students, and the interested public. Blogs can be used to advance the objectives of education in the following ways:

- They can provide mechanism for the dissemination of education related news.

- They can also be used by education/school authorities to engage parents and the stakeholders in dialogues as it will enable community members to offer their opinions/feedback on education issues.

- They can be used as repository for organizing various education issues (learning, budgeting, reforms, and test development, teacher education) to present and advance and collate community thinking on these issues.

- They can be used at the instructional level to allow collaboration between students and teachers, between students and students, and between teachers and teachers. Blogs can be used to showcase the works of children. To share is to learn.

Wikis

Wikis are a type of Web site that can be edited on a platform by a multiple numbers of users (Tapscott, 2008). Wikis allow individuals from varying backgrounds to work together to accumulate knowledge and offer opinions on the issues being discussed. Participant comments are shared collectively and are subject to group revision. Wiki interactions depend on a type of collaboration known as crowdsourcing, which suggests that knowledge can best be created by vetting discrete bits of information through crowds and taking their collective judgment as the best wisdom.

Wikis can be used for educational purposes. Here are some potential ways that schools can explore wikis.

- They can be useful for distance education and social work courses, especially with group projects that involve sharing views regarding applied practice experiences.

- They can be used to foster collaboration of students on research projects in the sciences, enabling them to share observations with one another and add comments and references to enrich reports.

- They can be used to organize the collaboration of teachers and students to assemble relevant course materials through a wiki interface.

Social Media and Mobile Devices

Social media refers to online services, mobile applications, and virtual communities that provide a way for people to connect and share user-generated content and to participate in conversation and learning.

There are many social media services on the Internet. The table below shows some of the main ones.

Table 1. Main types of social media services

Name	Function
Facebook	This is one of the largest social networks globally. Registered users create a personal profile and add other users as friends. They can exchange messages and photos. It allows people to recommend books, videos, and media articles to other people.
Pinterest	This is a very fast-growing social media. It is a virtual pinboard where you can individually (or as part of a team) post notes that contain text, links, and embedded video. Users download the application free on their computer, tablet, or smartphone and start adding "boards" or similar folders.
Twitter	This is a microblogging platform where users send 140 character messages to each other.
Instagram	This is a social network based on sharing pictures and fifteen-second videos, which can be posted to other social media sites.
Wikipedia	This is an online encyclopedia that anyone can edit.
YouTube	This is a video-sharing Web site.
Evernote	This is a digital tool for content creation. It can be used by students to save a collaborative collection of Web sites, pictures, references, and any other digital assets in organized folders.

In the spirit of global village, these platforms have the capacity to knit together discreet individuals and enable them to communicate recommendations, reactions, or remedies to others who have signed up at that site. This has great potential for transforming education. There are no better tools for students than the connectivity and interactivity of these kinds of outlets provided by the social media. They enjoy the opportunity to become part of specialized networks based on shared interests. This is especially true in educational settings. Generally, it seems clear that social media has the following characteristics: (a) challenges traditional models, (b) allows people to communicate, (c) allows people to collaborate, (d) gives people an audience, and (e) is open and transparent. These characteristics have enhanced many countries' current education reform objectives as they deal with the survival skills of the twenty-first century. It could be argued that by embracing the use of social media tools for learning and teaching, we can start to build a culture that may help contribute to the reform of our school systems.

Some of the specific uses of social media and mobile devices to education are:

- Educators can use Facebook, which has become a major source of interactive discussion, to interact with students on key classroom ideas. Facebook and other social networking tools make it possible to extend conversations virtually and reach large numbers of individuals.

- Twitter used in the classroom can help to engage students and educators in the learning process. This media could serve as a "back channel" for instructors after class. The tool can be used to share resources, promote brainstorming, extend class discussion, and promote in students a sense of community. An educator may use Twitter to create and update their class reading or news lists.

- Educators can use YouTube to reach many children with the services of best teachers from anywhere in the world. The lectures can be repeated several times until there is full understanding. Students can come back to the lessons at the time of revision for examinations. On YouTube students can be assigned for video homework.

Student Assessment

Technology can help immensely in improving educational assessment. In many African countries, educational assessment has historically been focused on annual student tests. At the end of the year or in specified grades, children take standardized tests measuring progress in the subjects. The results of these exams are compiled and released to parents and school administrators. These exams represent useful ways to evaluate student performance. But evaluation based on standardized testing is seriously flawed (Ravitch, 2010). In addition, standardized tests are not linked to any particular educational materials, so it is hard to know which instructional sets produced particular test results (Loveless, 2011). Educators want to know what made the difference; is it the curriculum or the teaching styles or classroom management?

Digital technologies offer opportunities for multiple measures of student performance. They can enable teachers to provide feedback at virtually every step of the learning process and use this regular evaluation to gauge progress. Through online means, teachers can look at not just what concepts students have mastered but also how much time they have spent on reading, solving math, and their contributions to the conversations in the social media on the instructional issues. Transformation of the assessment process could be a major way to improve learning. The virtue of the digital technology is that it enables assessment of each piece of the learning process. Through online devices, it is possible to increase the range of skills and concepts assessed and the manner and frequency by which these evaluations are undertaken.

With the aid of digital technologies, the district or local government area or schools can create data-sharing networks where information regarding students' performance and learning are stored and analyzed. Rather than rely on periodic student assessments or test performance, instructors can analyze what students know and how they master particular concepts. By using online databases, teachers can analyze learning in real time and in ways that are useful for the students.

Evaluating Teachers

Technology has increased the possibilities of teachers to be evaluated in ways that have more direct links to academic achievement. No longer are administrators limited to focusing on seniority, professional credentials, and head teachers' ratings. Instead, with the aid of technology, they can draw on information systems that link teacher performance to student test scores, engagement, and participation.

Teacher quality is one of the most important variables of children educational achievement. Teachers who (a) know their subjects and how to interact with children, (b) are committed to the profession, (c) and care about the learning of children will make a positive difference in education outcomes. But evaluating teachers is very complicated. There is still no consensus of what constitutes good teaching performance. What is currently being practiced is that teacher rewards are tied to professional credentials and seniority. But neither professional credentials nor seniority correlate highly with teaching and learning effectiveness.

With digital technology, it is possible to consider new and much more comprehensive approaches to teacher evaluation. Digital technology has the potential to enhance the ability to aggregate data, merge information, track student activities, and consequently monitor performance. As indicated earlier in the social media section, it will be possible to compile data on student engagement and participation in the learning process. The various social media devices can observe and record how long students devote to electronic devices, how they collaborate with other students, how long it takes them to master key concepts, how many valuable comments they submit to online discussions, and the extent to which they participate in team projects. By focusing on both process and outcome measures, administrators enhance their assessment metrics and incorporate more nuanced indicators of how individual teachers have contributed to student education (West, 2012). Technology enables school officials to measure a range of different factors that are important to student performance. They can look not just at test results but also at various instructional activities that may be linked to strong academic performance.

Distance Learning

Distance learning is the education that takes place via electronic media, linking instructors and students who are not together in the classroom. Distance learning offers the potential to reduce regional disparities and promote greater educational opportunity among underserved populations (Christenson, Horn, and Johnson, 2008). Distance learning is a relatively recent development, and so there is much about it that is not well understood.

There are various types of distance learning: (a) those that are Web-facilitated, (b) blended or hybrid (classroom plus online), (c) fully online. At the moment, most institutions in industrialized countries mainly use the blended approach, and most schools in Africa are still stock with the traditional brick-and-mortar model. In all societies the delivery of education has not moved away from the traditional model. What we currently have are various degrees of mixed approaches. Many schools are most comfortable injecting technology into the traditional instructional approaches.

It is very important to note that even with the restraints, distance learning (as indicated in the case studies) is gradually coming through in both the developed and developing countries. Let us look at university education. Six of the best colleges and universities in the United States have combined to post their courses on line for free (https://www.edx.org): Harvard, Massachusetts Institute of Technology (MIT), Berkeley, Georgetown, Wellesley, and University of Texas (Paul, 2013). Several universities in Africa and other parts of the globe are moving in this direction of online education. This points to the future of what will happen to classroom education.

One of the key questions about distance learning concerns its impact on student learning. Many studies tend to show that students in online learning conditions perform better than those receiving face-to-face instruction because online education will have additional learning time that facilitates student instruction (Means, 2010). On the cost side, online learning will reduce the cost of educating students to levels undreamed of some years ago. Just look at what is happening to all the post offices globally. They have all succumbed to e-mail, Facebook, etc. Schooling as we know it today will follow the same path.

Using Technology to Close Educational Inequalities

Throughout most history, good quality education has been the preserve of the few. As a result, regrettably, only a small part of our collective potential has been put to use. Today's rapid social, economic, and technological developments have brought us to a situation where quality education should no longer be seen as a privilege but as a right for everyone.

Despite recent global educational improvements, various forms of inequalities in education persist, particularly in developing countries. At the macro regional level, education inequalities cluster regionally in sub-Saharan Africa. While Southern Africa has relatively small differences, a belt stretching from West Africa, across Central Africa, and into East Africa, specifically Kenya, displays extensive ethnic, religious, and rural/urban education inequalities. At the micro level, in most countries in Africa, there are vast inequalities in education between ethnic and religious groups and rural and urban populations. In a recent USAID report (2014), across the thirty countries in Africa studied, urban residents have on the average one more year of education compared to rural residents. Among the four focus countries studied—the Democratic Republic of Congo (DRC), Liberia, Mali, and Nigeria—the greatest relative disparity in any group comparison is found in Mali, where women in rural areas have an average of only 0.6 years of schooling, compared to 3 years in urban areas. In Liberia, the average years of education among urban women are three times higher than among rural women (5.1 years compared to 1.7 years).

Overall, Nigeria has the most severe inequalities across all the different group comparisons (USAID, 2014). Among Hausas, the largest ethnic group in Nigeria, women have on average 1.6 years of schooling as compared to the Yoruba women, who have on average more than 9 years of schooling. A similar disparity exist for Nigeria's other large ethnic groups: Igbo women also have on average more than 9 years of schooling while Fulani women have on average only 0.8 years.

To a large extent this reflects a difference between Christians and Muslims across the West African subregion. In Liberia and Mali, Christian women have more than double the average years of education of Muslim women, while in Nigeria the average years of education is more than three times higher among Christians when compared to that of Muslims. Some of the conflicts in these places to a large extent could be related to the intergroup and intragroup inequalities in education in societies where the level of education determines who gets what.

How can technology help in closing the educational gaps? Recent years have seen extraordinary and accelerating developments in pedagogical potential of ICT, to improve traditional school teaching and learning methods at all levels, and to offer greater diversity in the delivery of instructions. Active participation in this process of continuous change is crucial for African countries where deep-seated social inequalities, deeply rooted in demographic, economic, cultural, and religious factors, still persist. ICT brings to education the capacity to reach massive audiences with consistent quality content, and to target groups with specialized needs. The use of the new technologies in Africa and especially Africa could contribute to solving traditional learning gaps and reduce the educational inequalities of the population and consolidating a national education system that offers quality services to all sections and sectors of society.

ICT could have positive impact on education in two ways: (a) ICT may help significantly to increase delivery and coverage of educational services to the different segments of the society, by offering more varied and flexible programmes and being able to respond to an increasing demand; (b) It could have considerable impact on the quality of education, in as much as it transforms the traditional teaching-learning process, to the point where a cognitive gap emerges between teachers and students with access to ICT and those without. It is important to look at some examples of how ICT has been used to universalize education

ICT for the Education of the Most Disadvantaged Children

Inequalities in education are further compounded by difficulties in delivering educational services to the disadvantaged children. Reaching the most disadvantaged children has become a national and international education priority as directed in the Education for All. The key question is: who are the most disadvantaged children? The most disadvantaged children can be categorized as follows (a) children with learning disability (b) those that have dropped out of school. One of the crucial indicators of a failed national system of education is when large numbers of children and youth including those who are disabled or from poor economic background enter school but leave before they complete their education.

Children with Disabilities

Most countries in Africa as part of the EFA implementation policy has passed policies that are designed to deal with the education of children with various aspects of disabilities. In most of the cases, the policies seek to integrate disabled students into the education mainstream rather than separate them from regular classes. The schools in Africa are not very equipped to deal with children having learning disabilities. Overall about 12% of students in African primary and secondary schools have one form of learning disability. According to the U.S. Department of Education, a learning disability is defined as "a disorder in one or more of the basic psychological processes involved in understanding or in using language, spoken or written, that may manifest itself in an imperfect ability to listen, think, read, write, spell, or do mathematical calculations" (NCES, 2011:32).

Technology offers the possibilities to resolve the long debate between mainstreaming and special-needs classes. Those who favor mainstreaming have argued that disabled students benefit from learning with fellow students. Those against claim that the needs of disabled students are best met in separate classes. But technology brings together the claims of the alternative approaches. Since technology allows for personalisation within the context of general educational programming and delivery, it enables those with special needs to be supported without taking them out of the classroom. It also creates opportunities for ongoing remedial support after school hours.

Technology offers the potential to aid these children through programs that allow students to tailor learning to their own requirements. Digital technology will enable teachers and administrators to tailor education to the pace and learning styles of individual students. As an example, educators have found that Apple's iPad tablet offers hope for autistic students who have difficulty expressing themselves verbally. Through an application called Proloquo2Go the student can tap on common screen images from the home, school, or store to request particular items (Howard, 2011). With this kind of user interface, students who have problems with verbal expression can expand their communication capabilities.

Out-of-school Youths

Africa has serious dropping-out and out-of-school problems. This is most serious in Nigeria. Despite the fact that primary education is officially free and compulsory, 8.7 million children of primary school age remain out of school—34.3% of primary age children (UNICEF and UNESCO 2012). Combined with junior secondary children out of school, 10.1 million children aged 5–14 are not in school. Education indicators for northern Nigeria are worse than for the rest of Nigeria, partly driven by demographics and the number of children who should be in school, partly by social attitudes towards "Western" education, and partly by the difficulties experienced by governments in ensuring provision in predominantly rural local government authorities (LGAs).

Examining the out-of-school population, children who are expected never to enter school are a bigger issue than children who drop out of school at the primary and junior secondary levels. These children form the largest portion of out-of-school children and are permanently excluded from school at all levels of the education system. Of primary-aged children, 74% are expected to never enter school while only 5% are classified as dropouts. At the junior secondary level, the situation is much the same with 76.9% or 2.2 million expected never to enter school while 22.1% or 625,000 drop out of school (UNICEF, 2012).

The issue is how digital technologies can be used to reach this group. In order to understand the potential that technology may have to reach disadvantaged learners, we must understand who these children and youth are and why they drop out of school. The dropout crisis is not just an education issue. It isn't just about schools, academics, curricula, and learning. It is also about context, communities, and families that collectively ensure that the needs of children are met.

Some Case Studies of Technology in Education

Technology has started to change instruction. Online courses and cyber schools are the most dramatic manifestations as they take students out of the traditional classrooms and brick-and-mortar schools. You can run schools from anywhere. Below are some examples of how technology is transforming schooling.

Advanced Academics in Bricktown District of Oklahoma, USA

The school in a small renovated warehouse in Bricktown District provides public education over the Internet to students in 29 states, 140 school districts, and 7 virtual schools from California to New Jersey and Alaska to Texas (Moe and Chubb, 2009). Sixty thousand high school students took courses there during the 2006–07 school year. The warehouse is home to a top-notch team of technologists. They create the platform on which courses are delivered and tests are administered and scored, and grades reported to state and local school systems. The warehouse is also home to about thirty teachers in all. Every course is supported by a teacher who is fully certified.

The teachers instruct their students as they work through digital lessons or complete assignments. Some of the instructions come through written "instant messages"; some occur via Whiteboard correspondence, with both teacher and student sketching ideas on the same electronic surface; some involve Internet phone calls. The teachers typically support four to five students at a time from their computers, providing a level of individual attention they could never offer to a regular high-school class.

Teachers like the informal atmosphere of the warehouse, where they easily interact all day long, sharing student challenges and brainstorming strategies. They enjoy the flexible hours. They can choose from a variety of hours, as students take their online courses around the clock, day and night.

The Khan Academy

The Khan Academy is a nonprofit educational Web site created in 2006 by educator Salman Khan, a graduate of MIT and Harvard Business School (Paul, 2013: 110). The stated mission is to provide "a free world-class education for anyone anywhere" (Khan, 2012: 7).

The Web site supplies a free online collection of more than four thousand micro lectures via video tutorials stored on YouTube teaching mathematics, history, health care, medicine, finance, physics, chemistry, biology, astronomy, economics, cosmology, organic chemistry, American civics, art history, macroeconomics, and computer science. Khan Academy has delivered over 240 million lessons; it is now the premier teaching site on the Web (Paul, 2013). The Khan Academy which started with one student—Nadia—the cousin of Salman Khan, who is now helping to educate more than six million unique students per month. The videos had been viewed over 140 million times, and students had done nearly half a billion exercises through the software (Khan, 2012).

Innovative School Access Program (SAP) in Nigeria

Nigeria historically has been offering free primary education for the past half century (Agu, 1986). But the country's education system has not been able to provide access to good quality education to all school-aged children. Only 44% of children in grades 7–12 attended school in the years 2007–2011. To address these challenges, Intel Education created the comprehensive School Access Program (SAP) education solution, which facilitates learning through modern digital technology.

SAP's objective is to entrench the use of ICT as a tool for teaching and learning, and to serve as a model for future programs that will also use technology to advance education within the country. The five main components of the program are:

1. Strengthening the technological infrastructure with supply of PCs and related hardware, as well as Internet connectivity

2. Enhancing schools' ICT readiness through refurbishing classrooms and deploying technology infrastructure, including electricity and other facilities

3. Creating relevant integrated software with local education content using e-learning solutions as well as sample lesson plans from Intel

4. Professional development of teachers through a five day training period

5. Training and support for suppliers and engineers to ensure proper delivery, installation, and ongoing maintenance

Intel first piloted SAP in a single Nigerian government school in 2007. The lessons learned from the pilot were used to guide the national rollout, which has reached three thousand schools in 2013. An assessment of the impact of SAP in 2012 showed positive outcomes on students and teachers. One of the most significant improvements seen in the schools was a rise in Senior Certificate Examination scores in English and physical sciences (Takang, 2012). The study also found that the SAP program had a strong impact on schools: (a) half the schools made additional investments in Intel Education Solution; (b) outside of class, 74% of teachers used the Intel Education Solution for research; and (c) teachers expressed confidence in their ability to use the Intel Education Solution.

Mexico: Television-Assisted Telesecundaria Programme

Telesecundaria was designed to meet the educational needs of hard-to-reach rural areas in Mexico, mostly communities of under 2,500 inhabitants. The main characteristics of Telesecundaria have always been: (a) using television to carry most of the teaching load, and (b) using one teacher to cover all subjects, rather than the subject matter specialists used in general secondary schools. At first it was offered in a few states (there are 31 plus the national capital), with a little over 6,000 students. In 2000, the programme was available at 13,851 locations nationwide, and served over 1,043,000 students and employed over 46,000 teachers.

Educational television has been the mainstay of the program throughout the years of operation. However, the mode of use of television has gone through three evolutions. The first stage involved a regular teacher delivering lectures through a television set installed in classrooms. Books and workbooks were provided to follow television program with exercises, revisions, applications, and formative evaluations. The second stage improved on this process and created programs with greater variety. The third stage, which began in 1995, deployed a satellite to beam the program throughout the country and used a wider range of delivery styles. Telesecundaria is now an integrated and comprehensive program providing a complete package of distance and in-person support to students and teachers. Telesecundaria has proved to be cost effective. The flow rates of Telesecundaria are found to be better than those of general secondary schools.

Interactive Radio Programme for Pastoralists in Nigeria

The nomadic population in Nigeria accounts for 9.4 million people, out of total population of 148,980,000 including 3.1 million school age population (UNESCO, 2008). The participation of the nomads in regular national education systems used to be extremely low. Access to radio and television as information and communication tools is very high in Nigeria, especially in Northern Nigeria. Through the National Communication for Nomadic Education, Interactive Radio Programmed (IRI) was launched in 1992 to provide open and distance education to pastoral nomads (Adeosun, 2010). Using the Federal Radio Corporation of Nigeria (FRCN), Kaduna, particular hours of the day are dedicated to air participatory instructions on basic functional literacy, numeracy, health and environmental education, modern techniques of animal husbandry, and civil responsibilities.

The objectives of using the radio for nomadic education were to (a) mobilize and sensitize the nomadic pastoralists to appreciate the value of modern education, (b) encourage nomads to contribute meaningfully towards the education of their children, (c) motivate nomads (both men and women) to enroll in adult literacy programmes, and (d) improve the quality of teaching and learning, particularly where performance is low and teachers are poorly trained.

The radio program is participatory, making it widely accepted by the nomads. They listen to this program which contains weekly news, interviews, discussion, music, drama, etc. There are also school-based IRI programs to improve quality of teaching and learning where performance is low and teachers are having challenges. As a result of the innovative strategies adopted by the commission, there have been great improvements in the quality of curriculum content delivery, with overall improvement in the learning achievement of nomadic schoolchildren and adults. The program is so successful that USAID adopted it as a strategy to improve literacy and numeracy skills of pupils in Lagos, Nasarawa, and Kano states in its Community Participation for Action in the Social Sector's (COMPASS) programme.

Digitizing Almajiri Schools in Nigeria

This is an initiative that is very current. Almajiri in the Islamic tradition arose from Alima Jirud, people who are moving from one part of the world to the other, looking for Islamic or Arabic knowledge. In Nigeria, this group of children has historically been not fulfilling their rights to the secular education opportunities. According to the Ministry of Education, it is estimated that there were 9.5 million almajiri children in the northern part of Nigeria in 2010 (Taiwo, 2013). From a rights perspective, it meant that about 9.5 million children mainly from Northern Nigeria have their rights to education being denied.

In April 2012, the Nigerian government formally signed into law Almajiri Education. The government set up a National Committee on implementation of Almajiri Education Programme and charged committee with integrating the almajiris into the UBE programme. Recently, the National Office for Technology Acquisition and Promotion (NOTAP) has begun the process of digitizing all the almajiri schools across the country. Building on tradition, the plan is to create a digital aloha. Aloha is the old almajiri slate. The digital slate will be embedded with different software programmes. Children could learn Arabic, math, English, healthy behaviours, etc. All forms of educative videos can be uploaded into the tablet for use. This has the potential of opening the global knowledge to the children. This could go a long way in liberating learning not only for the almajiris but for all. Successful implementation of digitizing almajiri education programme and enabling these children to acquire twenty-first-century skills could provide the model for Nigeria and other countries in similar educational situation to not only achieve universal basic education but also universal secondary education.

A Framework for Implementing the Integration of ICT in Education in Nigeria

All countries in Africa including Nigeria are signatories indicating commitment to the World Declarations on education (Education for All and the Millennium Development Goals). The NEPAD e-schools Initiative led by the e-African Commission has this as the key objective: "Ensuring that young Africans participate actively in the global information society and knowledge economy."

The Nigerian National Policy on Education National Policy on Education (2004) recognized education as instrument *par excellence* for affecting national development, through the acquisition of appropriate skills, abilities, and competencies, mental and physical, as necessary for the individual to live in and contribute to the development of his/her society (Federal Government of Nigeria, or FGN, 2004: 7). It stipulates that education has to be tailored towards self-realisation, individual and national efficiency, national unity as well as social, cultural, economic, political, scientific and technological progress (NPE, 2004: 7). The National Policy specified that ICT be integrated at all levels of the Nigerian education system. As examples, the policy stipulates that:

1. All states, teachers' resource centres, university institutes of education and other professional bodies in education shall belong to the network of ICT (FGN, 2004: 53).

2. Government shall provide facilities and necessary infrastructure for the promotion of ICT and its use as learning tools at all levels of education (FGN, 2004: 53).

The universal basic education (UBE) policy also supported the integration of ICT in the educational process. On the use of ICT to improve quality education and universal knowledge, Article 28 of the UBE policy had this to say: "UBE is also an opportunity for Nigeria to confront head-on the challenges of and to take full advantage of the possibilities offered by new information and communication technologies for improving the quality of education. The information age is also the age of knowledge. No school system can afford to stay outside the knowledge age while serving world that is now run by knowledge. The way out of the dilemma is the integration of computer appreciation, computer literacy, computer applications into UBE (Federal Government of Nigeria, 2003: 28).

Furthermore, the National Policy on Teacher Education (2007) envisioned the use of ICT in teacher development to: "produce quality, highly skilled, knowledgeable and creative teachers based on explicit performance standards through preservice and in-service programs to raise a generation of students who can compete globally" (Federal Ministry of Education, 2007: 6). Prior to the education policies endorsements of the use of ICT to improve the educational process, the National Policy for Information technology (2001) in its mission statement recognized the need to use IT for education (Federal Ministry of Science and Technology, 2001: iii). In addition, on the general objectives of the policy, 3 out of 31 focused on integrating ICT into the mainstream education and training.

The entire above are indicative of strong Nigerian governmental commitment to the use of ICT in improving the educational process. What is needed to take the process forward is a comprehensive implementation policy for ICT in education. There are some best practices nationally and globally as indicated in the case studies above. Other national examples that could be examined for consideration are: (a) School Net Nigeria; (b) Computers in-Schools Project; (c) One-laptop-per-child (OLPC).

Constraints to the Use of Technology for Improving Education Quality

There is no question of the use of technology to improve the quality of education in an information age. There is almost a consensus that the use of new technologies in the classroom is essential for providing opportunities for students to learn to operate in the global information age. We will argue that countries that do not incorporate the use of new technologies in schools cannot seriously claim to be preparing their students for life in the twenty-first century. However, getting a school system to adjust and change in the direction of ICT integration with schooling processes, like any change, has been and will be difficult. There are a lot of constraints with respect especially to the integration of ICT into teaching and learning. The constraints could be categorized into two: (a) teacher-level constraints and (b) school-level constraints.

Teacher-Level Constraints

These constraints are capability issues and are specific to the teachers. They will include: (a) lack of teacher competence, (b) lack of teacher confidence, and (c) teachers' resistance to change.

Lack of Teacher Competence

Many teachers especially in developing countries lack the knowledge and skills to use computers. A worldwide survey conducted by Pelgrum (2001), of nationally representative samples of schools from twenty-seven countries, found that teachers' lack of knowledge and skills is a serious constraint to using ICT in primary and secondary schools. Findings from a survey carried out in twenty-seven European countries show that teachers who do not use computers in classrooms claim that "lack of skills" are a constraining factor preventing teachers from using ICT for teaching (Empirica, 2006). Teachers' competence is directly related to teacher confidence, as explained below.

Lack of Teacher Confidence

A key constraint that prevents teachers from using ICT in their teaching is lack of confidence. Lack of teachers' confidence is complex and has many interlocking explanations: (a) fear of failure by teachers could be a contributory cause (Beggs, 2000); (b) limitations in teachers' ICT knowledge make them very anxious about using ICT in the classroom of children who perhaps know more than they do and so they are not confident to use it in their teaching (Balanskat et al., 2006).

Teachers' Resistance to Change

Teachers' attitudes and an inherent resistance to change have been found to be significant constraints (Schoepp, 2005). The reasons for resistance by teachers vary. One key area of teachers' attitudes towards the use of technologies is their understanding of how these technologies will benefit their teaching and their students' learning. Another area could be teachers' belief of not being supported, guided or adequately rewarded in the integration of technology into their teaching (Schoepp, 2005). We will like to argue that teachers' resistance to change could be an indication that something is wrong. The most important thing is to find out what is wrong by talking to the teachers.

School-Level Constraints

The school-level constraints are institutional in nature including: (a) lack of time, (b) lack of effective training, (c) lack of accessibility, and (d) lack of technical support.

Lack of Time

Time is a big constraint for teachers as the time for schooling is already allocated. How time impacts on teachers' efficient and effective use of computers in teaching and learning depends on the stage of ICT education integration. For the school systems that have computers in schools, the issue could be the difficulty in scheduling enough computer time for classes and the time for teachers to explore the different Internet sites or look at various aspects of educational software. For the teachers in the school systems that are at the basic stage, the time issue could be time to learn and understand the concepts and techniques that accompany the use of technologies. Use of technology in education will require initial time allocation to understand, plan, and interact with the program.

Lack of Effective Training

Lack of effective training is a constraint. The success of ICT integration in the schooling process depends on the adequacy of the training. The issue of training is certainly complex because it is important to consider several components to ensure the effectiveness of training. For the trainings to be adequate, they should include: (a) training in digital literacy; (b) pedagogic and didactic training in how to use ICT in the classroom. According to Newhouse (2002), teachers need training in technology education (focusing on the study of technologies themselves) and educational technology (support for teaching in the classroom). Where training is ineffective, teachers may not be able to access ICT resources.

Lack of Accessibility

Inadequate access to resources in schools, including home access, is another complex factor constraining teachers from integrating new technologies into education. Inaccessibility of ICT resources could be as a result of (a) unavailability of the hardware and software or other ICT materials within the school and home for a many of the developing countries and (b) poor organisation of resources, inappropriate software, and poor-quality hardware. Accessibility of ICT resources may not guarantee its successful implementation in most African schools because of the main issue of adequate electricity supply to power the utilisation of the resources.

Lack of Technical Support

One of the top barriers to ICT use in education is lack of technical assistance (Pelgrum, 2001). The new technologies have quite some naughty technical problems such as: Web sites not opening when needed, failing to connect to the Internet, printers not printing, malfunctioning computers. These problems are constantly being faced by most organisations including the international organisations working in most African countries. Consequently, these organisations tend to employ technical computer experts to assist in solving these endless reoccurring problems. ICT integration in schooling will definitely encounter these problems. These barriers have impeded the use of ICT in schools that have been trying it and will increasingly constrain its introduction in many African countries given the state of the maintenance culture in Africa which is low. This will be compounded by the fact that teachers in Africa will definitely have to work with donated computers which in most cases will be old.

Conclusion

Technology for the African Schools of the Future

In a very short time, the world changed, becoming flat and increasingly changing very fast, changing the skills that used to be important for our survival. There is no question that the socialisation agencies—family, schools—should adapt very quickly to be able to cope with the new realities. With respect to schools, we argue that the methods of teaching and learning must adapt to these changes: (a) All students need skills to thrive in a global knowledge economy; (b) In the age of the Internet, using new information to solve new problems matter more than recalling old information; (c) Today's children and youth are differently motivated when we compare them to previous generations.

Over the next twenty years, the African continent will experience huge growth in the number of young people in its society. These young people will need to be healthy and educated in order to be competitive and productive. Investment in quality education that targets the poor and hard to reach groups will be very critical in influencing the direction the continent will take. In Africa, almost everyone believes education provides hope. It is a huge achievement that in the continent of Africa, almost all children of primary age will soon have a place at a school, except the countries having internal conflict. But they need to be armed with twenty-first century skills to be productive.

However, to meet the aspiration of making the twenty-first century an African one, schools in Africa have to be innovative in dealing with the double achievement gaps that the African educational systems have to contend with. These are (a) the gap between the quality of schooling that most children from the middle class get and the quality of schooling available for most poor and vulnerable children and the consequent disparity in outcomes, and (b) the global achievement gap, which is the gap between what even our best schools are teaching and testing versus what all students will need to succeed as learners, workers, and citizens in today's global economy.

African schools for the future should be able to deal with the challenges posed by (a) and (b) simultaneously. This will entail a wider range of strategies that will improve the quality outcomes of schooling, at the same time providing the twenty-first-century skills and capabilities people will need to work in the increasingly knowledge driven economy. In Africa it will take many decades to improve the existing public school system to an acceptable level, if we follow the traditional school improvement models. As a conclusive argument and suggestion, the most relevant strategies could be to exploit the opportunities of new technologies, particularly the mobile phone, to make learning possible in new ways. In that way Africa will be teaching today's students for tomorrow and not yesterday (according to Dewey). This will entail Africa adopting the policy of *"Going beyond Education for All* to *Education Above All"* as recommended by Professor Emeritus Pai Obanya.

References

Adeosun, O. (2010) "Quality basic education development in Nigeria: Imperative for use of ICT" *Journal of International Cooperation in Education,* Vol.13, No.2 pp. 193–211 (CICE Hiroshima University).

Agu, A. O. (1986): The Implementation of universal primary education in Nigeria: Nation states and schools. Unpublished Doctoral Thesis, Harvard University Cambridge, USA.

Balanskat, A., Blamire, R., and Kefala, S. (2006). *A Review of Studies of ICT Impact on Schools in Europe.* European Schoonet.

Beggs, T. (2000). "Influences and barriers to adoption of instructional technology" Paper presented at the Proceedings of the Mid-South Instructional Technology Conference, Murfreesboro, TN. April 9–11 2000.

Castells, M. (2006) "The network society: From knowledge to policy" In Castells, M. and Cardoso, G. (eds.) *the Network Society: From Knowledge to Policy.* Washington DC: John Hopkins Centre for Transatlantic Relations pp. 3–22.

Chakravartty, P., and Sarikakis, K. (2006). *Media policy and globalization.* Edinburgh: Edinburgh University Press.

Christensen, C., Horn, M., Johnson, C. (2008) *Disrupting class: How disruptive innovation will change the way the world learns.* New York: McGraw-Hill.

Dale, R. (2005) "Globalization, knowledge economy and comparative education." *Comparative Education,* 41, (2) pp. 117–149.

Dewey, J. (1944). *Democracy and education* New York: Macmillan.

Empirica (2006). *Benchmarking access and use of ICT in European schools 2006: Final report from head teacher and classroom teacher surveys in 27 European countries.* Germany: European Commission.

Federal Government of Nigeria (FGN) (2004). *National policy on education* Lagos: NEDRC.

Federal Government of Nigeria (FGN) (2003). *Universal basic education act* Lagos: NEDRC.

Federal Ministry of Education (FME), (2007) "Current policy reforms for teacher education," Paper presented at the Federal Ministry of Education/NUC National Workshop on *Tertiary Education Financing: Which Way Forward* Lagos: University of Lagos 23–24 April 2007.

Federal Ministry of Science and Technology (2001). *National policy for information technology*

http://www.nitda.gov/docs/policy/ngitpolicy.pdf.

Howard, J. (2011). "App Called Proloquo2Go. Whets Appetite of Non Verbal Child," *Montreal Gazette,* April 12.

Khan, S. (2012). *The one world school house: Education re-imagined.* New York: Twelve Hachette Book Group.

Kline, D., and Burstein, D. (2005) *Blog: How the newest media revolution is changing politics, Business and Culture* Weston, Conn.: Squibnocket Partners.

Loveless, T. (2011). *How well are American students learning? The 2010 Brown Centre Report on American Education* New York: Brookings Press.

Means, B. (2010). "Evaluation of evidence-based practices in on line learning: A meta-analysis and review of online learning studies" Office of Planning, Evaluation, and Policy Development, U.S. Department of Education.

Meyer, J., and Jepperson, R. (2000). "The actors of modern science: The cultural construction of social agency." *Sociological Theory* 18 (1): p. 100–120.

Meyer, J., Ramirez, F., and Soysal, Y. (1992) "World expansion of mass education, 1870–1980." *Sociology of Education* 65 (2), p. 128–149.

Moe, T., and Chubb, J. (2009). *Liberating learning: Technology, politics and future of American education.* San Francisco CA: Jossey Bass.

Newhouse, P. (2002). *Literature Review: The impact of ICT on learning and teaching.* Perth, Western Australia: Department of Education.

Obanya, P. (2015). "An African perspective in humanistic education." *Educationeering, no.1*

Paul, R. (2013). *The school revolution: A new answer for broken education system* New York: Grand Central Publishing.

Pelgrum, W. (2001). "Obstacles to the integration of ICT in education: Results from a worldwide educational assessment" *Computers and Education,* 37, p. 163–178.

Ravitch, D. (2010). *The death and life of the great American school system: How testing and choice are undermining education.* New York: Basic Books.

Rychen, D., and Salganik, L. (2003). "A holistic model of competence." In Rychen, D., and Salganik, L. *Key Competences for a successful life and a well-functioning society.* Gottengen, Germany: Hogrefe and Huber.

Sachar, H. (1990). *The course of modern Jewish history.* New York: Vintage Books.

Schoepp, K. (2005). "Barriers to technology integration in a technology rich environment. *Learning and Teaching in Higher Education: Gulf Perspectives,* 2(1), p. 1–24.

Selwyn, N. (2013). *Education in a Digital World: Global Perspectives on Technology and Education.* New York and London: Routledge.

Taiwo, F. (2013). "Transforming the Almajiri education for the benefit of Nigerian society" *Journal of Educational and Social Research* Vol. 3, No. 9 November p. 67–72. (MCSER Publishing, Rome Italy).

Takang, A. (2012). "Intel EMPG Summary Report: Nigeria Academic Impact Assessment Report" December.

Tapscott, D. (2008). *Wicinomics: How mass collaboration changes everything.* New York: Portfolio Hardcover.

Wilmarth, S. (2010). "Five socio-technology trends that change everything in learning and teaching" In Jacobs, H. (ed.) *Curriculum 21: Essential Education for a Changing World* Alexandria USA: ASCD p. 80–96.

West, D. (2012). *Digital Schools: How technology can transform education* Washington DC: Brookings Institution Press.

National Centre for Educational Statistics (NCES) (2011) "The condition of education, 2011" (U.S. Department of Education).

USAID (2014). "Conflict and educational inequality: Evidence from 30 countries in sub-Saharan Africa" (Final Report) prepared by the Aguirre Division JBS International.

UNESCO (2010) *EFA Monitoring Report, MDGs 2010* Canada: UNESCO Institute for Statistics.

UNESCO (1990). *World declaration on education for All: Meeting basic learning needs.* Paris: UNESCO.

UNICEF and UNESCO (2011). "Imperative for quality education for All in Africa: Ensuring equity and enhancing teaching quality" Background Paper Prepared for the ECOSOC Annual Ministerial Review (AMR) Regional Preparatory Meeting for Africa, Lome Togo, 12 April 2011.

Wagner, T. (2008). *The global achievement gap: Why even our best schools don't teach the new survival skills our children need and what we can do about it.* New York: Basic Books.

West, D. (2012). *Digital schools: How technology can transform education.* Washington D.C.: Brookings Institution Press.

Chapter 3

Integration of Technologies in Education: A Study on University of Benin Mass Communication Undergraduates

Ofomegbe Daniel, Ekhareafo,
B. O. J. Omatseye
&
Blessed Friedrick, Ngonso

Introduction

Developments in information and communication technologies (ICTs) have brought fundamental changes to different phases of human life. In the area of education, there appears to be a transition of emphasis from an analogue educational instruction to that of a digital knowledge-based education. It has become very imperative in the area of media education. Nzewi (2009) and Umoren (2006) observed that recent developments in ICT have drastically affected educational procedure for improved quality of education offered to students. ICT resources in instructional delivery in schools will serve a dual purpose and result in more efficient classroom instruction. The government's desire for a robust education that is in line with the global trend in ICTs led the federal government to review the National Policy on Education (1998) to the current one (NPE, 2004). This is with the intent to accommodate the introduction of ICTs into the school system and bring both the schools and their products to the realities of the changes occasioned by technologies.

Given this background, it is expected that students at all levels of education ought to be exposed to ICTs. With particular reference to students in mass communication, they ought to be provided with adequate ICTs which can offer them quality higher education. In addition, students need access to such technologies which could facilitate their learning experiences and prepare them for the world of work. This can only be attained when it is drastically integrated into the instructional process and the practical sessions the students need to deepen their knowledge of the nuances of the discipline. According to Jude and Dankaro (2012), productive instructional delivery enhances learners' creative and intellectual development through the use of ICT resources, for instance, in the use of multimedia images, graphics, audio, text, and motion for high-quality learning.

There is no gainsaying that ICT in media education has the potential to transform the way students learn about the current trends in mass communication since modern media practice is highly digitalised. This can be better realised through a well-developed media instruction and the necessary pedagogical competencies for instructional delivery through ICT resource utilisation.

Idemili and Sambe (2007) refer to ICT as the electromagnetic technology bewildering an array of interconnected forms. This includes micro-circuiting, micro-graphics, holographic memory, micro-electronics, optic fibre satellites, video discs, telex, view data, digital broadcast systems, teletext, facsimiles, videophones, computers, and microprocessors. On the other hand, Umoren (2003) from an education perspective, sees ICT resources as instructional delivery tools used to explore, investigate, solve problems, interact, reflect, reason, and learn concepts in the classroom. This innovation permits alternative types of educational patterns for facilitating the teaching and learning process. It is in this regard that Umoren conceived the ICT resources as the e-learning that uses an information network through the Internet.

In a similar vein, Jude and Dankaro (2012) noted that virtual teaching is another type of ICT resource which entails instructional delivery through teleconferencing. Web-based instruction uses the Internet and the World Wide Web (WWW) as the major components of learning materials and instructional resources for effective instructional delivery. Audio media (instructional slides and tutorials) are teaching and learning aids made and written into compact disks, graphics, and texts. Through the PowerPoint, instructional delivery is impactful to slow and fast learners.

Apart from the above, when we talk of ICT in media education, it is largely predicated upon the use of digital computers, the Internet, and other portable digital devices like mobile phones, iPods, MP3 players, and the digitalisation of communication. It is the awareness and use of ICTs or new media technologies for communication purposes. According to Olley (2009), citing Croteau and Hoynes (2003) who acknowledged Neuman (1999), "We are witnessing the evolution of a universal interconnected network of audio, video, and electronic text communications that will blur the distinction between interpersonal and mass communication and between public and private communication." They further argued that new media will:

- alter the meaning of geographic distance
- allow for a huge increase in the volume of communication
- provide the possibility of increasing the speed of communication
- provide opportunities for interactive communication
- allow forms of communication that were previously separate to overlap and interconnect

Several studies(Olley ,2009;   Croteau & Hoynes ,2003) have been conducted on the influence of ICTs in teacher education, broadcasting, and publishing; however, the focus in this study is specifically, the impact and relevance of ICTs in mass communication in terms of students' awareness, utilisation, availability, adequacy and accessibility, and barriers to their use. The case study focused on University of Benin mass communication students.

Statement of the Problem

Modern societies are now technology driven. The impact of ICTs can best be seen from the ease at which human activities are now being

61

carried out. As an area of academic discourse, studies on ICTs have sought to understand the genealogies of new media platforms and texts; tracing the distinct pasts of digital computers and the media, and understanding how these paths came to intersect and what role they play in modern communication and other human activities and endeavours.

An exegesis into the history of educational expansion in Nigeria suggests that development in the quality of education provided leaves much to be desired. Although the curriculum of media education may appear laudable, the defective implementation of the content often creates a wide gap between theory and practice. This can only be handled when media educators and students are provided with adequate ICTs that can facilitate the teaching and learning process, just as knowledge can only be enhanced through adequate utilisation of ICT resources that will lead to quality instructional delivery during lessons and student own self-practice.

Ordinarily, one would have thought that with the growing influence of ICT in today's society, adequate provision would be made in the training of future users of the technologies in the university programme of action for realising its strategic plan. Where this is the case, the university would have contributed in raising a generation of students who are not only acquainted with ICT but would have mastered the skills necessary to fit into the competitive world of work.

In the University of Benin, especially in the Mass Communication programme, it is difficult to determine the level of ICTs availability, access, and usage by the students, perhaps because one is not certain if the students have a media centre where the various ICTs could be housed for students learning experiences. Certain questions become pertinent in this regard: Do the students of mass communication use ICTs to support their studies and courses? Do the students have the requisite knowledge and skills needed for usage? Where they are available, and are they adequate? What is the greatest barrier to using ICTs by the mass communication students?

The challenges of new media technologies in media education cannot be overemphasised. In fact, it is the lack of exposure of students to the technologies that seem to hinder their capacity to fit easily into the industries. There is no doubt that Nigeria universities are still trailing behind in the race of acquisition of new media technologies for the purpose of education. Iroh, as cited in Agba (2001) and Olley (2009), lamented the pace of ICT development in Nigeria when compared to the rest of the world.

The nonprovision of adequate necessary ICTs in Nigeria universities which could facilitate access and usage of the technologies is far from encouraging, especially when compared to some other developing countries like South Africa, India, and other South American countries. This may be responsible for the sharp information imbalance between students trained with ICTs and those not trained. It is against this backdrop that this study seeks to answer the question, what is the role of ICT in education especially in the University of Benin Mass Communication Department?

Purpose

In a broader perspective, this study aims at determining the level of ICT resource utilisation, availability and accessibility by mass communication students in the University of Benin. Specifically, it seeks to achieve the following objectives: Find out if technologies are integrated in the teaching and learning of students of mass communication. The use of ICTs to support education is now the current trend in the world. For the purpose, the study considered the level of awareness of ICT resource by mass communication Students in the University of Benin. It determined the extent of ICTs resource utilisation by mass communication students in the University of Benin. It also tried to find out the influence of the integrated technologies on mass communication students, as well as determine the greatest barrier to using technologies by the mass communication students.

Research Questions

The following were the research questions formulated to guide this study:

1. Do the students of mass communication use ICTs to support their studies/courses?

2. What is the level of awareness of ICT resource by mass communication students in the University of Benin?

3. What is the extent of ICTs resource utilisation by mass communication students in the University of Benin?

4. What is the influence of ICT availability on mass communication students or the department?

5. What is the greatest barrier to using ICTs by the mass communication students?

Review of Literature

There is no gain saying the fact that the field of education has been affected by ICTs. This influence has in strong ways affected pedagogical issues in learning, teaching, and research. Jhuree (2005) contends that much has been said and reported about the impact of technology, especially computers in education. This is why Yusuf (2005), citing Lenke and Coughlin (1998), observed that ICTs have the potential to innovate, accelerate, enrich, and deepen skills; to motivate and engage students; to help relate school experience to work practices; to create economic liability for tomorrow's workers, as well as strengthening teaching and helping schools change.

The notable influence could be seen in the area of research; where the volume of research literatures is not made available on the Internet. Schools, which hitherto had old books in stocks now create ritual libraries connected to content providers regularly, update their research centres. Thus, researchers and students now find the Internet a valuable research resource with the volume of information that could be accessed from it. Young (2002) captures this position more appropriately; with the help of ICT, students can now browse through e-books; sample examination papers and previous years' papers and can also have easy access to resource persons, mentors, experts, researchers, professionals, and peers all over the world. This flexibility has heightened the availability of just-in-time learning and provided learning opportunities for many more learners who previously were constrained by other commitments.

Cabero (2001) noted that the flexible time-space accounted for by the integration of ICT into teaching and learning processes contributes to an increase in the interaction and reception of information. Such possibilities suggest changes in the communication models and the teaching and learning methods used by teachers, giving way to new scenarios which favour both individual and collaborative learning.

The integration of ICTs in education has been successfully described by Amin (2008) citing Older (2000) when he noted that conventional teaching has emphasised content, and for many years, courses have been taught with the use of ICT Teachers have taught through lectures and presentations interspersed with tutorials and learning activities designed to consolidate and rehearse the content. He observed that contemporary settings are now favouring curricula that promote competency and performance. Curricula are starting to emphasise capabilities and to be concerned more with how the information will be used than with what the information is. Contemporary ICTs are able to provide strong support for all these requirements, and there are now many outstanding examples of world-class settings for competency and performance-based curricula that make sound use of the affordance of these technologies.

Young (2002) also noted that mobile technologies and seamless communications technologies support 24/7 teaching and learning. He however observed that choosing how much time will be used within the 24/7 envelope and what periods of time are challenges that will face the educators of the future. Towing the same line of thought to a degree, McGorry (2002) noted that ICT has the potential to remove the barriers that are causing the problems of low rate of education in any country. It can be used as a tool to overcome the issues of cost, less number of teachers, and poor quality of education as well as to overcome time and distance barriers.

New Media Consortium (2007) asserts that ICT presents an entirely new learning environment for students, thus requiring a different skill set to be successful. Critical thinking, research, and evaluation skills are growing in importance as students have increasing volumes of information from a variety of sources to sort through. Idoko and Ademu (2010) also investigated the challenges of information and communication technology for teaching and learning as perceived by agricultural science teachers in secondary schools in Kogi State. Both found that ICT availability was one of the most important obstacles to technology adoption and integration in learning. They indicated that there is urgent need for more computers if a country is to successfully integrate ICT in public secondary schools.

In the same vein, Bottins (2003) asserts that the use of ICT can improve performance, teaching, administration, and develop relevant skills in the disadvantaged communities. Yoen, Law, and Wony (2003) noted that ICT improves the quality of education by facilitating learning by doing, real-time conversation, delayed-time conversation, directed instruction, self-learning, problem solving, information seeking and analysis, and critical thinking, as well as the ability to communicate, collaborate and learn. Amin (2008) surmises the place of ICTs in education, when he noted that ICTs such as videos, television and multimedia computer software that combine text, sound and colourful moving images can be used to provide challenging and authentic content that will engage the student in the learning process. Interactive radio likewise makes use of sound effects, songs, dramatisation, comic skits, and other performance conventions to compel the students to listen and become more involved in the lessons being developed.

Becker (2000) noted that ICT increases student engagement, which leads to an increased amount of time students spend working outside class. Valasidor and Bousiou (2005) observed that ICT helps students in their learning by improving the communication between them and the instructors. Jude and Dankaro (2012) investigated the utilisation of ICT in the instructional mix by teacher educators in College of Education (COE) Katsina-Ala, Benue state, Nigeria. The findings revealed that ICT resources were not available, and for that reason, teacher educators could not access them for instructional development purposes.

The efficacy of ICT in higher education has been noted by Umoren (2006). She contends that ICTs have been known to enhance educational opportunities of individuals and groups constrained from attending traditional universities as well as the use of computers as tutors for drills and practise as well as instructional delivery. Tella (2011) investigated the level of availability and use of ICT in some southwestern Nigeria colleges of education. The results of the survey on the college of education staff on the level of availability, use of, and perception of the impact of ICT on teacher education in Nigeria revealed a low level of usage of ICT gadgets; nonavailability of ICT equipment and that the respondents were disgruntled with the sluggish use and integration of ICT.

Ezeoba (2007) carried out an investigation of ICT availability in schools in Onitsha on one hundred nursery school teachers which revealed that the media availability average was less than 20% of fifty. It also found out that the degree of utilisation in instructional delivery was that teachers used mostly books and over 60% did not use ICT resources at all. Fakeye (2010) also investigated English language teachers' knowledge and use of ICT in Ibadan Southwest LGA of Oyo State and found that availability of computers and their connectivity to the Internet was non-existent in virtually all the schools studied. Since utilisation is premised on availability, and because availability is poor, the obvious result was that usability was also poor.

Integration of technologies into classroom learning has impacts on the practice of teachers, in particular when ICT is conceptualized as a tool that supports a real change in the pedagogical approach. Newhouse (2002) identified the following impact:

- The balance of roles they play with a perceived risk of reduced influence
- Providing greater access to information, leading to increased interest in teaching and experimentation (Gather and Bridgforth, 2002)
- Requiring more collaboration and more communication with teachers, administrators, and parents
- Providing more time to engage with students, leading to greater productivity

Information and communication technology (ICT) has helped transform educational research from an analogue educational research to that based on digital knowledge based on technological development in education. This has made it one of the basic building blocks of modern society. Jude and Dankaro (2012) noted that recent developments in ICT have drastically affected educational procedure for improved quality of education offered to students. They go further to say that productive instructional delivery enhances learners' creative and intellectual development through the use of ICT resources, for instance, in the use of multimedia images, graphics, audio, text, and motion for high-quality learning.

Nzewi (2009) and Umoren (2006) believed that ICT resources in instructional delivery in schools will serve a dual purpose and more efficient classroom instruction. This innovation permits alternative types of educational patterns for facilitating the teaching and learning process. She conceptualized the ICT resources as the e-learning which is most commonly associated with higher education and corporate training that uses an information network through the Internet, an intranet (LAN) or extranet (WAN). Electronic learning (e-learning) is used both in informal and formal educational setting for facilitation, instruction, interaction, and instructional delivery. Web-based learning is also a subset of e-learning. Another type of ICT resource is the virtual teaching which entails instructional delivery through teleconferencing the video conferencing technique.

Okpoko (2011) did an assessment of Information and Communication Technologies utilisation in the University of Nigeria. Her findings indicated that 86% of the respondents have computers in their offices but only 28% were networked. She further discovered that majority of the computers were essentially for typing (35%) and storing (26%) of office documents, 14% for sending and receiving mails. ICTs were also used for Internet browsing (13%) and 10% for sending and receiving files from colleagues. The findings further revealed that although 98% of the respondents were aware of the school official Web site only 26% use it effectively, 48% use it irregularly while; 18% do not use it at all.

The use of the Internet for official communication stood at 5.5% unlike written mails which stood at 86%. In terms of the factors responsible for ICT usage, power failure, narrow distribution of ICTs and low competence, and funding were acknowledged as hindrances. She recommended Internet and intranet penetration and a decentralised management information system to faculties in order to stimulate access and use amongst others.

The advent of ICTs with their convergence will deepen mass communication students' media literacy skills. The beauty of media convergence according to Oyero (2007) is that

convergence has brought the delivery of radio directly to individual listeners over the internet. Traditional, over-the-air stations also have their web based stations with differences in their programmes. An example in Nigeria is Radio Lagos FM 93.7found on www. radiolagosfm.com. Others are web-based only and they permit the simultaneous downloading and accessing of their radio files. Also, the convergence of television and the internet, just underway, holds the potential to reinvent both media, particularly because of the promise of fuller interactivity.

The convergence of media brought about by the new media technologies is changing the flow of information from a linear to a three dimensional form of information: Mass (one-to-many), Interpersonal (one-to-one) and computing (one-to-many). The present introduction according to Oyero (2007) is the many-to-many model of communication as illustrated by the diagram below.

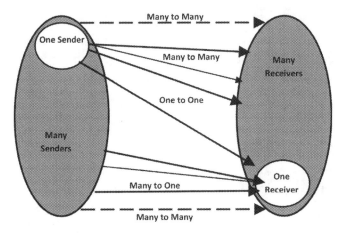

Figure 1: Model of communication
Source: Oyero (2007)

In addition, new media did not only help to achieve contact with the targeted audience (participants) but also sustain this contact, thereby aiding one of the fundamental aims of communication, which is making communication a participatory affair. Moreover, media messages would not have been able to have the kind of impact it has on people today if not for the new media technologies. ICTs give room for the improvement of teaching. For example, using the Internet, students can offer suggestions for improvement as quick as possible through blogs created by their teachers. Writing from a media perspective, McQuail (2005) rightly noted that new media transgresses the limit of the print and broadcasting media as it enables many-to-many conversations and providing instantaneous global contact.

Furthermore, new media facilitate access to up-to-date information to support real-time decisions and bring out accuracy in learning. For instance, the mobile telephone, as presented by Elegbeleye (2005), is able to make people communicate on real-time basis, saving time and money among other conveniences like enabling people to download information or even listen to live broadcast messages. In addition, the WebTV turns a television set into a computer screen, permitting access to the Internet where recent and upcoming programmes are displayed. As such, it provides not only access to the Web but also to several sites of its own, including one for programme schedules. The WebTV also offers several features that allow the Internet to enhance television viewing. For example, it provides a program listing feature that takes viewers from a renewable screen schedule directly to a show with a simple click (Baran, 2009). Media students can take advantage of these features and benefits and become versatile, current, and productive in their learning careers.

The impact of these literatures to the study lies in the fact that today's education is now technology-driven. The adoption of ICTs by school authorities will prepare students not only for the world of work, but will also enable the students to fit into the modern ways of learning.

Theoretical Framework

This research work is anchored on the technology determinism theory propounded by Marshall McLuhan in the 1960s. This theory states that technologies shape how individuals in a society think, feel, and act and how society operates as we move from one technological age to another and that the linking of electronic information media would create an interconnected global village. Dominick (2009, p. 57) asserts that the thrust of the theory is technology drives historical change. The theory explains that all social, political, economic, and cultural changes are influenced by advances in technological innovations. Williams (1990) aligns with this position when he noted that the assumption lies in the inevitable power of technologies to cause widespread social change. Mcluhan traces the change concept from the invention of the printing press to satellite communication.

In relation to this study, there is no doubt that the adoption of technologies for learning affect the nature of learning and the ways of learning. For instance, the library, which was hitherto the main place of research for students, is gradually been replaced by the virtual library and the Internet. Technology has also made it possible for online learning and supervision. All the possibilities technology creates that have implication for teaching and learning rest with the technological determinism theory.

Significance of the Study

This study is significant in the following ways. The study would expand the frontiers of research in ICT, especially with the use of new media technologies by mass communication students. This way, it will bring to the understanding of many the level of acceptability, application, and operation of ICTs in a mass communications department. University management would know the full benefits and barriers to ICT usage in the Department Of Mass Communications in the University of Benin; thereby facilitating a home-grown and functional ICT policy that will place the Nigerian mass communication student in a position to compete well in the industry in line with world standards. It would give direction as to how to gradually transform from the current inadequate state to more adequate ICTs for students access and usage. It would also add to the body of academic literatures on ICTs and provide empirical data for future researchers interested in this area of research.

Research Design

The research design for the study is a survey. The choice of survey is anchored on the fact that the study sought to find out the place of technology integration in education with a focus on mass communication undergraduates. Hence, the study sought their opinion on the level of access, extent of usage, availability, and barriers to ICT usage in the mass communications programme. Surveys are useful in the measurement of public opinion, attitudes, and orientations that are dominant among a large population at a particular period. The total population of full-time mass communication students in the University of Benin is 611 students. A breakdown of the population shows the following: 400 level = 151, 300 level = 197, 200 level = 120, 100 level = 143.

Sample Size and Sampling Technique

Given the nature and size of the population, a sample of 400 respondents from the four levels under survey was selected as respondents using Taro Yamane's statistical formula. The choice of

400 is informed by Wilmer and Dominic (2005) who contended that in a survey, 100 sample is poor, 200 fair, 300 good, 500 very good, and 1,000 excellent. The sample represents 67% of the study population of 611. The choice of Taro Yamane's formula is informed by the fact that the study has a heterogeneous and finite population. The limitation of the respondents in this sample size was to ensure that the sample is manageable, bearing in mind factors such as the total population, time, and the financial resources at the researcher's disposal. Again, this is in line with what Wimmer and Dominick (2005) advocated when they contend that the size of the sample required for a study depends on at least one or more of the following seven factors: project type, project purpose, project complexity, amount of errors tolerated, and time constraints. Others are financial constraints and previous research in the area.

To determine and select the four hundred respondents, the researcher adopted each class level as a stratum. As a result of the population imbalances and demographic characteristics of the study population, respondents were first stratified along class levels, and respondents were then selected, using the simple random sampling technique, to pick the actual respondents. The table below shows a breakdown of the class levels and actual sample from each class based on Taro Yamane formula.

Table 1: Class level and sample size

S/N	Stratum	Population	Sample size
1	**400 level**	**151**	**98**
2	**300 level**	197	128
3	**200 level**	120	78
4	**100 level**	140	93

Research Instrument

The instrument used in the collection of data was the questionnaire. The questionnaire was designed for the respondents to tick only the options that best agreed with the items in the survey.

The research instrument was validated by instructional technology experts and educationist to ensure that it addresses the content of the topic under study. In addition, it was to ensure that the items contained in the instrument are as much as possible coherent, sequential, comprehensive, and therefore capable of testing what the study was set out to test.

To test the reliability of the instrument, the researchers used the split-half method to establish reliability. The researchers used twenty samples of the questionnaire to establish reliability using Spearman-Brown prediction formula. The scores of the two halves were correlated, and the coefficient obtained stood at the value of 0.56, which is high. Therefore, the instrument was considered reliable. The researchers with the aid of class coordinators randomly distributed the validated questionnaire to the students before general classes.

Data Analysis

In this study, data collected were analysed using the simple percentages and were presented with the aid of tables. The data measured were in terms of the frequency, mean and percentage distribution of the different categories of variables that were displayed in the tables. The analysis was done using Statistical Package for the Social Sciences (SPSS).

The study primarily focuses on the integration of technologies in education particularly amongst University of Benin Mass Communication students undergraduates. Although 400 respondents were randomly drawn to represent the population of study, 375 of the administered questionnaire on the respondents were returned, thus representing 93.75% return rate. This was tangible enough for analysis. In the same vein, out of the 375 questionnaire returned, 208 of the respondents were females while 167 of them were males, representing 55.47% and 44.53% respectively. The dominant age category of the respondents was 20–25 years.

Table 2: Technologies support education

Ns	Option	SA	A	Neutral	D	SD	Mean
1.	Students are interested to use ICTs in education.	132 (35.2%)	96 (25.6)%	79 (13.86)%	45 (4%)	23 (6.13%)	2.90
2.	The education system in Uniben is sufficient to support ICTs in mass comm.	155 (41.33%)	114 (30.4)	99 (26.4%)	7 (1.86)	-	3.03
3.	The use of ICTs in mass comm. Makes its training and learning effective.	177 (47.2%)	164 (43.73%)	34 (9.06%)	-	-	2.78
4.	ICT has made education easier	199 (53.06%)	136 (36.26%)	40 (10.66%)	-	-	2.66
5.	ICTs are invented to promote education.	108 (28.8%)	124 (33.06%)	119 (31.73%)	23 (6.13)	1 (0.26%)	2.84
				Grand mean = 14.21			

As revealed in table 1, the mean scores of variables 1, 2, 3, 4, and 5, are: 2.90, 3.03, 2.78, 2.66, and 2.84 respectively. The calculated grand mean on whether student of mass communication use ICT to support their studies is 14.21 this is higher than the assumed 2.5 benchmark. The manifest data from table 1 show that students are interested in using ICTs in their education. This is represented by 132 respondents, representing 35.2% who strongly agreed with this view. They also strongly agree with the position that the university is interested in using ICT to support education. It also shows that the 177 respondents, representing 47.2%, strongly agreed that the use of ICTs in mass communication makes its training and learning effective. The respondents also strongly agreed that ICT has made education easier. In the same vein, 124 respondents agreed that ICTs are invented to promote education.

Table 3. Use of ICT

Ns.	Option	SA	A	Neutral	D	SD	Mean
6.	ICT is a necessary part of mass communication	157 (41.86%)	142 (37.86%)	52 (13.86%)	15 (4%)	9 (2.4%)	2.54
7.	Students training is essential, apart from using ICTS in education	65 (17.3%)	52 (13.86%)	76 (20.26%)	68 (18.1%)	114 (30.4%)	2.81
8.	ICT can assist students of mass comm.	109 (29.06%)	142 (37.86%)	84 (22.4%)	26 (6.93%)	14 (3.73%)	2.63
9.	ICTs make mass communication students independent and self-learners.	128 (34.13%)	107 (28.53%)	62 (16.53%)	54 (14.4%)	24 (6.4%)	2.81
10.	Most people think that ICTs are limited to computer and Internet.	84 (22.4%)	153 (40.8%)	71 (18.93%)	45 (12%)	22 (5.8%)	2.69
						Grand mean = 13.48	

Table 3 revealed the mean scores on what the extent of which ICTs resource utilisation by mass communication students in Uniben. Item 6 had a mean of 2.54, item 7 had a mean of 2.81, item 8, had a mean of 2.63, item 9, had a mean of 2.81, and item 10 had a mean of 2.69. The grand mean of 13.48 was calculated. This means that the grand mean is higher than the assumed 2.5 benchmark. From the data above table, 156 respondents, representing 86% strongly agreed that ICT is a necessary part of mass communication. They however disagreed with the assumption that students training is essential, apart from using ICTS in education

This is obvious from 114 respondents, representing 30.4% of the total respondents. Conversely, they agreed that ICTs make mass communication students independent and self-learners. In the same vein, 153 respondents, representing 40.8% agreed that most people think that ICTs is limited to computer and Internet. The import of the data based on the fact that adequate training of mass communication students rest on how well ICTs are deployed and used.

Table 4. ICT awareness

	Option	SA	A	N/TRAL	D	SD	Mean
11.	Students are well aware of ICT.	123 (32.8%)	107 (28.53%)	88 (23.46%)	44 (11.73%)	13 (3.46%)	2.66
12.	Existing methods of teaching are enough to support ICT.	85 (22.66%)	42 (11.2%)	73 (19.46%)	99 (26.4%)	76 (20.26%)	2.69
13.	ICT in education creates awareness for teaching and learning.	128 (34.13%)	171 (45.6%)	7 2 (19.2%)	4 (1.06%)	-	2.81
14.	There is a need to improve the awareness of ICT among mass com students.	174 (46.4%)	149 (39.73%)	52 (13.86%)	-	-	2.66
15.	The Mass Communication Department of Uniben is trying to create awareness about the importance of using ICT in mass comm.	101 (26.93%)	94 (25.06%)	143 (38.13%)	32 (8.53%)	5 (1.33%)	2.69
						Grand mean = 13.51	

Data from table 4 show the mean scores of variables 11, 12, 13, 14, and 15, are 2.66, 2.69, 2.81, 2.66 and 2.69 respectively. The calculated grand mean what is the level of awareness of ICT resources by mass communication Student in the University is 13.51 this is higher than the assumed 2.5 benchmark. Thus, the manifest data from the table indicate that 123 respondents, representing 32.8% are aware of ICTs. However, 99 respondents, representing 26.4% disagreed with the assumption that existing methods of teaching support

ICT. Nevertheless, 171 respondents, representing 45.6% agreed that ICT in education creates awareness for teaching and learning. The respondents strongly agreed that there is improve the awareness for teaching and learning ICT. This is indicated by 174 respondents, representing 46.4%. The respondents were however indifferent on the assumption the department is trying to create awareness about the importance of using ICT in mass communication.

Table 5: Difficulties facing ICTs usage

	Options	SA	A	N/TRAL	D	SD	Mean
16	Teacher factors	116 (30.93%)	102 (27.2%)	82 (21.86%)	30 (8%)	45 (12%)	2.78
17	Lecturer hall factor	94 (25.06%)	152 (40.53%)	70 (18.66%)	34 (9.06%)	25 (6.66%)	2.90
18	Energy related problem (PHCN)	252 (67.2%)	81 (21.6%)	42 (11.2%)	-	-	2.75
19	Inadequate Internet connectivity	117 (31.2%)	168 (44.8%)	56 (14.93%)	9 (2.4%)	25 (6.66%)	3
20	Knowledge of computer usage	52 (13.86%)	37 (9.86%)	98 (26.13%)	127 (33.86%)	61 (16..26)	2.93
21	Availability of computers	43 (11.46%)	24 (6.4%)	18 (4.8%)	169 (45.06%)	121 (32.26%)	2.81
22	Student's attitude towards computer.	135	69	103	45	23	2.59
						Grand mean = 19.76	

In table 5, the mean scores of variables 16, 17, 18, 19, 20, 21, and 22 are 2.78, 2.90, 2.75, 3, 2.93, 2.81, and 2.59 respectively. The calculated grand mean on the level of awareness of ICT resources by mass communication student in University of Benin is **19.76**. This is higher than the assumed 2.5 benchmark. The table indicates that teacher factors, Internet connectivity, and lecture hall affect ICT usage. It also reveals that power supply also affect ICT usage. This is represented by 252 respondents, representing 67.2% who strongly agree with this position. However, 169 respondents, representing 45.06%, disagreed with the assumption that availability of computers is germane to usage, while 135 respondents opined that students' attitude pose a difficulty to usage.

Table 6. Availability of ICT tools

Ns	Option	SA	A	NEUTRAL	D	SD	Mean
23	Usage of ICTs depend on the following tools	119 (31.73%)	59 (15.73%)	123 (32.8%)	52 (18.86%)	22 (5.86%)	2.75
24	Lack of ICT tools affect the use of ICTs in mass comm.	187 (49.86%)	141 (37.6%)	43 (11.46%)	4 (1.06%)	-	2.63
25	The availability of ICTs tool will ensure the development of ICT infrastructure.	111 (29.6%)	89 (23.73%)	132 (35.2%)	17 (4.53%)	26 (6.93%)	2.27
26	Students with computer knowledge facilitates the use of ICT in mass comm.	187 (49.86%)	112 (29.86%)	57 (15.2%)	14 (3.73%)	5 (1.33%)	2.60
					Grand Mean = 10.25		

Table 6 revealed the mean scores on what is the influence of ICT availability on Mass Communication Department. Item 23 had a mean of 2.75, item 24 had a mean of 2.63, item 25, had a mean of 2.27, and item 26, had a mean of 2.60. The grand mean of **2.56** was calculated. This means that the grand mean is higher than the 2.5 benchmark. The implications of the data from the above table indicate that the respondents strongly agree with the assumption that the availability of ICTs tool is crucial to use. It also means that students with computer knowledge will facilitate ICT usage. The data also suggest that the respondents were indifferent on the assumption that ICT availability will facilitate ICT development. They were also indifferent on the assumption that availability of ICT tools will motivate usage.

Table 7. Problems of access to electronic information in the institution

Options	Number of Response	Percentage (%)	Mean
Adequate number of personal computer	152	40.53%	2.30
No computer laboratory	181	48.26%	1.66

No campus computer network	26	6.93%	2.0
No Internet connectivity	-	-	-
Lack of support of its staff	16	4.26%	2.85

The table above shows that the greatest problem of access to electronic information in the University is the absence of computer laboratories. This is represented by 181 respondents, representing 48.26% of the total respondents. This is closely followed by 152 respondents, representing 40.53% who see the lack of adequate number of PC as the problem of access to electronic information.

The import of the data from table 8 suggests that the problem of access to electronic information stems primary from the nonavailability of computers. In other words, when computers are made availability, access to ICTs and the various software used in their operations would have been addressed.

Table 8. The greatest barrier to using ICT

Option	Number of responses	Percentage (%)	Mean
Time	106	28.26%	2.15
Cost	184	49.06%	3.03
Training	55	14.66%	2.06
Technical support	30	8	2.39

The data in table 9 show that the greatest barrier to using ICT in the department of mass communication is the high cost of acquiring the hardware and software of ICT. This inferred from 184 respondents, representing 49.06% who ticked this. It is closely followed by 106 respondents who see time as the greatest barrier. While 55 and 30 respondents, representing 14.66% and 8% respectively, ticked training and technical support. This means that the students lack the financial resource necessary for them to acquire and use ICTs.

4.3 Discussion

The answers to the questions are based on the analysis provided in tables 1–8.

Research Question 1: Do students of mass communications use technologies to support their studies?

The analysis in table 1 indicates that the calculated grand mean for the sets of questions probing to know whether students of mass communication use ICT to support their studies was 2.84 and which is higher than the 2.5 benchmark. This suggests that mass communication students use ICTs to support their studies. The manifest data from table 1 show that students are interested in using ICTs in their education. This is represented by 132 respondents, representing 35.2% who strongly agreed with this view. They also strongly agree with the position that the university is interested in using ICT to support education. It also shows that the 177 respondents, representing 47.2% strongly agreed that the use of ICTs in Mass communication makes its training and learning effective. The respondents also strongly agreed that ICT has made education easier. In the same vein, 124 respondents agreed that ICTs are invented to promote education.

The import of the analysis lies in the fact that students have come to understand the place of ICTS and they are integrated as part of the teaching and learning. The extent to which they bring the interest to bear depends on the curriculum provision for their usage and how well the university authorities create avenues for ICTs to properly support their course of study. Students who subscribed to the idea that ICTs made teaching and learning effective probably based on their positions on the few courses that offered them leverage in using ICT either for research or the teacher's ability to use the tools for effective teaching. The finding also aligns with the value of ICTs in education as highlighted by Jude and Dankaro (2012), when they noted that recent developments in ICT have drastically affected educational procedure for improved quality of education offered to students. They go further to say that productive instructional delivery enhances learners' creative and intellectual development through the use of ICT resources, for instance, in the use of multimedia images, graphics, audio, text and motion for high quality learning.

Research Question 2: What is the extent of ICTs resource utilisation by mass communication students in the University of Benin?

Table 2 provides the answer to this question. The table revealed the mean scores on what is the extent of ICTs resource utilisation by mass communication Students in the University of Benin. Item 6 had a mean of 2.54, item 7 had a mean of 2.81, item 8, had mean of 2.62, item 9, had mean of 2.81 and item 10 had mean of 2.69. The grand mean of 2.69 was calculated. This means that the grand mean is higher than the 2.5 benchmark. This means the level of utilisation is high. The analysis indicates that 156 respondents, representing 86% strongly agreed that ICT is a necessary part of mass communication; they however disagreed with the assumption that students' training is essential apart from using ICTs in education. Conversely, they agreed that ICTs make mass communication students independent and self-learners. In the same vein, 153 respondents, representing 40.8% agreed that most people think that ICTs is limited to computer and Internet.

This means that mass communication students are not only using the computer and Internet for learning but other valuable ICTs necessary for their course of study. It also suggests that mass communication students are using ICTs regular. This is manifested from respondents who assert that they use ICTs regularly. The results of this finding differ considerably from an earlier study by Okpoko (2011), who did an assessment of information and communication technologies utilisation in the University of Nigeria. While this finding holds promise for the department in terms of usage, it is important for the school to make room for other ICTs apart from computers and Internet access; this will go a long way to deepen the students' knowledge and skills in ICTs.

Research Question 3: What is the level of awareness of ICT resources by mass communication student in University of Benin?

The mean scores of the variables in table 4 and 5 provide answer to this question. The calculated grand mean on the level of awareness of ICT resource by mass communication student in University of Benin is 2.70 and 2.82 respectively; this is higher than the 2.5 benchmark.

This means there is a high level of awareness on ICT resource by mass communication students.

This is supported from the manifest data in the tables which indicate that 123 respondents, representing 32.8% are aware of ICTs. However, 99 respondents, representing 26.4% disagreed with the assumption that existing methods of teaching support ICT. Nevertheless, 171 respondents, representing 45.6%, agreed that ICT in education creates awareness for teaching and learning. The respondents strongly agreed that there is improved awareness for teaching and learning ICT. This is indicated by 174 respondents, representing 46.4%. The respondents were however indifferent on the assumption that the department is trying to create awareness about the importance of using ICT in mass communication.

The implication is that awareness on ICT in education is relatively high amongst mass communication students. The number who seems not aware might be connected with those who see ICTs from the perspective of high calibre software and hardware. The department therefore needs to sensitize the students on what constitutes ICT especially from the general view and especially the perspective of media education.

Research Question 4: What is the influence of ICTs availability on Mass Communication Department?

The answer to this question stems from table 3. The grand mean scores on what is the influence of ICT availability on Mass Communication Department is 2.56, which is higher than the 2.5 benchmark. This means that available ICTs have a relatively low level of influence on the department. The respondents strongly agree with the assumption that the availability of ICTs tool is crucial to use. It also means that students with computer knowledge will facilitate ICT usage. The data also suggest that the respondents were indifferent on the assumption that ICT availability will facilitate ICT development. They were also indifferent on the assumption that availability of ICT tools will motivate usage.

While it is correct to say that ICTs have influence in education, the available ICTs have not had a strong influence on the department or the course of study. This is because, the available ICTs are inadequate and are incapable of making a substantial influence on students, this creates room for students to show lukewarm attitude to ICT. As noted in the literature review, the state of ICT infrastructure and resource in the mass communication programme is far from the ideal. Although there is a media studio with the different studios, only the radio studio and the control room and the photolaboratory can boast of some digital equipment which is the hub of modern broadcasting and photography.

The lighting facilities are obsolete and not in tune with modern digital lighting equipment. The department does not have a computer laboratory where students can learn typing, editing, and desktop publishing. The lack of these shortchanges the students, as it deprives them of the capacity for online news selection, graphic design, and typing skills. This seeming lack of infrastructure renders the influence of ICT availability on the course less impactful.

Research Question 5: What is the greatest barrier to using ICTs by mass communication students?

The answer to this question finds support from table 7 and 8. The result from the tables shows that the greatest problem of access to electronic information in the Mass Communication Department of the University of Benin is the absence of computer laboratories. This is represented by 181 respondents, representing 48.26% of the total respondents. This is closely followed by 152 respondents, representing 40.53% of the total respondents who see the lack of adequate number of PC as the problem of access to electronic information.

Apart from computers, the finding also shows that the greatest barrier to using ICT in the department of mass communication is the high cost of acquiring the hardware and software of ICT. This is inferred from 184 respondents, representing 49.06% who reasoned in this light. This means that the students lack the financial resource necessary for them to acquire and use ICTs. It suggests that the level of funding of the programme is not encouraging since the basic requirements for a robust education that integrates ICT are not provided. Where this is the case as noted from the analysis, it will be difficult for the students to acquire a modern education that is in tune with global trends.

Conclusion

Based on the findings, the following conclusions are reached.

Students have come to understand the place of ICTs in education. The extent to which they bring their interest to bear depends on the curriculum provision for their usage and how well the university authorities create avenues for ICTs to properly support their course of study. What is obvious is that students' ICTs usage is self-directed and provided.

A good number of mass communication students are using ICTs regularly but a sizable number do not use ICTs. This is because the department has not created the level of awareness that could stimulate their interest. Apart from the issue of awareness, the absence of computer laboratories that could create access and the necessary availability for usage is a barrier.

While finding indicates that ICTs have influenced education, the available ICTs have not had a strong influence on the department or the course of study. This is because the available ICTs are inadequate and are incapable of making a substantial influence on students. This creates room for students to show lukewarm attitude to ICT. This means that the students lack the financial resources necessary for them to acquire and use ICTs.

ICT education is not properly funded in the department of mass communication. What is dominant is that students acquire their own ICTs for personal use. Since there are no laboratories, the teachers are not in a position to teach the students in this area. Besides, many of the students are poor; some have knowledge of the concepts but lack access since they have no resources to acquire the tools that will stimulate their usage.

Recommendations

No doubt, a research involves an investigation geared towards increasing knowledge and providing ideas to solving problems. From this fact, coupled with an enthusiastic desire to ensure confirmatory evidence on this study, particularly in the aspect of achieving a greater feeling of certainty for the likely purpose of making generalisations in the future, the researcher therefore deems it fit and necessary to make the following recommendations:

1. Government needs to increase funding to education, so that university authorities can provide the necessary ICTs tools, laboratories, and studios required for teaching and learning.

2. The level of adequacy of ICTs in the department of mass communication is not encouraging. Management needs to prioritize ICTs acquisition with a particular focus on those that ICTs that have direct impact on daily teaching and learning.

3. Students who use their own personal ICTs should be encouraged, and universities should organize workshops and seminars that could deepen their self-knowledge.

4. The department needs to sensitize students on the various types of ICTs use in mass communication so that students who can acquire them can start well ahead of time.

5. Teachers and instructors need training on ICTs to facilitate students' knowledge and usage of ICTs.

References

Agba, P. C. (2001). *Electronic reporting: Heart of the new communication age*. Enugu: Snaap Press Ltd.

Amin, S. N. (2008). An effective use of ICT for education and learning by drawing on worldwide knowledge, research and experience:

ICT as a change agent for education: A literature review. Available at http//:www.nmc/horizanreport.pdf. Accessed 9/2/2015.

Croteau, D. and Hoynes, W. (2003). *Media Society: Industries, images and audiences.* Thousand Oaks: Pine Forge Press.

Elegbeleye, O. S. (2005). Prevalent use of global system of mobile phone (GSM) for communication an interactional enhancement or a drawback? *Nordic Journal of African Studies,* vol. 2, No. 14. pp. 25–37.

Ezeoba, K. O. (2007). Instructional media. An assessment of the availability, utilization and production by nursery school teachers. *Journal of Applied Literacy and Reading.*3 (Special Edition) pp. 33–38.

Fakeye, D. O. (2010). Assessment of English language teachers' knowledge and use of Information and Communication Technology (ICT) in Ibadan Southwest Local Government of Oyo State. *American- Eurasian Journal of Scientific Research.*5 (4). pp. 56–59.

Idemili, S. O. and Sanbe, S. A. (2007). The Nigerian media and ICTs implications and challenges. In Nwosu, I. E and Soola, O. E. (Eds.) *Communication in Global ICTs and Ecosystem perspectives. Insights from Nigeria.* pp. 181–188. Enugu precision Publishers Ltd.

Idoko, J. A. and Ademu, A. (2010). The challenges of Information and Communication Technology for teaching—Learning as perceived by Agricultural Science teachers in secondary schools in Kogi State. *Journal of Educational Innovators.3(2).* pp. 43–49.

McGorry, S. Y. (2002). From distance education to online education. The Internet and HigherEducation (3) 2, pp. 163–174.

Morris, M. and Ogan, C. (2002). The Internet as a mass medium. In McQuail's D.(Ed.).

McQuail's Reader in Mass Communication. London: Sage.

New Media Consortium (2007). Horizon Report, Available at www. nmc.org/pdf/2007/horizonreport.pdf. Accessed on 9/2/15.

Newhouse, P. (2002). *The Impact of ICT on learning and teaching: a Literature Review:* Western Australia Department of education.

Nzewi, U. (2009). *Information and Communication Technology (ICT) in Teaching and Learning. Curriculum and Practice.* Abuja; Curriculum Organization of Nigeria.

Okpoko, C. (2011). An Assessment of Information and Communication in the University of Nigeria, Nsukka. *International Journal of Communication.* No. 13. pp. 73–85.

Oyero, O. S. (2007). The new media technologies: Prospects and Challenges for Development in Africa. In Mojaye, E. M. V., Salawu, A., and Oyewo, O. O. (Eds). *Ebenezer Soola Conference on Communication: Proceedings.* Ibadan: pp. 168–175. Ebenezer Soola Conference on Communication.

Umoren, G. (2006). Information and Communication Technology and Curriculum. N*igerian Journal of Curriculum Studie*s. Calabar Chapter. 2 (1), pp. 57–83.

Valasidou, A., Sidiropoulos. D., Hatzis, T.and Bousiou-Makindou, D. (2005) Guidelines for the design and implementation of E-learning programme, proceedings of the IADIS. International Conference of IADIS e-society, 27–30 June, Qawra, Malta.

Olley, W. (2009) New Media Technologies in Broadcasting: An unpublished Master of Arts dissertation in Mass Communication, University of Nigeria Nsukka.

Young, J. (2002). The 24-hour Professor. *The Chronicle of Higher Education*, 48, 38. pp. 31–33.

Yuen, A., Law, N. and Wong, K. (2003). ICT implementation and school leadership case studies of ICT integration in teaching and learning. *Journal of educational administration* 41 (2), pp. 158–170.

SECTION 2

Utilisation of Educational Technology Tools for Teaching and Learning

Chapter 4

Availability and Utilisation of Information and Communication Technology Resources in Secondary Schools in Central Senatorial District of Edo State, Nigeria

Eimuhi, Justina Onojerena
and
Ikhioya, Grace Olohiomereu

Introduction

Secondary education in Nigeria has been seen as the major avenue for producing school leavers for useful living in society and for preparing them for higher or tertiary education. In an attempt to meet the need of fulfilling the strategies of National Economic Empowerment and Development Strategy in education, the government has become conscious of introducing technology at all educational levels. This is expected to be in line with the world driven communication technology agenda of the United Nations Educational Scientific and Cultural Organization (UNESCO). The UNESCO (2002) strategic objectives in education include improving the quality of education through the diversification of contents and methods and promoting experimentation, innovation, the diffusion and sharing of information and best practices as well as policy dialogue. Some of the strategies of National Economic Empowerment and Development Strategy (NEEDS) in education are highlighted below:

- Ensuring that 80% of secondary school leavers are computer literate.
- Ensuring that 30% of secondary schools have functional facilities.
- Ensuring that 30% of secondary schools have functional information and communication technology (ICT) facilities.
- Ensuring that 50% of teachers at all levels are trained in computer skills.

Yusuf (2005) noted the specific application objectives of the National Policy on Education on Information and Communication Technology to be the following:

- To develop a pool of IT engineers, scientists, technicians, and software developers;
- To increase the availability of trained personnel;
- To provide attractive career opportunities; and
- To develop requisite skills in various aspects of IT.

Information and communication technology (ICT) has become a key tool and has a revolutionary impact on how we see the world and how we live in it (Nwachukwu, 2006). The world today is becoming a global village through the use of information superhighways. ICT is having an impact on the methods of teaching and learning globally. However, this revolution does not seem to be widespread, and many need to be strengthened to reach a large percentage of the population. In a complex society like Nigeria, many factors affect the utilisation of ICT, so an interdisciplinary and integrated approach is very necessary to ensure the successful development of Nigeria's economy and society (Mac-Ikemenjima, 2005).

Information and communication technology is the use of electronic equipment, especially computer, for sorting, analyzing and distributing information of all kinds including words, numbers, and pictures. It is the application of modern electronic technologies to data processing and communication. According to Ngurukwem (2005), it covers the task performed, procedures followed, and devices employed in gathering, manipulation, transformation, storage, retrieval, and dissemination of data and information. No sector of the national economy can grow effectively without the incorporation of ICT in its mode of operation. Information and communication technology (ICTs) are indispensable and have been accepted as part of the contemporary world especially in the industrialized societies (Nwachukwu, 2006). In fact, cultures and societies have adjusted to meet the challenges of the knowledge age. The pervasiveness of ICT has brought about rapid changes in technology, social, political, and global economic transformation. However, the field of education has not been unaffected by the penetrating influence of information and communication technology. Undoubtedly, ICT has impacted the quality and quantity of teaching, learning, and research in teacher education. Hence, ICT provides opportunities for students, student teachers, and academic and nonacademic staff to communicate with one another more effectively during formal and informal teaching and learning (Yusuf, 2005).

The academic landscape in Nigeria includes the teaching and learning process along with the educational programmes and courses and the pedagogy or methodology of teaching; the research process, including dissemination and publication, libraries and information services, including higher education management and administration (Beebe, 2004). The relationship between education and government is dependent on policy formulation and implementation. The Nigerian government acknowledges the importance of ICT in its national policy on education in section 4, subsection 19(m): "In recognition of the prominent role of Information and Communication Technology in advancing knowledge and skills necessary for effective functioning in the modern world, there is urgent need to integrate information and Communication Technology (ICT) into education in Nigeria." Also, in section 5, subsection 30(t); it states that "Government shall provide necessary infrastructure and training for the integration of ICT in the school system in recognition of the role of ICT in advancing knowledge and skills in the modern world" (FRN, 2004).

The national policy on education, or NPE (1998) in Ademola (2005) give a special focus on vocational and technical education in the hope of producing graduates both for servicing the industries and self-empowerment. As contained in the NPE, technical and vocational education makes clear the government's intention to build Nigeria into a technologically developed and self-reliant nation (FRN, 2004). As Ngurukwem (2005) rightly puts it, the computer is the heart of the present-day ICT. With skills in computer training technology, a vocation is created. All that is needed is for students to learn entrepreneurial skills (the ability to set up a business) to manage and sustain small computer businesses. This no doubt will reduce the unemployment problems in Nigeria. When personal self-reliant effort is put in place, it will transform into national self-reliance because when the individual is gainfully employed, the gross national product (GNP) is enhanced and the income per capital is also increased.

The Federal Executive Council of Nigeria approved the nation ICT policy in March 2001, and the implementation started in April 2001 with the establishment of the National Information Technology Development Agency (NITDA), which was charged with the implementation responsibility. The policy, however, recognized the private sector as the driving engine of the IT sector. Consequently, government inaugurated the Nigeria Action Plan Committee to develop a new ICT globally and in Nigeria the review of the initial policy became necessary.

It is one thing to succeed in getting our research finding translated into national policies; it is another ball game altogether to get them implemented. This may often necessitate further follow-up and intensive lobbying. Part of the problem has to do with incessant changes within the government. Lack of understanding of the power of well-formulated policy and diligent implementation, which may be affecting educational and national development, may apparently account for this.

It is evident that the telecommunication industry has achieved much within a period of five years, but not much seemingly in the education sector of our economy. Obviously, since the year 2001 when the communication industry was deregulated, so many jobs have been created for graduates and other skilled and unskilled individuals. The graduates are taking jobs with the telecommunication operator; the skilled are out there unblocking lines and unlocking phones while the unskilled are selling phones, lines, and rechargeable cards. This is wealth creation that will certainly stop youths or the highly skilled from seeking greener pastures abroad—resulting to brain drain. Information and communication technology in education is relatively low in terms of application and availability of facilities. Utilisation of ICT facilities in education in Nigeria is less than 5% despite the emphasis on implementation of ICT in the curriculum especially in the universities in Nigeria. This implies that only a few schools, especially universities, in Nigeria have access to ICT education (*Guardian Newspaper*, 2007).

There are developments in the Nigerian education sector which indicate some level of ICT application in secondary schools. The Federal Government of Nigeria recognizes in the National Policy on Education (FRN, 2004) the prominent role of ICTs in the modern world, and has integrated ICTs into her educational programmes. To actualize this goal, the document stated that government would provide basic infrastructure and training at the primary school. At the junior and secondary school, computer education has been made a prevocational elective. It should be noted that the year 2004 was not the first attempt the Nigerian government made to introduce computer education in schools. In 1988, the Nigerian government enacted a policy on computer education. The plan was to establish pilot schools and diffuse computer education innovation first to all secondary schools, and then to primary schools; unfortunately, the project did not really take off beyond the distribution and installation of personal computers (Okebukola, 1997; cited by Aduwa-Ogiegbean and Iyamu, 2005). Okebukola (1997), cited by Aduwa-Ogiegbean and Iyamu (2005), concludes that the computer is not part of classroom technology in more than 90% of Nigerian public schools. This implies that the chalkboard and textbooks continue to dominate classroom activities in most Nigerian secondary schools.

Knowledge in ICT on the part of job seekers is a criterion for job acquisition in the twenty-first century. When ICT curriculum is not there in the schools, how would applicants be relevant in this digital era? Every student is expected to have access to a computer in school and live with one at home to acquire skills. However, the techniques and skills of doing something can be learned informally through apprenticeship, or on the jobs, or formally in training institutions (Amawwhule, 1998 in Ademola, 2005). Many youths or school leavers today are roaming the streets because of unemployment in the nation, and because they lack entrepreneurship education in ICT. The fact is that, while the developed nations continue to increase their IT budget, in developing countries, including Nigeria, the budget seems to continue to fall, further widening the digital divide. Assessment of the level of implementation of ICT in schools therefore cannot be underestimated.

According to WEF's global evaluation, new technologies have not been able to increase significantly the access to basic education. Instead, to some extent it has increased the gap between the haves and the have-nots with regard to access to quality education (Peraton and Greed, 2000). The potential of ICT in widening the access to education has not been fully utilized; of course one can ask why this has not been in focus when researching into ICT and education. If ICT is the solution, what is the problem? Improved secondary education is essential to the creation of effective human capital in any country (Envoh, 2007). The ability to use computers effectively has become an essential part of everyone's education. Skills such as bookkeeping, clerical and administrative work, stocktaking, and so forth now constitute a set of computerized practices that form the core IT skills package: spreadsheets, word processors, and database (Reffell and Whiteworth, 2002).

The demand for computer and ICT literacy is increasing in Nigeria, because employees have realized that computers and other ICT facilities can enhance efficiency. They have also realized that lack of skills in computer application can be a threat to other jobs, and the only way to enhance job security is to be computer literate. With the high demand for computer literacy, the teaching and learning of these skills is a concern among professionals (Oduroye, n.d.). New instructional techniques that use ICTs provide a different modality of instruments. For the student, ICT use allows for increased individualisation of learning. In schools where new technologies are used, students have access to tools that adjust to their attention span and provide valuable and immediate feedback for literacy enhancement, which is currently not fully implemented in the Nigerian education system (Emuku and Emuku, 2000).

There are developments in the Nigerian educational sector, which indicate some level of ICT application in secondary schools. The Federal Government of Nigeria, in the National Policy on Education (FRN, 2004), recognizes the prominent role of ICTs in the modern world, and has integrated ICTs into education in Nigeria. To actualize this goal, the document stated that government would provide basic infrastructure and training at the primary school.

The Federal Ministry of Education has launched an ICT—driven project known as School Net (www.snng.org) (FRN, 2006; Adonai, 2005; Okebukola, 2004), which is intended to equip all schools in Nigeria with computers and communications technologies. In June, 2003 at the African Summit of the World Economic Forum held in Durban, South Africa, the New Partnership for Africa Development (NEPAD) launched the e-school initiative, which is intended to equip all schools (high schools) in Africa with ICT equipment, scanners, digital cameras, and copiers among others. It is also meant to connect African students to the Internet. The NEPAD initiative will be executed over a ten-year period with the high school component being completed in the first five years. Three phases are envisaged, with fifteen to twenty countries in each phase. The phases are to be staggered, and an estimated 600, 100 schools are to benefit. The aim of the initiative is to impact ICT skills to young Africans in primary and secondary schools, and to harness ICT to improve, enrich and expand education in African countries (Aginam, 2006).

If we look at the recent development in the education sector globally, we can summarize the implications and demands of global information society in the education system as follows:

- Demand for widening the access to education.
- Continuous lifelong learning (e.g., fading the boundaries between preset and inset, formal education and working life).
- Global versus local cultural developments.

- Creation of new educational networked organisations (e.g., global virtual universities, virtual schools, and multinational educational consortiums).
- Changing of educational management from hierarchical institutions to equal distributions of network organisations, from commanding to negotiating
- Demand for more flexible and general skills (e.g., mental skills such as problem solving, searching, information, learning skills etc.).

There are several impediments to the successful use of information and communication technology in secondary schools in Nigeria. These are: cost, weak infrastructure, lack of skills, lack of relevant software, and limited access to the Internet. New instructional techniques that use ICTs provide a different modality of instruments. For the student, ICT use allows for increased individualisation of learning. In schools where new technologies are used, students have access to tools that adjust to their attention span and provide valuable and immediate feedback for literacy enhancement, which is currently not fully implemented in the Nigerian education system (Emuku and Emuku, 2000). It is against this backdrop this study was set to investigate the utilisation of ICT resources in schools. The study specifically looked at the level of implementation of ICT resources in schools, ascertain the proportion of students who are computer literate and to examine student computer ratio in schools.

Research Questions

The following questions are raised to guide the inquiry

1. What is the level of ICT utilisation in secondary schools in Edo Central Senatorial District of Nigeria?
2. What proportion of students is computer literate?
3. What is the student computer ratio in schools?

Design of the Study

The research method adopted for this study was descriptive design based on survey research design. A survey research design is one in which a group of people or items is studied by collecting and analyzing data from only a few people or items considered to be representative of the entire group. This method was chosen because it was considered to be more economical and has accurate assessment of the characteristics of the whole population through the study of a sample considered to be representative of the entire population.

Population of the Study

The population for this study includes all students in secondary schools in Central Senatorial District of Edo State of Nigeria. It comprises all public and private students in all the Secondary Schools in the study area. The population was made up of 3,500 students comprised 1,950 public secondary school students and 1,550 private secondary school students respectively.

Sample and Sampling Technique

Sampling is the process of selecting a number of individuals to represent the large group from which they were collected (Nworgu, 1991). A sample of 350 students was used for this study using the simple random sampling technique after the population had been stratified according to local government areas. A total of 50 schools comprising 25 secondary schools and 25 private secondary schools drawn across five Local Government Areas, representing 10% of secondary schools in Central Senatorial District of Edo State was studied.

Instrument of the Study

The research instrument was a checklist of twenty items carefully drawn to elicit responses from students. The checklist comprised two sections—A and B. Section A deals with personal data of the respondents while Section B comprises 12 items designed to elicit responses from students on the level of utilisation of ICT in secondary schools.

Validity of the Instrument

The researcher, after carefully designing the checklist items, presented the items to the supervisor, who is an expert in educational administration who carefully reviewed the checklist items to ensure content validity.

Reliability of the Instrument

Since a checklist was used, the researcher did not carry out a reliability test.

Administration of Instrument

Copies of the checklist were administered personally by the researcher. Each of the secondary students was visited by the researcher. A brief section was held with students of each school with the assistance of the staff on the importance of the research and the need to be honest in their responses before the items were finally administered. The researcher, after distributing the checklist, gave some time for the filling, and retrieved on a second visit. However, for each school, the researcher placed a staff that monitored the students and retrieved copies of filled instrument.

Method of Data Analysis

The statistical techniques used to analyze data in this project were percentages and ratio.

Results

What is the level of ICT utilisation in secondary schools in Edo Central Senatorial District of Nigeria?

Table 1. Level of ICT utilisation in secondary schools in Edo Central Senatorial District of Nigeria

Data field	High	Medium	Low	Very low	Total
You are active in the use of computer in the classroom	53	85	163	49	350
You can connect the computer and its peripherals	99	157	73	21	350
You can locate and run application program e.g., Microsoft Word	38	97	116	99	350
You can use spreadsheet package very well	29	80	157	84	350
You can download files from the Internet	78	110	105	57	350
You can communicate online with other students on homework/assignment	51	79	140	80	350
You can use Web search engines (Google, Mamma, Devilfinder, AltaVista etc.) very well	86	100	113	51	350
You can chat on the Internet using instant messages (Yahoo, MSN, Hotmail, Gmail, etc.).	28	82	182	58	350
You can sent or receive e-mail messages.	31	70	125	124	350
You can move files between drives (e.g., C: to A:).	61	65	130	94	350
There are practical classes for computer studies in my school.	59	171	70	50	350
There is a functional Internet facility in my school that is accessible to all students.	65	155	91	39	350
Total	678 = 16.1%	1251 = 29.8%	1465 = 34.9%	806 = 19.2%	**4200**
	1929=46%		2271=54%		

This table shows the level of usage of ICT by students in secondary schools. In the analysis, 678 (16.1%) shows high level, 1,251 (29.8%) shows medium level, 1,465 (34.9%) shows low level and 806 (19.2%) shows very low level of utilisation of ICT in secondary schools. The table further revealed that high and medium usage of ICT resources by students amounted to 46% while low and very low amounted to 54%. Therefore, the level of usage of ICT in secondary schools by students is low.

What percentage of the students is computer literate?

Table 2. Percentage of computer literate students in secondary schools

Ownership of school	Number of students	Number of computer-literate students	Percentages (%)	Remark
Public	195	73	37%	Low proportion from public schools
Private	155	71	46%	Moderate proportion in private schools
Total	**350**	**144**	**41%**	**Proportion of computer literate students is low.**

This table presents the proportion of computer literate of students in secondary schools. The analysis showed that 41% of students are computer literate which a low proportion is. It also shows that that the proportion of computer literate students in private secondary schools is fairly higher than that of students of public secondary schools.

What is the student computer ratio in schools?

Table 3. Student-computer ratio in secondary schools

Ownership of school	Number of computers	Students	Mean	Ratio
Public	38	640	17	1:17
Private	52	520	10	1:10
Total	**90**	**1160**	**13**	**1:13**

From the analysis of table 3, it is observed that the ratio of computer to a student is 1:17 in public schools and 1:10 in private schools. On the average, student computer ratio is 1:13. This shows that the number of computer per student in secondary schools is inadequate.

Discussion of Results

It was found that the level of usage of ICT in secondary schools by students is low. This shows that many students neither have access to ICT tools nor have adequate knowledge of their usages. The Federal Government of Nigeria, in the National Policy on Education (FRN, 2004), recognizes the prominent role of ICTs in the modern world, and has integrated ICTs into education in Nigeria. To actualize this goal, the document stated that government would provide basic infrastructure and training at the primary school. At the junior secondary school, computer education has been made a prevocational elective, and is a vocational elective at the senior secondary school. Okebukola (1997), cited by Aduwa-Ogiegbean and Iyamu (2005), concludes that the computer is not part of classroom technology in more than 90% of Nigerian public schools. This implies that the chalkboard and textbooks continue to dominate classroom activities in most Nigerian secondary schools.

The study also found out that the percentage of computer-literate students in secondary schools is low. It is as low as 41%. It was also discovered that the proportion of computer-literate students in private secondary schools was fairly higher than that of students of public secondary schools. This could be as a result of inadequacy of computers in schools especially government-owned secondary schools where the computers available are hardly put to use due to low level of manpower, electricity problems, and poor maintenance culture. There is therefore the need to increase the level of students computer literacy since the world has turn a global village and only students equipped with computer skills may be able to compete favourably in the competitive market.

The study equally found that the ratio of students to computers was very low and unencouraging. This shows that the number of computers per student in secondary schools is inadequate. UNESCO strategic objectives in education (UNESCO, 2002) highlighted some of the National Economic Empowerment and Development Strategy (NEEDS) as follows: Ensuring that 80% of secondary school leavers are computer literate, ensuring that 30% of secondary schools have functional facilities, ensuring that 30% of secondary schools have functional information and communication technology (ICT) facilities. But in secondary schools in Nigeria, these are not realizable. It imperative therefore that school owners should increase the infrastructural facilities and computer equipment in schools to increase the level of ICT awareness in secondary schools.

Recommendations

Based on the findings of this study, the following recommendations were made:

1. The government should provide adequate computers and its accessories to secondary schools and ensure that students have access to them.

2. Government should train more manpower in ICT that would train the students on the use of computers and other ICT resources for national development.

3. All stakeholders of the education industry in Nigeria have to be directly involved in the business of education especially in the donation of instructional materials and equipment especially computers to public schools to enhance student-computer ratio.

4. Parents of students should organize out-of-school computer tutorials for their children especially during the holidays to improve the literacy rate of students.

Conclusion

This chapter has showed that information and communication technologies (ICTs) are necessary technological tools, which include the use of computers and peripherals, the use of the Internet and online library and laboratory facilities, and visual and audiovisual aids like projectors and films in the pedagogical preparation and delivery of lessons for effective teaching and learning in the classroom. Schools in Nigeria, particularly public tertiary institutions, lack the necessary technology tools for effective teaching and learning.

It is the belief of the researcher that provision of ICT facilities in schools in their right quantity and quality and engaging the right caliber of staff with ICT knowledge will help in the realisation of the objective of ICT in education in Nigeria. Full implementation of ICT in education in Nigeria will also aid in the development of the country and enable the products of the education system (graduands) to compete favourably in the world of work especially at the international level.

References

Aduwa-Ogiegbaen, S. E., and Iyamu, E. O. S. (2005). Using information and communication technology in secondary schools in Nigeria: Problems and prospects. *Educational Technology and Society*, 8 (1), 104–112.

Aginam, E. (2006). NEPAD scores students' ICT education in Africa Low. *Vanguard.*

Beebe, M. A. (2004). Impact of ICT revolution on the Africa landscape. *CODESRIA Conference on Electronic Publishing and Dissemination*, Dakar, Senegal.

Emuku, U. A. and Emuku, O. (2000). Breaking down the walls: computer application in correctional/prison education. *Benin Journal of Education Studies*, 3(4), 34–46.

Evoh, C. J. (2007). Policy networks and the transformation of secondary education through ICT in Africa, the challenges of NEPAD e-schools initiative. *International Journal of Education Using Information and Communication Technology* (IJEDICT), 3(1), 64–84.

Federal Republic of Nigeria (1994). *National Policy on Education (4ᵗʰ ed)*. Lagos: NERDC Press.

Mac-Ikemejima, D. (2005). E-education in Nigeria: Challenges and prospects. Paper presented at the 8th UNICT Tsak Force Meeting, Dublin, 13–15.

Ngurukwem, J. (2005). Information and communication technologies and teacher education in Australia. Technology. *Pedagogy and Education*, 12 (1), 39–58.

Nwachukwu, P. O. (2006). Appraising the relationship between ICT usage and integration and standard of teacher education programs in developing economy. *International Journal of Education and Development, 2(3), 70–85.*

Nworgu, B. G. (1991). *Educational research: basic issues and methodology.* Ibadan: Wilson Publishers Ltd.

Peraton, E. and Greed, U. (2000). *Integrating ICT into learning worldwide.* London: Macmillan Press Ltd.

Reffell, P. and Whitworth, A. (2002). Information fluency: critically examining IT education. *New Library World*, 103 (182), 427–435.

Okwudishu, C. H. (2005). *Awareness and use of information and communication technology (ICT) among village secondary school teachers in Aniocha South Local Government Area of Delta State.* Abraka: Delta State University. Unpublished B.Sc. (LIS) project.

Yusuf, M. O. and Onasanya, D. (2005). "Information and communication technologies and education: Analyzing the Nigerian national policy for information technology." *International Education Journal* 6 (3), 316–321.

Chapter 5

Utilisation of Educational Technology as Correlates of students' Academic Performance in Ekiti State Technical Colleges

Ogunlade, B. O.
and
Babalola, J. O.

Introduction

In any nation, development can be in terms of sociocultural or socioeconomic advancement that later transforms into enterprises of technology. The ingredient that leads to such development is the ability of man to utilize and explore satisfying relevant needs and aspirations of the society. Education is the only means of developing the natural potential of an individual which many people do not recognize. In other words, educational technology is the design and application of teaching tools like whiteboard and pen, bulletin board, computer, projector, radio, television, and telelecture in order to improve the teaching and learning process.

Educational technology can be regarded as devices, objects, or things that are used by teachers to transmit, transfer, and share their lessons with their students. In the process, the teacher must facilitate the learning to be effective, in order to provoke curiosity, capture the attention, and arouse and sustain the interest of the learners. To achieve this is to formulate and to initiate such lesson with educational technology which should be designed, produced, and presented in a systematic manner (Ibe-Bassey, 2004).

Okoro (2007) supported the above assertion that if a teacher introduces novelty into teaching, the ability and interest of students will be aroused. There is therefore the need to present the contents of the subject using educational technology. Cox (2000) found out that utilisation of educational technology in technical colleges makes the lesson more exciting and interesting for teachers as well as students. More so, the use of educational technology in the classroom could add to the understanding of cognitive and affective capabilities of students. This can equally help students to interact with each other, argue, and think well and to make decisions. Teachers are indispensable in the utilisation of educational tools in achieving better academic performance in a school system. Because of the major role they play and the position they occupy in the implementation of the curriculum, they remain the facilitators of instructional activities in the school system. Hence, more attention should be paid to the uses of the available materials or tools that make for connectivity between students and teacher's education. Research has shown that there is a very strong relationship between students' performance and utilisation of educational technology (Sanders and Rivers, 1996).

Ogundare (2003) equally subscribed to the idea that educational resources utilisation increases the rate of learning and simultaneously gives relative freedom to the instructors to use more time to access gainful activities. Ajayi (1999) and Yoloye (2004) found that inadequate and inconsistent in the use of educational technology has been responsible for students' poor performance in technical education. Ajayi (2004) stressed further that success recorded especially in language education can be attributed to the proper utilisation of educational technology which undoubtedly leads to improved and efficient teaching and learning activities in the technical schools.

Yusuf (2005) in his observation noted that combining educational technology with teaching and learning enhance better performance of students who use it when compared with their counterparts who did not. It is essential to develop a lifelong education and to incorporate the required educational technology. Akpan and Essien (1992) stated that there is consensus on the opinions that the use of educational tools to complement conventional instruction obviously improved students' achievement. Harbor-Peter (2001) is of the opinion that poor performance of students in technical subjects can be attributed to poor and ineffective instructional skills and methodologies employed by the facilitators. In the same vein, Ogunlade (2011) explained that poor performance is as a result of not utilizing necessary educational technology in solving technical subjects' problems.

Technical education plays vital roles in human resources development of a country because it creates skilled manpower, enhances industrial productivity, and improves the quality of her citizen (Ogunna, 2009). In order to find permanent solutions to some of the problems of technical education in Nigeria, the challenges posed by use of educational technology and modernisation must be rigorously pursued in line with internationally acceptable educational technology standards as a shortcut to fortifying the Nigerian education system. Hence, proper utilisation of available educational technology can promote academic performance and give to the learners increased conceptualisation and understanding of vital language skills as available in different school subjects as verbal explanations (Akude and Anulobi, 2014). The utilisation of educational technology becomes effective when the users know how to identify and select the right medium.

Ogunsheye (2001) has the opinion that effective utilisation of educational technology in this global comunity requires the training of human resources for a Nigerian learner. At any rate, the learner should not be left out in the use and application of this new technology. Access to educational technology is a major concern of technical educators and administrators because the level of instructors' awareness of educational technology is still low (Ogunlade, 2011).

Statement of the Problem

It is observed that the utilisation of educational technology seems to be declining among instructors and students in Technical Colleges in Ekiti State, which may be responsible for the poor performance of students in internal and external examinations in the state. If educationally relevant academic tools are not well attended to, it may negatively affect academic performance of students. The inability of instructors to properly utilize and appropriately match specific learning objectives to ensure understanding of educational and technology is the concern of the researchers. Consequently, the study investigated the utilisation of educational technology and its relationship with students' academic performance in technical colleges in Ekiti State.

Purpose of the Study

The study is aimed at investigating utilisation of educational technology and their relationship with students' academic performance with particular reference to Ekiti State technical colleges. The study examined the academic performance of students using educational technology with those that are not using it. This study is designed to investigate whether educational technologies are effectively utilized or not.

General Question

A general question was raised to further address the problem of the study: Do the instructors in technical colleges utilize educational technology or not?

Research Hypotheses

The following null hypotheses were generated to further address the problem of the study.

1. There is no significant relationship between utilisation of educational technology and students' academic performance.

2. There is no significant relationship between instructors' qualifications and utilisation of educational technology.

Significance of the Study

The findings of this study will be useful to students, instructors, principals, parents, curriculum planners, policy makers, society, government, and organisations that are involved in the creation of awareness for the improvement of teaching and learning through educational technology. The information generated in the study will assist those designing the curriculum of technical education programmes which may enhance a better change in behaviour among instructors and students in general.

Methodology

The research design employed was a descriptive survey study. The study was carried out in Ekiti State. There are five public technical colleges. The target population of this study consisted of 307 instructors and 1,270 students in Ekiti State technical colleges. Purposive sampling technique was used to select the instructors in all technical colleges while multistage and simple random sampling were used to select students in identified technical colleges. Twenty hundred students and fifty instructors were finally drawn from the colleges.

Research Instruments

The researchers utilized two self-designed sets of instruments for this study, Personal Information Characteristics Utilization of Educational Technology (PICUET) and Students Performance Test (SPT). The "PICUET" has two parts; part A has sections, personal information and characteristics of instructors while section B elicited information on educational technology tools that are used by the respondents. Part B consisted of forty questions from four general and compulsory subjects, ten questions elicited information on General Science, ten questions from technical drawing, mathematics, and English Language.

The method used in validating the instrument were face and content validity procedures. The items constructed were presented to specialists in test and measurement, technical education, educational technology, and English language in the faculty of education, Ekiti State University for the determination of the appropriateness of the instrument. Expert judgments were used in determining the content validity. However, the experts took time to check the extent to which the items of the instrument were representative of the content being measured. Based on their suggestions, a corrected copy was tested for reliability. Copies of the questionnaire and test were administered twice (two weeks interval) to equivalent respondents who were not part of the final sample. Responses from the instruments were compiled, scored, and analyzed using test-retest method and reliability coefficients of 0.75 for PICUET and 0.63 for SPT were obtained, and these were considered significant at 0.05 level.

Administration of the instruments

The instruments were personally administered by the researchers. The personal contacts with the respondents facilitated the speedy and effective response to the test and the questionnaire.

Data Analysis

Both descriptive and inferential statistical methods were used to analyze the data collected. The descriptive statistics such as percentage, mean, and standard deviation were used while Pearson Moment Correlation Coefficient was used for the inferential statistics and all the hypotheses were tested at 0.05 level of significance.

Results and Discussion of Findings

A total of fifty and two hundred questionnaire copies were administered on instructors and students.

General Question 1: Does the instructors in technical colleges utilize educational technology or not respectively?

Table 1. Frequency and percentage showing utilisation of educational technology by the instructors

		Utilised		Not utilised	
S/N	Variables	F	%	f	%
1	Whiteboard and pen	28	56	22	44
2	Bulletin board	50	100	-	-
3	E-mail	10	20	40	80
4	World Wide Web	11	22	39	78
5	Computer	40	80	10	20
6	Projector	20	40	30	60
7	Films and videotape	30	60	20	40
8	Movies	10	20	40	80
9	Radio	20	40	30	60
10	Overhead transparency	-	-	50	100
11	Game and simulation	1	2	49	98
12	Slide	10	20	40	80
13	Telelecture	5	10	45	90
14	Television	50	100	-	-
	Overall	285		415	
	Overall Percentage		42.1%		57.8%

Table 1 shows the extent of utilisation of educational technology tools in technical colleges by the instructors. Fourteen educational tools were listed and examined if they are used for teaching. The analysis shows that 56% of the respondents used whiteboard and pen while 44% did not use them. All of the respondents utilised the bulletin board. Twenty percent of the respondents made use of electronic mail while 80% did not use it. Twenty percent utilised World Wide Web while 78% indicated they did not use it. Eighty percent of the respondents indicated that computers were used while 20% percent did not make use of it. Forty percent of the respondents made use of projector while 60% did not. Sixty percent utilised films and videotape while forty percent did not utilise it. Sixty percent of the respondents indicated that they made use of radio while 40% did not utilise it. Overhead transparencies were not used. Ninety-eight percent did not utilise game and simulation while only 1-2% of the respondents used it. Twenty percent utilised slides and 80% of the respondents did not. Ninety percent did not make use of telelecture while 10% utilised it. All the respondents (100%) utilised television for instructional purpose.

Table 1 revealed that fourteen items were generated, which attracted a total score of 700 for utilisation of educational technology facilities, while 415 were recorded as nonutilisation. The results therefore indicate that educational technology facilities were not adequately utilised in the technical colleges.

Research Hypotheses

1. There is no significant relationship between utilisation of educational technology and students' academic performance.

Table 2. Summary of relationship between utilisation of educational technology and students' academic performance

Variables	N	Mean	SD	r-cal	p-value	Comment
Utilisation of education technology	50	7.020	1.449	0.134	0.195	N.S
Students' academic performance	200	16.400	3.687			

$P<0.05$

Table 2 shows that r-cal (0.134) is less than P-value (0.195) at 0.05 level of significance. The null hypothesis is accepted. Therefore, there is no significant relationship between utilisation of educational technology and students' academic performance.

2. There is no significant relationship between teaching qualification and utilisation of educational technology

Table 3. Summary of teaching qualification and utilisation of educational technology

Variables	N	Mean	SD	r-cal	p-value	Comment
Teaching qualification	50	6.10	2.401	0.055	0.195	N.S
Utilisation of educational technology	50	8.720	3.602			

$P<0.05$

Table 3 shows that r-cal (0.055) is less than P-value (0.195) at 0.05 level of significance. The null hypothesis is accepted.

Findings and Discussion

The findings indicated that educational technology facilities were not properly utilised in the technical colleges because of infrastructure problems. These findings agree with the position of Akude and Anulobi (2014), who discovered that public schools encounter problems in the utilisation of instructional materials in the areas of power failure, and lack of in-service trainings for the teachers. Nonutilisation of educational technology also affects education as supported by Obibuaku (2011) who indicated that poor instructional facilities are the causes of decline in the quality of education.

Table 2 revealed that the relationship between teaching qualification and utilisation of educational technology is positive, weak, and not significant. This implies that as much as teaching qualification increases, the utilisation of educational technology will also increase. The findings on the relationship between teaching qualification and utilisation of educational technology also agree with the view of Doug (1996) who confirmed that teachers do not use their instructional medium to yield maximum productivity in the teaching and learning process.

This research finding showed that the use of educational technology is not in line with the qualification of instructors. A lot of issues can hinder qualified instructors from performing well, such as supply of the material, lack of skills and technical know-how on the part of the instructors, lack of in-service trainings for instructors to update their knowledge on educational technology, lack of incentives for instructors to encourage improvisation, lack of maintenance culture, and inadequate power supply. This is not in line with the finding of Kadzera (2006) that additional qualification has a relationship with the use of instructional technology in the secondary schools.

Implications of the above findings are that educational technology, if effectively used can improve student learning and raised their performance in both internal and external examinations. Government, nongovernmental organisations, education stakeholders should give more attention to the procurement and utilisation of educational technology materials.

Conclusion and Recommendations

This study has shown that the use of educational technology such as projectors, computers, television, videotape, and other Internet facilities by instructors in technical colleges in Ekiti State is still very low. It further reveals that the utilisation of educational technology is not taken seriously in our schools by the instructors even though educational technologies are the materials that can increase standard and quality of education. Negligence of the utilisation of these tools can cause a serious decline in the quality of education made available to the learners. Qualification of instructors is an important ingredient to foster excellent teaching and learning processes, but other factors have to be present in order to produce better learning outcome. Utilisation of educational technology materials will improve if the following recommendations can be taken into consideration:

1. More attention should be given to all technical colleges in terms of provision and utilisation of educational technology materials.

2. Educational administrators should make efforts in monitoring instructors in the use of educational technology.

3. Learning resource centres should be established in technical colleges for training, seminars, and workshops regularly for improvement.

4. Adequate and alternative electricity (solar power) should be provided in all technical colleges.

References

Ajayi, D. T. (2004). *Educational technology methods, materials, machines.* Jos: University Press Ltd.

Akpan, A. P., and Essien, E. F. (1992). The effect of instructional visual on achievement and retention of concepts in secondary school students, Published Research Thesis University of Calabar.

Akude, I. and Anulobi, J. C. (2014). Utilization of educational technology materials in teaching and learning in public and private primary school in Owerri, Imo State *Journal of Educational Media and Technology* 18 (1): 37–47.

Cox, M. J. (2000). Information and communication technologies: Their roles and value for science educator in Monik M. (Eds); Good Practice Science Teaching What research has to say? U.S.A. Open University Press.

Doug, J. (1996). Evaluating the impact of technology. The less sample answer, *Educational Technology Journal.* http://www.isd771k12:mn.us/staffdir/staff2/johnsonDoug.html.

Ibe-Bassey, G. S. (2004). Principle and principle of instructional communications. Uyo: Dor Publishers.

Kadzera, C. M. (2006). Use of instructional technologies in teacher training colleges in Malawi. Unpublished PhD Dissertation, Virginia Polytechnic Institute and State University, USA.

Obibuaku, L. O. (2011). Free education in Imo State: Challenges and possibilities. Owerri. Assumpta Press, 14–21.

Ogunojemite, G. B. (2006). *Distance education in perspective*, Lagos. Bifocal Publisher.

Ogundare, S. E. (2003). Fundamentals of teaching social studies. Oyo: Immaculate Publishers.

Ogunlade, B. O. (2011). Availability and utilization of instructional technology as correlates of students' academic performance in Ekiti State Technical Colleges. Unpublished Master's Thesis, Ekiti State. University, Ado-Ekiti.

Ogunsheye, F. A. (2001). *Syllabuses for library use, education programme in secondary schools and teachers' colleges,* Ibadan. University of Ibadan, Abadina Media Resources Centre.

Ogunna, J. T. (2013) Applying educational technology in a developing economy for motivational learning of Home Management Education, *Journal for Technology Education* 9(1): 248–255.

Okoro, C. C. (2007). Basic concepts in educational psychology. Uyo: Abaam Publishing Company.

Sanders, W. L. and Rivers, J. C. (1996). Cumulative and residual effects of teachers on future academic achievement. Knoxville: *University of Tennessee Value-Added Research and Assessment Centre.*

Yusuf, M. O. (2005). Information and communication technology and education: Analysis of Nigerian national policy for information technology. *International Educational Journal.* Adelaide, Australia 5 (3): 319–321.

Chapter 6

Design and Utilisation of Educational Technology Resources for Improving Teaching and Learning of Technical Vocational Education and Training Programmes in Nigerian Schools

Olabiyi, Oladiran Stephen

Introduction

In Nigeria, a lot of opportunities are available to ambitious and progressive technical vocational education and training (TVET) teachers to learn various techniques that will make their outputs more effective, while at the same time make their jobs more interesting and pleasurable. One such avenue exists to train, procure, and produce teachers that will be able to impart the required knowledge and skills to the technical vocational education students. Therefore, efforts should be made to equip the schools with technology tools, human, materials, and financial resources so as to achieve the objectives of TVET programs.

Educational technology has been variously defined by different authors; however, three major areas of perception of the concept include hardware, software, and system approach. The hardware approach, which is the oldest concept, sees it as mechanizing or automating the process of teaching with devices. On the other hand the software approach looks at it from the point of view of learning theories and reinforcement principles in the design and presentation of stimulus materials. The systems approach sees it as systematic way of designing, presenting, and evaluating instruction. The Association for Educational Communication and Technology (AECT, 1982) gave a comprehensive definition as a systematic way of designing, carrying out, and evaluating the total process of teaching and learning in terms of specific objectives based upon research in human learning and communication and employing a combination of both human and nonhuman resources to bring about a more effective instruction. An important aspect of educational technology is instructional design, which according to Siemens (2002) is the systematic process of translating general principles of learning and instruction into plans for instructional materials and learning. This should involve real application of educational technology resources competencies in the teaching process. The Educational Resource and Information Centre (ERIC, 2008) defined educational technology as the study and ethical practice of facilitating human learning and improving teachers performance by creating, using, and managing appropriate technological resources and process.

Educational technology is aimed at the interaction process among the teacher, media, and learner, upon which all other systems are dependent. All technical education programmes planning, management, and curriculum development have in the end to justify themselves by the effectiveness of learning outcomes. The effectiveness of teaching and learning of technical vocational education programs has to meet the needs of both the individual and the society through the use of multimedia. One of the most current and crucial moves towards better educating students is the application of technology tools in all areas of studies (Albion, 2001). Multimedia has been considered as an effective media for enhancing teaching and learning process. These tools are commonly found in many schools, from elementary schools to universities. However, the world of multimedia is relatively new in some places and can be defined as the combination of different media (Howard, 1994), such as text, sound, image, and video, to convey information. To be more precise, Marshall (1999) states that multimedia is the field concerned with the computer-controlled integration of text, graphics, drawings, still, and moving images (video), animation, audio, and any other media where every type of information can be represented, stored, transmitted, and processed digitally.

The common use of a technology tool is due to its functionality that brought the fundamental change for people in sending and receiving and interpreting information. Pea (1991) states that multimedia help students in many ways. It provides the context for the abstract word or explanation such as a different emotion on a human face, and to fulfill individual needs and preferences so that students can learn the material accordingly. Therefore, Teoh and Neo (2007) state that multimedia offers an alternative way of instruction to the current or traditional learning process. The new function of multimedia has led programmers to develop new software for teaching and learning process in technical vocational education and training (TVET).

Technical Vocational Education and Training (TVET)

Technical vocational education and training according to Thomson (2012) is defined as an instructional program which includes general studies, practical training for the development of skills required by the chosen occupation and related theory, in which the proportion of these components vary considerably depending on the program (vocational agriculture, vocational business, vocational home economies, and vocational technical), but emphasis is usually on practical training. By implication, vocational education helps in developing an individual's psychomotor, cognitive, and affective skills, so that they can take their rightful place in society. Vocational education is more all-embracing than technical education. The latter is seen as a postsecondary vocational training whose primary purpose is to produce self-reliant technicians who are technical driven with affective skills in different occupations. Technical vocational education and training (TVET) is one of the recognized and effective processes by which quality, up-to-date, information, literate, and knowledgeable workers are prepared and trained for sustainable society. Both of them required educational technology tools.

The Federal Republic of Nigeria (2004) describes TVET as a comprehensive term referring to those aspects of the educational process involving, in addition to general education, the study of technologies and related sciences, the acquisition of practical skills, attitudes, understanding, and knowledge relating to occupations in various sectors of economic and social life. In a nutshell, TVET prepares human resources for the ever-changing world of work. The reason for this is that, for effective participation in the world of work, the study of technologies and related sciences as reflected in the description of TVET is of paramount significance that can be realized with adequate educational technology resources arrangement in TVET institutions. Practical skills can be delivered virtually via a well-organized technology tools setup; gone are the days when practical skills are taught using only hands-on learning. Programmed instruction in the form of software and interactive video made it easy for practical skills to be taught using technology tools. TVET incorporates the total learning experiences offered in educational ideas and abilities to make mature judgments and be in a position to create goods and services in the area of business education, industrial technical education, computer education, home economics education, agricultural education, and fine and applied arts education.

In a nutshell, TVET prepares human resources for the ever-changing world of work. For effective participation in the world of work, the study of technologies and related sciences is of paramount significance, and this can be achieved through educational technology resources. TVET in the views of Okorie (2001) has a good potential of creating jobs for the unemployed graduates, and reducing poverty level of the people since those who have undergone training in this area can establish their own business thereby getting income to take care of their families, and making people stand on their own economically without depending on others (self-reliance). Productive, competent, and flexible personnel are a prerequisite for further economic development, TVET though effective utilisation of education technology resources will increase expertise and capacity of technical institutions to provide effective work skills to the populace.

A quality TVET programme plays an essential role in promoting a country's economic growth and contributing to poverty reduction as well as ensuring the social and economic inclusion of marginalized communities. UNESCO views TVET as the master key to poverty alleviation and social cohesion and as central to promotion of sustainable development. An educational technology resource has the potentials to simplify and facilitate the quality of instructional processes in our schools. Cavas, Cavas, Karaoglan, and Kisla (2009) explained that ICT supports teaching and learning processes by providing new opportunities for interaction between students and knowledge; provides students easy access to information; offers the potential to meet the learning needs of individual students to promote equal opportunity, and also promote interdependence of learning among students. Additionally, educational technology tools support holistic learning, collaborative grouping, problem-oriented activities and integrated thematic units. Teachers wishing to teach in this way will be both more efficient and effective if they employ tools to reach their goals (Dellit, 2002).

Training Resources Required for Teaching Technical Vocational Education and Training

The acquisition of practical and applied skills as well as the basic scientific knowledge that would facilitate efficient occupational efficiency requires good manipulation of skill-oriented training resources in a favorable environment. Such a learning situation can be created through effective utilisation of training resources. Furthermore, an important component in teaching TVET is training facilities. It enables an instructor or teacher to promote multiple senses of skill acquisition in TVET students, as we have seen from learning theories, the more sense stimulated in the teaching and learning process, the easier it is for students to (recall) remember what they have learned. TVET teachers and instructors have to continually bear in mind that learners learn through their senses. Some learn better by one sense or some other senses; to some, seeing is believing. To others, the sense of learning by touch, smell, and taste dominate in technical vocational education. Technology tools provide many new opportunities to address issues such as learning styles and student-centred instruction and promote higher-level thinking. One of the possible means of acclimatizing TVET to develop human resources for the ever-dynamic world of work is to focus its investment in design and utilisation of educational technology resources in the curriculum implementation process (teaching and learning). The aim of TVET is to prepare people for self-employment and to be a medium of evolution for people to the world of work, by helping individual to have a sense of belonging in their communities. Consequently, TVET is seen as an instrument for reducing extreme poverty (Hollander and Mar, 2009). These distinctive features of TVET make technology tool application a mandatory component that can aid in achieving a sustainable and globally recognized workforce.

Within TVET learning, software such as computer-assisted instruction, Energy Performance Design System (EDPS), many programs for improving drawing, spatial ability, creativity, and writing skills are easily found within the Internet. A lot of resources can be accessed to get these materials; the most common one is CAD, while TVET teachers can adapt these tools or design his own to use in teaching.

Also, the concept of training facilities is perceived by various authors. Nwachukwu (2001) posited that skill acquisition that would facilitate occupational efficiency requires performance and a skill oriented situation. He explained that training facilities are all practical and skill acquisition resources that would facilitate the process of teaching and learning and evaluation of technical vocational education skills. He also described it as the electronic system, tools, equipment, and other resource materials that could be utilised in directing and controlling technical vocational education operations and for reinforcing the teaching and learning. He further explained that training facilities are devices developed and acquired to assist technical vocational teachers in transmitting organized knowledge, skills, and attitudes to learners within an instructional situation directed towards learning and acquisition for work. Ogwo and Oranu (2006) described training facilities as any device employed by the instructor or teacher to transmit facts, facilitate skill/knowledge acquisition, and improve an attitude, understanding of learners. The authors emphasised that training facilities include models, objects, drawings, graphs, and charts, pictures, films and specimens. And all these are found in the domain of educational technology.

Awotu-Efebo (2002) explained that training facilities comprises all available and accessible theoretical, practical, and skill-oriented resources which facilitate the learning, acquisition, and evaluation of technical vocational education teachers in transmitting the facts, skills, attitudes, and knowledge to the learners within the instructional system and as may be applied to the world of work. Awotu-Efebo asserted that they are resources or equipment which are essential in the teaching process to achieve the objective of teaching. This is in line with Olaitan (1996), who cautioned that without the use of some materials, tools, or equipment in teaching skill-acquisition programmes, certain skills that might be required for entry into some vocational occupational areas might not be imparted. It therefore means that educational technology tools are necessary for some vocational skills to be learned by TVET students.

Olowu (2002) classified training resources into human and nonhuman resources. Human resources, according to Olowu, include people who have various educational backgrounds, experiences, skills, position, and status and are engaged in the teaching and learning relevant skills. These include people who are subject specialists, such as woodworkers, home economists, auto mechanics, professionals such as doctors, engineers, nurses, and teachers or instructors among others. Nonprofessionals according to Olowu are classified as administrative support staff, cleaners, clerical officers, and semiprofessionals. Semiskilled workers include auxiliary teachers and nurse ward maids, copy typists, and laboratory attendants, among others. These categories of human instructional resources are invited as resource persons in any teaching and learning situations. Gerlac and Ely (2000) regarded instructional resources as including a wide range of materials and equipment such as motion pictures, television programmed instruction models, demonstrations, books, audiotapes, and computer-assisted stimulatory printed materials, equipment, and combination of them.

Principal tools, equipment, facilities, machines, and materials used for instructional purposes in TVET programs include the following functional library, which comprises basic reference materials, such as encyclopedias, dictionaries, modern textbooks on various subjects, a library built with facilities such as recording studies where books are recorded on cassettes, Internet facilities, and audiovisual equipment. Olaitan, Nwachukwu, Igbo, Onyemachi, and Ekong (1999) explained that in vocational technical education, acquisition of diverse technical skills calls for utilisation of diverse training resources. This forms the basis for categorizing training materials in vocational technical education into three groups, namely: tools and equipment, fixed facilities, and consumable materials.

Tools and equipment describe all the portable and heavy instruments or mechanical devices useful for performing special operations in TVET teaching and learning situations. They are the instruments that can be handled easily while carrying out special operations as well as instructional or learning activities. Tools are usually used in transmitting knowledge in the workshop to the learners. They are used for demonstrations, for practices in the process of learning skills, and for testing skills in specified vocational areas. Equipment, on the other hand, refers to portable or heavy mechanical devices usually used for TVET operations in workshops. Equipment is more sophisticated than tools. The use of equipment is mainly on technical specialized skills, practices in the instructional and learning environment. Tools and equipment used for technical education are classified in specialized areas: automobile, building, electrical and electronics, metalwork, and woodwork.

Fixed facilities are the facilities positioned at a particular place for the performance of specified and specialized operations for providing required services. An example of a fixed facility is a building needed for numerous purposes. A workshop is meant for woodwork, metalwork, automobile, electrical and electronics operations with fixed equipment, schools, farmlands, orchards, plantations, fishponds, different livestock structures and buildings, and processing storage facilities. Consumables are materials that are utilised for machines as components of the production of observable job outcomes. Consumables are the basic materials requirement for facilitating skill development activities and practices. The special areas of TVET have a need for different consumables; for instance in woodwork, the consumables are: wood, nails, glues, wood products, paints, fittings, and fixtures among others, metalwork requires steel, electrodes, and sheet metal. Electrical operation requires consumables such as cables, nails, wire, sockets, masking tape among others (Olaitan, et al., 1999).

Application of educational technology tools in teaching both science and technology subjects according to Odumosu and Keshinro (2000) is the integration of learning aids, simulation, modeling, Internet services, lesson presentation, AutoCAD, ArchiCAD, drill and practice and computer-assisted instruction (CAI). In addition, Adewoyin (2006) indicated that ICT can be applied in producing smart drawing, semantic, GIF construction, videotape, relay chart, tool book, graphic design, and animation.

Design of Educational Technology Teaching Resources in TVET

Design is the human beginning of every manmade product in existence, there has to be practical or scientific reasons for this involving exercise. Design, therefore, is an intentional planning or inventing and making of an article for a particular use. According to Walton (1974) the term *design* refers to the article itself as well as the planning of its construction, operation and appearance. To Olabiyi (2005) design is viewed as a creative process of using available resources to provide what people need or want, which can be made by either hand or machine to meet people's specific need. From this definition, design must satisfy the felt need. It is purposeful creation activity resulting from creative thinking. The end result is usually a new product, either as an original creation or modification of an already existing product. If products are designed from the standpoint of function, materials, appearance, and construction technique, it is believed that those products will ever be anything but contemporary. Design as applied to educational technology in the production of teaching materials, resources, and devices, is purposeful creation or modification of existing teaching resources or materials consequent upon which new products are produced to satisfy felt instruction needs. Design is generally concerned with how things work (technology), how things look (aesthetic), and how things feel (emotions) (Olumba, 1996).

Stages in Designing and Selecting Educational Technology Resources in TVET Programs

Educational technology resources are being designed from the moment they come to mind. When thinking about the product, ask some questions such as How would the product serve the purpose? How will it look like? Are there materials and tools or equipment to produce the product? How much will it cost to make the product? Because of these questions, it is essential to develop and record the mental picture and identify the stages involved in designing educational technology resources in technical vocational education and training resources. Specifically the following need to be considered:

- Determine your students' needs and the purpose of the resources.
- Generate a design or develop ideas, develop several sketches of the product, develop a model and do some experiments, develop the sketch and model into final drawing.
- Select the materials and construct the resources using good workmanship.
- Judge the product to see it fits your students' needs.

Educational technology resources either commercially produced or teacher-made are designed, produced and utilised for executing different types and levels of instructional objectives. It could be for a specific lesson, unit, and modules or for an entire course of study. For whatever purpose, the responsibility of selecting the most appropriate resources rests on the teacher. Superficially, making the choice could be adjudged a very simple task. In fact, it transcends ordinary selection from the abundance of competing educational resources. To succeed in this important instructional decision, the technical teacher requires good judgment, intelligence, and unique competency. These qualities are desirable because selection and utilisation of an unsuitable educational resource exert far-reaching adverse effects on the learners. On one hand, it could constitute a great distraction to learning rather than attracting the users. In the same vein, hearing may be blocked instead of being facilitated. Apart from the above effect, there are other factors that are important and worthy of consideration in selecting educational resources for instructional purposes. These, among others factors, include attributes of the instructional resources, individual differences among the learners, instructional objectives, class size, availability of infrastructural amenities, teachers' capability in handling the teaching resources, learner's cultural background, nature of the learning task, and the enabling task. Other factors include durability, portability, currency, availability, maintenance and service, cost, ease of operation, safety, aesthetic, and availability of support personnel. In addition to the above variables, a requisite knowledge of the psychology of human learning is an essential criterion for selecting instructional resources.

The Learner. The conventional classroom or workshop comprises learners with distinct and diverse biological, social, cultural, religious, and economic background. As a result, these learners possess different abilities, interests, attitudes, motivations, and rates of learning. Despite these observable differences, learners are grouped in a class or workshop under the guidance of the teacher. For teaching to be considered effective, teachers must work for the benefit of all. The teacher provides a conducive learning environment, a relevant and interesting classroom or workshop experience motivating enough for each of the learner. Thus, the consideration of the learner in instructional resources selection behooves the teacher to identify inherent differences among the learners and select the most suitable educational technology resources that will beneficial to all. Thus, the educational resources selected should be tailored to suit the nature, interest, need, and aspiration of the learners.

Instructional Objectives. Instructional objectives are the expected end results of an activity which directs the behaviour and activities involved in teaching technical education programs. Instructional objectives emphasised what teacher should prepare to achieve. Instructional objectives indicate the changes expected in the learners after completing a learning experience. Effective teaching is based upon well-defined and specific objectives stated in behavioural forms. Because learning requires active efforts of learners, the objectives must be stated in terms of activities that will best facilitate students' learning and what they should be able to do at the end of the teaching experience. Instructional objectives guide in the selection of instructional resources. As a matter of importance, the instructional objectives should be decided on the basis of available instructional resources and the teaching method to be adopted.

Furthermore, the instructional objectives suggest the activities expected of the learners and the materials to be used. Therefore, instructional objectives (affective, cognitive, and psychomotor) should guide teachers in the selection of instructional materials for instruction. In a situation where instructional materials bear no relationship on stated instructional objectives, the attainment of the objectives may prove difficult. Marrying instructional objectives with instructional materials demands thorough analysis of the lesson content to identify the important, concepts, principles, and tasks. Selection of learning contents involves listing the knowing.

Types of Learning and Tasks. A teacher's knowledge of the types of learning and the relevant activities to facilitate them equips him better in his choice of instructional materials. This is important because a teacher in a single lesson may have to promote different types of learning from simple to complex types as concepts, principle, and problem solving. For instance, in a wood workshop, the teacher wants to teach the students how to construct mortise and tenon joints; hand tools required for making mortise and tenon joint, steps in wood preparation to the required sizes, methods of marking mortise and tenon, cutting of tenon and chiseling of mortise, trial fitting of joints. Content analysis of the lesson shows that three types of learning are involved. In hierarchical order they are: chaining, concept learning, and principle learning and problem solving. These types of learning are facilitated through discussion, demonstration, observation, and the questioning method. Therefore, instructional resources capable of enhancing the methods will also promote the various hierarchies of learning identified

Teacher Capability in Handling Instructional Resources. The ultimate goal of utilising instructional resources in technical education is to promote or facilitate teaching and learning. This presuppose that the teacher possess the requisite competencies for effective utilisation of the instructional resources and management of instruction, oftentimes, instructional resources, especially the hardware constitute nuisance in the class resulting from the teacher's lack of skill or inexperience in their operation. Instead of facilitating learning, such resources tend to block it and the teacher thrown into state of dilemma, confusion, and even ridicule. The panacea is for the teacher to always strive to select instructional materials and devices which they possess enough skill and knowledge to operate. For sophisticated hardware, last-minute rehearsal prior to actual demonstration in class is important. The teacher thus evaluates his competency and certifies the device functionality. Where necessary the services of a knowledgeable technician should be employed.

Class Size. The Federal Republic of Nigeria (FRN, 2004) recommended that the class size should be 1:40; in reality a laboratory should enroll an average of twenty students. The FRN (2004) emphasised that for effective participation of students in practical work, the teacher-students ratio shall be kept at 1:20. This consideration relates to the number of learners in the class and the physical dimensions of the classroom. Some instructional materials have the potency for large and small group instruction while others are best used for small groups. For instance, in a large lecture theatre that accommodates about five hundred to one thousand students, projected aid such as film, film strip, transparences, and slides are suitable. In such a large class, multiple screen can be used. Others like chalkboards and other types of boards, flip charts, specimens, among others are ideal for small groups. For visual materials, visibility and clarity should predicate their use.

Cost. Before deciding on the use of any instructional material or device for classroom or workshop instruction, the financial involvements should be ascertained. Instructional resources should be the type that can be procured or accommodated within the limits of the funds available whether commercially produced or improvised. Where there are several alternatives, their cost effectiveness relative to performance needs to be determined. There is nothing wrong in using a cheaper instructional resource, provided it is capable of achieving the objectives of the lesson.

Learner's Cultural Background. Learners in the classroom come from different cultures. Diverse as the culture may be, they exert distinct and enormous influences on the learners which subtly or overtly manifest in the class. The influences shape learners' perceptions, thinking, values, and interests, methods of learning, outlook in life, motivation, and belief systems. As in the consideration for individual differences, instructional resources selected to enhance learning should consider the learner's cultural background and belief systems especially. Instructional resources that constitute a taboo in a given geographical area should be avoided. Attempt at using such materials may conceive as deliberate effort at ridiculing the people and their culture.

Currency of the Educational Resources. Newness of the educational instructional resources is another important pragmatic criterion that should guide the instructor or teacher in making his choice. This brings to focus the teacher currency on the substantial number of new educational resources available in the market. Knowledge of this nature is acquired through reading professional journals, attending conferences and workshops. The benefits of using current educational technology resources are innumerable. More powerful, efficient, and result-oriented methods of storing, retrieval, and dissemination of educational information in the classroom may be acquired on one hand. On the other hand, learners may acquire skills on improved methods of learning for better results. It is important that teachers and instructors should use sound judgment and intelligence devoid of prejudice in selecting the instructional resources according to need and peculiar nature of the learning content, learners and local environment.

Qualities Required of Competent TVET Teachers in Utilising Instructional Resources

Skillful instructors and teachers are vital to every dynamic, successful technical vocational education program. Facilities, instructional materials, equipment, and personnel with specialized technical knowledge are also needed, but without instructors and teachers competent in the art of teaching, no educational program can be completely successful.

Competent in the Subject Contents and Instructional Materials. There is no substitute for experience and detailed knowledge and skill in the subject being taught. The competent instructor and teacher will do a better job of studying one lesson ahead of the class and will be more at ease with students than an inexperienced instructor or teacher. The instructor or teacher should be thoroughly competent in the skills to be taught as well as in the use of the suitable learning resources that will facilitate the skills and knowledge, and related information. This is particularly true where students are being prepared for positions involving specific and specialized tasks. Students are usually alert, capable, and are quick to appraise their instructors. The competent instructor earns their respect.

Mastery of the Techniques of Instructor and use of Instructional Materials. The competent instructor or teacher prepares each lesson with the best use of student's time, in accordance with the instructional objectives of the lesson, ensuring that planning of the lesson is flexible enough to meet the needs of individuals in the class. The techniques of presentation of content and resources include being able to speak clearly without shouting; organize instruction and resources according to students learning capability; repeat and emphasise key materials in such a way that it stands the best chance of being remembered; conduct a demonstration skillfully using suitable instructional materials, and provide practice sessions and performance tests in such a way as to promote and develop desirable skills and attitudes.

Resourcefulness and Creativity. Only the incompetent instructor or teacher uses the same instructional resources and method all the time. The resources and methods that work well for one individual or for one class or for one lesson may not be satisfactory in another learning situation. The good instructor is alert to even the slightest evidence that confusion, misunderstanding, or lack of interest is present among the students, and is able to adjust his or her approach instantly to correct the situation. One of the reasons for varying instructional resources and procedures with different classes or individuals is that individual differ to a marked degree in native capacity, in background of experience, and in learning pattern. The rapidity with which a particular individual learns a particular subject depends to a large extent on how well he adopts his learning pattern to the method by which the instructor or teacher presents materials. Resourcefulness is demonstrated when the instructor or teacher designs a new instructional resource to help illustrate a principle, uses a current event to emphasise a concept, builds an advanced project to develop his own competence in the subject he teaches, or discovers a more effective way of measuring the progress of each student.

Knowledge and Application of Evaluation Procedure. The good instructor or teacher is like the good cook who keeps testing the food to see if the flavor is right, or like the craftsman who uses the senses of sight, smell, and touch to indicate when the power tool is cutting properly and safely. The instructor or teacher must be sensitive to the way the students are responding. There must be a constant desire to find out the extent to which students are learning. This can be done periodically by examinations if the questions are designed to find out whether or not the students have achieved the objectives specified for instruction. The good instructor or teacher makes every effort to tell by the expression on the faces of the students, by the questions asked, and by other indications of the extent to which a particular idea, process, or skill is being assimilated by the learner.

Desire to Teach with Instructional Resources. The attitude of the instructor or teacher towards use of instructional resources is very important. Few occupations make such demands on the emotional and mental composition of the individual as teaching. The instructor or teacher must always be projected into the thinking of others, and must do this not in the sense of command—of ordering students to do things—but as a sympathetic and understanding guide. The instructor or teacher must show, tell, and guide patiently until the beginner has acquired the necessary competency. He must learn to transfer his interest from the subject to the students. He must make central in his thinking not what he is teaching, but what students are learning.

Ability to Develop Good Personal Relations. The instructor or teacher must learn to get along with students, supervisors, and administrators. Emotional factors influence learning. Favorable attitudes such as a feeling of confidence tend to increase the level of student achievement, while unfavorable attitudes or strong emotions such as the fear of failures may block learning entirely. It must be remembered that the instructor who has good interpersonal relations with his students can require, and get, much more achievement from them in the long run than one who is disliked, resented, or not respected for his ability. The instructor or teacher must work in harmony with his fellow instructors and with his supervisors. Demonstrating willingness to do more than is required is a great help in earning the respect of one's associates and supervisors. In summary, the instructor or teacher needs to be friendly, cooperate with other staff, take part in social activities, complement the achievements of others, use tact and consideration, maintain good personal appearance, and be a professional—taking pride in the role and demonstrating a high degree of ethical behaviour.

Principles of Utilising of Training Resources in TVET

The utilisation of training resources involves the teacher manipulating tools, equipment, fixed facilities, and consumable materials to facilitate the teaching and learning process. Vocational and technical education skill development in students is facilitated by the use of training resources which, when properly used, serve the following: demonstrating specific skills, carrying out manually operated functions, providing supportive functions of used tools and equipment, performing mechanically operated activities, aiding students' skill-development activities; promoting students' memory development and recall, aiding the construction and production purposes and evaluating the success of skill acquisition (Olaitan et al., 1999, and Nwachukwu, 2002).

Availability of good quality educational education resources is not a sure guarantee of successful and efficient utilisation in the classroom or workshop. The instructor or teacher's personality and background in instructional resources utilisation are vital determinants to successful presentation. A poorly presented instructional resource may constitute a great hindrance in the teaching and learning process. Inadvertently, the instructor or teacher's diligent plans for an interesting and successful instruction would have been shattered. Therefore, the instructor or teacher needs to be guided by some fundamental principles in utilising educational technology resources.

Adequate Preparation. Effective utilisation of educational technology resources in TVET brings to scrutiny (and often times ridicule) the professional competencies of teachers/instructors. The reason is simple. Good presentation indicates teacher's/instructor's level of preparedness, knowledge of the subject matter, classroom management and organisational skills, techniques of evaluation, and dexterity in handling the instructional resources and equipment. Amidst these relevant professional competencies to be exhibited, the instructor or teacher is inadvertently on test before the learners. To pass the noncredit examination, he needs to adequately prepare himself, the class, the environment and the educational technology resources and equipment. It is required of technical teacher to master the operational guidelines of the equipment or instructional resources and the best method of presentation. This is what instructor or teacher should rehearse in his mind and in the laboratory using the intended instructional materials before the actual presentation in the classroom or workshop. Where an assistant is needed, he should be available before the commencement of the lesson. Combining good techniques for presentation with the above preparations, ultimately, will improve the utilisation and utility of educational technology resources.

Educational Technology Resources Must Relevant to Instructional Objectives Stated. The major purpose of utilising any educational technology resources is to facilitate teaching and improving learning in order to realize the expected outcomes of lesson. A strong relationship should therefore exist between the instructional resources and instructional objectives. For instance, a behavioural objective which expects the learner to describe or sketch a hand plane should prompt the use of charts, maps and diagram which illustrates the features. In the same token, instructional resources such as slides, transparencies, motion picture, and television are effective in teaching task identification and procedural knowledge and concepts.

Adapt Instructional Resources to Learner's Instructional Differences. Learners differ from one another. Thus, in some sense, they always require special attention. There are several approaches to cope with the problem. One common approach is to group the learners. Another approach is to apply techniques and measures that lead to individualized instruction to teach each student's needs. Adapting instruction to meet individual needs may require utilising special methods and instructional resources. The instructional resources should be interesting and motivating to all; to enable each child to progress at his pace and ability level. Fast learners and slow learners alike should benefit ultimately from the use of instructional resources. Also, factors related to individual differences that must be considered when utilising instructional resources are the learners' background or experiences, emotional needs, and physical and psychological needs. No particular instructional resources in realities possess the potentialities for satisfying the immediate needs of all learners at any given time. Combined resource is a way out (multimedia). Multimedia approach implies the uses of more than one type of instructional resources in a given lesson, instructor or teacher needs to acquire skills on how to combine suitably different categories of instructional resources for teaching.

Educational Technology Resources Should Be Utilized When Necessary and Appropriate. There is no law that stipulates when the instructional resources should be introduced during classroom teaching or workshop practice, the point emphasised here is that instructional materials should be introduced in a lesson at the most appropriate time and when necessary provided it will achieve its objective as considered relevant in the lesson context. Instructional resources could be introduced at the beginning, middle, or end of the lesson. The choice is that of the instructor or teacher and according to his plan. Nevertheless, the most effective instructional resources are the one introduced in "the nick of time." Experience and common sense will guide the practicing instructor or teacher more than research evidence.

Educational Technology Resources Should Not Substitute the Teacher. There may be a need to warn TVET teachers in effective utilisation of instructional resources. On no account should instructional resources should substitute the instructor or teacher; rather it should serve as a good supplement. This is a common mistake among technical teachers, even the highly experienced. With this understanding, teachers will appreciate their primary goal of guiding learning in audiovisual programs (i.e., educational television and instructional television) and audio programs (i.e., radio broadcast); there must be a point when the teacher's explanation or illustration is important to students.

Summary

Technical vocational education and training has been defined as the process of teaching individuals the systematic skills, knowledge, and attitude involved in the production of specific products or services, while educational technology is the systematic, planned approach of integrating suitable instructional materials and devices into teaching-learning plan for successful attainment of clearly stated technical vocational education goals or instructional objectives. Educational technology tools are widely utilised in teaching for example, instructional usage of involve such activities as preparing for courses, searching the Internet for course contents, making presentation in the class, carrying out laboratories experiments, preparing e-mail for sending course materials and lesson contents to students.

References

Adewoyin, J. A. (2006). The place of Information and Communication Technology in designing and utilizing instructional materials. A Book of Proceeding on a One Day Train the Trainer Open Workshop on Understanding New Technologies in Instructional Media Materials Utilization.

Albion J.D. (2001). Using technology in teaching and learning. London, Kogan.

Cavas, B., Cavas, P., Karaoglan, B., and Kisla, T. (2009). A study on science teachers' attitudes Toward Information and Communication Technologies in Education. Retrieved May 10, 2012 from *http://www.tojet.net/articles/v8i2/822.pdf.*

Dellit, J. (2002). Using ICT for quality in teaching-learning evaluation processes. In UNESCO Using ICT for quality teaching, learning and effective management. Retrieved May 10, 2012 from *http:// www. unesdoc.unesco.org/images/0012/001285/128513eo.pdf.*

Educational Resource and Information Centre (ERIC, 2008). Toward a New Consensus among Social Foundations Educators: Draft

Position Paper of the American Educational Studies Association Committee on Academic Standards and Accreditation. *Educational Foundations, 7* (4). (pp. 362–369).

Federal Republic of Nigeria (2004) *National Policy of Education.* NERDC, Press, Lagos Gerlac, O. S. and Ely, D. (2000). Teaching and Media: A Systematic approach. New Jersey, Prentice Hall Inc.

Hollander, A. and Mar N. Y. (2009). Towards Achieving TVET for All. In R. Maclean and D. Wilson (eds). International Handbook of Education for the Changing World of Work: Springer Science + Business Media BV, pp. 1863–1877.

Howard, D. (1994). Toward a theory of educational technology. In Willis, J., Robin, B., and Willis, D. A. (Eds). *Technology and Teacher Education* pp 393–397.Charlottesville, VA: Association for the Advancement of Computing in Education.

Marshall, D (1999). What is Multimedia? http://www.cs.cf.ac.uk/Dave/ MM/OLD_BSC/node10.html (accessed July 1, 2013).

Odumosu, A. I. O., and Keshinro, O. (2000). *Effective science teaching and improvisation in the classroom* Lagos: Obaroh and Ogbinaka Publishers.

Ogwo, B. A. and Oranu, R. N. (2006). *Methodology in formal and non-formal technical and technical education*: Ijejas Printer and Publishers.

Okorie, J. U. (2001). *Developing Nigeria workforce. Calabar:* Page Environs Publishers

Olabiyi, O. S. (2005). *Effective teaching method in vocational and technical education principles and practice.* Lagos: Raytel communications Ltd.

Olaitan, S. O., Igbo, C. A. Nwachukwu, C. E. and Ekong, A. O. (1999). *Curriculum*

Development and Management in Vocational Technical Education. Onitsha; Cape Publishers

Olaitan, S. O. (1996). *Vocational and Technical Education in Nigeria (Issues and Analysis).* Onitsha: Noble Graphic Publishers.

Olowu, F. A (2002). Instructional Resources in Social Studies. In Ajetunmobi and Omoladi, L. A. (Eds) *Themes in Social Studies Education and Culture*, Lagos: Raytel Publisher.

Olumba, B. A. (1996). Educational media and instruction In B. A. Ogwo (Ed) *curriculum Development and Educational Technology* (pp. 182–243) Markudi: Onaivi Printing & Publishing Co. Ltd.

Pea, R. D. (1991). Learning through multimedia. III Computer Graphics and Application, 11 (4), 58–66.

Siemens, K. C. (2002) Students perception of multimedia technology integrated in classroom Learning. *International Journal of Humanities and Social Sciences*, 2(11), 67–70.

Teoh, B. S. and Neo T. (2007). Interactive multimedia learning: Students' attitudes and learning impact in an animation course. *The Turkish Online Journal of Educational Technology*, 6(4), 28–37.

The Association for Educational Communication and Technology (AECT), 1982). *Instructional Resources and Materials* pp 39–50.

Thomson, O. P. (2012) Rethinking Teaching with Information and Communication Technologies (ICTs) in Architectural Education. Teach. Teacher. Education, 25: (1132–1140).

Walton. (1974). *Woodwork Theory and Practice.* England: Longman Group Limited.

SECTION 3

Adoption and Integration of Educational Technology Tools in the Classroom

Chapter 7

Teen Perceptions of Cellular Phones as a Communication Tool

Denise Jonas, Gary L. Schnellert, and Attia Noor

Introduction

Technological change has been a catalyst for discussion and debate for centuries. The emergence of new systems or inventions have influenced society and changed the way people learn, work, and communicate (Andersen, 2009) From system theories on crop rotation to the invention of the computer, technological change can be conflicting as early adopters embrace its presence while laggards are skeptical of its use. In essence, this mixture of resistance and acceptance in the evolution of technology contributes to a "love-hate relationship" for its users. Cellular phones are a phenonmena in the evolution of technology and are at the forefront of discussion and debate of whether users should resist or adopt.

Research reveals the cellular phone is the most popular technology on the market today. A report by the Pew Research Center noted 91% of American adults own a cellular phone, Blackberry, iPhone, or other related device. Teens, ages twelve to seventeen are the fastest-growing market of cellular phones users with 75% owning a cellular phone (Commonsense Media, 2009), revealing an increase from 2007 when two-thirds of United States teens ages thirteen to nineteen owned or carried a cellular phone (Cotten, Shank, & Andersen, 2014). Wireless connectivity and advanced cellular phone features make cellular phones the "must-have" gadget for teens today as about one-third of teens access Internet via cellular phones (Cyberbullyalert.com, 2008).

Teens embrace cellular phones for a variety of opportunities. Instant communication anytime anywhere contributes to an underlying feeling of safety (eSchool News Staff, 2006). Phone features such as text messaging, phone cameras, organisation tools, and entertainment applications have changed the way teens entertain themselves and communicate (Harris Interactive, 2008). High-tech cellular phone features and software applications have turned mobile phones into mini-PCs with Internet resources, instant messaging, and video capabilities Kolb, 2010). The cellular phone is a "boombox" for this generation with its capacity to store thousands of music selections. Due to these advanced features, cellular phones have emerged as an educational tool, to enable, engage, and empower teens for learning (KVOA.com, 2009).

Despite cellular phone opportunities, there are perceived distractions associated with this device and how it is used for communication. Technological change naturally contributes to resistance due to loss of control, fear of misuse, or legal ramifications (Lenhart, 2010). Disruption behaviours by teen cellular phone users have parents and schools questioning whether cellular phones are "toys or tools" (Madden, Duggan, Cortesi, &Gassser, 2013). Research suggests schools and parents are uncertain the benefits of cellular phones outweigh the disruptive behaviours associated with them, such as cheating, classroom disruptions, sexting, and harassment (OMNI Inc., 2004).

Administrators have been cautious to embrace cellular phones in schools as districts have faced legal action in court for enforcing school cell phone policies and lost. For example, in *Foster v. Raspberry (Parker, 2007)* and *Klump v. Nazareth Area School Dist. (Pew Research Centerfor the people and the Press, 2010),* the courts upheld a student's Fourth Amendment right to be secure against unreasonable searches following a cell phone violation at school. In *Miller v. Skumanick* (Project Tomorrow, 2010), the school was found in violation of a student's First Amendment rights of freedom of speech when penalized for sharing a suggestive photo. The debate between opportunities and distractions has schools across the nation conflicted as to whether to ban cellular phones to avoid problems or whether to embrace cellular phones has a learning tool to reduce problems.

Method

A. Research Questions

The purpose of this study was to determine teen perceptions of cellular phones as a communication tool. The following two research questions guided this study:

1. What is the current level of cellular phone ownership and usage by middle-level teens in a Midwest suburban community?

2. What are teen perceptions of how they use cellular phones to communicate in their everyday lives?

B. Participants

A total of 243 middle school students were identified which were enrolled in a technology course and chosen for the study. The school was a Midwest suburban school with a total population of 1394 students. Participants were drawn from grades 6, 7, and 8 from a growing Midwest suburban community. The sample consisted of the following groups:

1. Students by grade level (Grades = 6, 7, and 8) attending the Midwest Suburban School District's Central Middle School (N = 1394). Students were identified through the school's database system.

2. Students by grade level (Grades = 6, 7, and 8) enrolled in a technology-based program (N = 243) in the same Midwest Suburban School District.

C. Instruments

The lack of a suitable instrument in the literature required the researcher to design a survey instrument and questions specifically for conducting the study. Surveys that influenced the design of this survey instrument and questions included: Pew Internet Research Center, Speak Up 2009, A Generation Unplugged, and Common Sense Media. Two survey instruments were developed for the study: (a) an open-ended survey was used to gather qualitative data from focus groups, and (b) a closed and open-ended survey was designed to gather quantitative and qualitative data from all the participants.

D. Procedures

Focus groups were identified as the method to gather qualitative data for the study. "A Toolkit for Conducting Focus Groups" was used as a framework for focus group procedures. Fourteen opened-ended questions were constructed for the focus group survey instrument. A pilot test of the instrument was completed to confirm reliability of the questions and face validity of the survey after comparing student responses. The final focus group survey instrument was utilised as a guide for all focus groups.

Three focus groups, representing Grades 6, 7, and 8, were used to collect data from the population to be studied. A method of stratified random sampling was used to select eight to ten participants, of varying demographics, for each grade level to participate in the study's focus groups. The twenty-five participants were representative of demographic subcategories for the CMS population to be studied. An unbiased college intern served as moderator to facilitate one-hour sessions for each group. During the data collection process, focus-group participants were coded for anonymity and to establish identity to compare recordings and notes for data analysis. A schematic was used to code participants. A number represented participants' grade level, and a letter differentiated between individual participants. For example: 6A, 6B, 6C identified individuals in Grade 6; 7A, 7B, 7C identified individuals in Grade 7; and 8A, 8B, 8C identified individuals in Grade 8, and so on. Following the focus group discussion, convenience sampling was used to pilot the Teen Cellular-Phone Survey (TCS) electronic survey instrument and questions. Focus group feedback was used to check for survey question clarity, vocabulary, and as a cross-check of reliability towards the final electronic survey instrument to be used in the quantitative data collection process.

The electronic survey instrument was comprised an opening page with information about the survey including: purpose for the study, population to be studied, directions, length, and a waiver of consent explaining anonymity of the results. After the electronic survey was completed in its on-line format, another pilot test was completed by the Central Middle School's technology department, administration, and school personnel to check for question clarity, vocabulary, spelling, format logic, and reliability of the electronic instrument in its final format. SurveyMonkey, a survey software and questionnaires template, was used as the medium for collection. Using SurveyMonkey software, the survey was designed for online collection of data and questions were manually entered. Participants were required to respond to all questions. A question loop was established to restrict non-cellular-phone users from answering questions only for cellular owners. Following the final pilot test, the electronic Teen Cellular Phone Survey (TCS) survey instrument was approved for use.

The study's two survey instruments were made available to the following participants:

1. The focus group instrument was utilised during focus group sessions at the Central Middle School's technology-based program. Twenty-five students participated in three focus groups, Grade 6 ($N = 9$), Grade 7 ($N = 8$), and Grade 8 ($N = 8$).

2. The electronic TCS survey instrument was made available to teens in Grades 6, 7, and 8, teens present in school the day the survey was given via the Central Middle School's Web site.

3. The electronic TCS survey instrument was made available to middle level teens enrolled in the technology-based program,

Grades 6, 7, and 8, present in school on the day the survey was given via the Central Middle School's Web site.

The electronic TCS survey instrument was used for individual student collection of the data. CMS teens were informed of the survey during the spring MAP (Measures of Academic Progress) assessment. A convenience sample created the best opportunity for students to access technology needed for the participants to access the online survey and generated the most efficient method for students to participate. A verbal explanation of the survey, its purpose, and how to access the survey was given to participants by the CMS technology specialist. A written explanation and overview of the survey, purpose, length, number of questions, and waiver of consent were given at the beginning of the survey. The survey was distributed to students using Central Middle School's Web site for fourteen days.

Findings

A. Research Question 1

What is the current level of cellular phone ownership and usage by middle level teens in a Midwest suburban community? Displayed in table 1 is data related to the question on popular communication devices used by families comparing traditional landlines versus cellular phone ownership. Respondents reported cellular phone ownership in 87.6% of their homes. Roughly, two in five respondents (37.6%) indicated their families only used cell phones (e.g., no landlines), while families that used traditional landline phone connections only (no cell phones) made up 2.5 of the sample). Responses were taken from the *Teen Cellular Phone Survey* question 6. "Valid percent" refers to those respondents who answered the items. It will always sum to 100.0 because the datum "ignored" nonrespondents.

Table 1. Cellular phone ownership vs. traditional landline ownership

Category	Frequency	Valid Percent
Cellular phones (ONLY).	366	37.6
Only a landline telephone (not a cell phone).	24	2.5
A landline telephone and cellular phones.	584	60.0
TOTAL	974	100.0

One of the study's primary questions was with respect to the level of teen cellular phone ownership by Central Middle School teens. Responses were taken from the *Teen Cellular Phone Survey* question 8. A majority (81.5%) of respondents reported owning cellular phones while 3.1% indicated they share a phone with their parents, and 15.4% stated they do not own a cellular phone. Responses to these questions are provided in table 2. For the remainder of the study, when inferential analyses was employed as to cell phone owners, the researcher used this variable to limit the analysis to students who either owned a cell phone or who shared one with a family member ($N = 824$); this eliminated respondents who identified themselves as a non-cellular-phone owner.

Table 2. Teen cellular phone ownership

Cellular Phone Ownership	Frequency	Valid Percent
Own cellular phone.	794	81.5
I do not own my own cellular phone but share one with my family.	30	3.1
Do not own a cellular phone.	150	15.4
TOTAL	974	100.0

To understand why participants reported themselves as not owning a cellular phone, the researcher included an open-ended comment box in question 8. Of the 15.4% of respondents who reported they did not own cellular phones, 62 respondents noted comments as to why. Categories emerged from responses which included: age, responsibility, cost, parent choice, parent sharing, safety, new to community, and no need.

"Age" was one of the most common responses as to why a respondent did not own a cellular phone. Seventeen (17) of 62 of the respondents

(27%) mentioned "age" as a reason they did not own a cell phone. This may be reflective of the fact Grade 6 (40.7%) was the largest group of survey respondents.

The category of "responsibility" also received 17 of 62 statements (27%) from respondents as to why the respondent was not a cellular phone owner. However, within this category, the researcher discovered respondents reporting themselves to be non-cellular-phone owners were reflective of students who may have owned a cellular phone in the past, but were currently without a cellular phone. Thus, responses for the category of "responsibility" could be further defined as the need to show responsibility to become a first-time cellular phone owner or the need to be responsible to continue ownership of a cellular phone.

"Parent choice" and "cost" were key factors reported by respondents as rationale for not owning a cellular phone. Twenty percent (13 of 62) of the respondents stated their parents, "don't think they need one" [cellular phone], "don't want them to have one," or "won't buy them one." Cost resonated as a subcategory of parent choice with 11% (7 of 62) of respondents reporting this as parent justification for not allowing them to own a cellular phone.

B. Teen Ownership of Popular Technology Devices

Responses were taken from question 5 of the TCS and self-reported by respondents. Table 3 was compiled to give insight into the survey respondent's perceptions of accessibility to today's most popular technology devices in their home.

Table 3. Self-reported frequency of popular technology devices

Technology Device	Frequency	Valid Percent
Cellular phone	853	87.60
Internet access	849	87.20
Xbox or Wii	793	81.40
Laptop	736	75.60
Desktop	630	64.70
MP3 player	599	61.50

iPod touch	445	45.70
Smartphone	298	30.60
E-reader (ex. Kindle)	168	17.20
iPad	156	16.00

Respondents were given the option to select multiple technology device categories. The valid percent represents the frequency of respondents who have access to technology in each category in comparison to the total number of survey respondents who actually ventured a response to the item ($N = 974$).

Cellular phones were the most popular technology of choice at 87.6% of respondents mentioning cell phones, with Internet access close behind at 87.2%. Gaming systems such as Xbox and Wii were reportedly used by 81.4% of the respondents, outweighing computer use. Survey respondents reported using laptops more often than desktops at 75.6% compared to 64.7%, respectively. Over half of the respondents (61.5%) reported having access to an MP3 player. Less than half of the respondents reported access to the following technologies: iPod touch (45.7%), smartphone (30.6%), e-reader (17.2%), and iPad (16.0%). The survey question also gave survey participants the option to contribute other technologies not on the list. Responses included television, cable net, iPod, PlayStation, Blu-ray, PSP, DS, DVD, and radio.

A cross-tabulation comparison of access to popular home technology devices as a function of ethnic categories is provided in table 4. Responses were taken from TCS questions 5 and 7. On average, Hispanics reported the highest percent of cellular phone ownership at 96.8% with Caucasians close behind at 90%. Smart-phones were most prevalent among Asian students at 37.2% and Native Americans were next in line for using smartphones at 33.3%.

Data indicated laptops are more popular than desktops in all ethnic categories. Hispanics reported the greatest percent of laptop ownership at 80.6% while Caucasians reported computer desktop access in their home to be at 66.1%. Technology devices for entertainment such as MP-3 Players were most popular with Hispanics at 67.7% while X-Box or Wii were used more by white students at a reported 85.2%. Native

American participants reported the highest access to the Internet at 91.7% while African Americans reported the lowest access at 65.6%. The iPad and e-Reader were most popular with Asian students at 25.6% and 18.6%, respectfully. Finally, iPod touch ownership was most prevalent among white students at 48.4%, while African Americans were less likely to own an iPod touch at 29.7%.

Table 4: Respondent Home Access to Popular Technology Devices by Ethnicity

Category	African American (N = 64)		Asian (N = 43)		Native American (N = 24)		Hispanic (N = 31)		White (N = 812)	
	Freq	%	Freq	%	Freq	%	Freq	%	Freq	%
Cell phone	43	67.2	30	69.8	19	79.2	30	96.8	731	90.0
Smartphone	14	21.9	16	37.2	8	33.3	6	19.4	254	31.3
Laptop	41	64.1	29	67.4	18	75.0	25	80.6	623	76.7
Desktop	37	57.8	27	62.8	14	58.3	15	48.4	537	66.1
MP3 player	28	43.8	22	51.2	14	58.3	21	67.7	514	63.3
Internet access	42	65.6	32	74.4	22	91.7	25	80.6	728	89.7
iPad	8	12.5	11	25.6	4	16.7	5	16.1	128	15.8
iPod touch	19	29.7	14	32.6	8	33.3	11	35.5	393	48.4
e-Reader	7	10.9	8	18.6	3	12.5	4	12.9	146	18.0
X-box or Wii	36	56.3	24	55.8	17	70.8	24	77.4	692	85.2

A cross-tabulation was used to further analyze access-to-technology-devices by gender and by race/ethnicity. However, due to small numbers in some of the racial or ethnic categories, results produced cell sizes with expected frequencies less than five, thus invalidating chi-square tests. In order to generate an appropriate group size for frequency analysis, the racial/ethnic variables were converted to a two-level (bivariate) variable. Because using all racial categories in cross-tabulations with gender and each "access" variable produced expected cell sizes less than five, the racial-ethnic variable was recoded as a bivariate distribution (white/Euro-American vs. students of colour). This allowed for the chi square analyses of the cell size problems, except where noted.

Table 5 compares access to popular devices in relation to gender and the bivariate group of white versus students of colour.

Table 5. Gender comparison bivariate distribution (whites vs. students of colour)

Device	Male				Female			
	White		Student of colour		White		Student of colour	
	Access	No access	Access	No access	Access	No access	Access	No access
Cell phone (N)	350	51	64	21	381	30	58	19
(%)	87.3	12.7	75.3	24.7	92.7	7.3	75.3	24.7
Smartphone (N)	138	263	30	55	116	295	14	63
(%)	34.4	65.6	35.3	64.7	28.2	71.8	18.2	81.8
Laptop (N)	303	98	66	19	320	91	47	30
(%)	75.6	24.4	77.6	22.4	77.9	22.1	61.0	39.0
Desktop (N)	264	137	53	32	273	138	40	37
(%)	65.8	34.2	62.4	37.6	66.4	33.6	51.9	48.1
MP-3 Player (N)	261	140	48	37	253	158	37	40
(%)	65.1	34.9	56.5	43.5	61.6	38.4	48.1	51.9
Internet Access (N) (%)	361	40	65	20	367	44	56	21
	90.0	10.0	76.5	23.5	89.3	10.7	72.7	27.3
I-Pad (N)	61	340	16	69	67	344	12	65
(%)	15.2	84.8	18.8	81.2	16.3	83.7	15.6	84.4
I-Touch (N)	201	200	33	52	192	219	19	58
(%)	51.1	49.9	38.8	61.2	46.7	53.3	24.7	75.3
e-Reader (N)	64	337	16	69	82	329	6	71
(%)	16.0	84.0	18.8	81.2	20.0	80.0	7.8	92.2
X-box or Wii (N) (%)	346	55	62	23	346	65	39	38
	86.3	13.7	72.9	27.1	84.2	15.8	50.6	49.4

No interaction existed between racial groups and access to cell phones among males, but with females the chi-square statistic was significant – ψ^2, 1 df = 21.7, p< .001. The significant difference was produced because many more white female students (92.7%) had cellular phones, while only 75.3% of females of colour indicated this level of access.

No difference between males or females on the variable of gender was noted in ownership of smart-phones. Thus, it can be inferred males, females, and students across racial and ethnic groups enjoyed similar levels of access to smart-phones. Further, no differences in access to laptops by racial groups were noted for males. However, the effect for females was significant (ψ^2, 1 df = 9.839, p = .002). The significant chi-square was produced by moderately higher access to laptops on the part of white females versus female students of colour (78% vs. 61%).

No statistically-significant differences were noted between ethnic groups for males with regard to the ownership of desktop computers. For females, however, a significant result accrued (ψ^2, 1 df = 5.91, p = .019). The significant inferential test resulted from the fact a significantly higher proportion of white females reported access to desktop computers than did their counterparts of colour (about 66% vs. 52%).

Noteworthy is the evidence of differences in Internet access amongst whites and students of colour for both males and females. About 90% of white students reported access while students of colour reported access at75%.

With new technologies such as iPad, iPod touch, and e-Reader, no differences in access to iPad devices were noted between genders or ethnic groups. All groups demonstrated access levels between 15% and 20%. Ownership of iPod touch technology was close to being an effect with males, but the difference was not significant. However, there were again differences between ethnic groups for females in the ownership of iPod touch technology (about 47% white vs. 25 % students of colour). Females of colour had less access than did white females. Electronic reading devices, such as e-Readers, were the least owned technology device by gender and between ethnicities with all reporting no access (about 80% to 92%).

Technology entertainment devices such as X-Box and Wii game systems were analyzed to compare perceived ownership by gender and ethnicity. Again, there were no observed differences between males within ethnic groups regarding access; however, there were again major differences between white females and female students of colour with game systems (about 84.2% vs. 50.6%). It is also noticed that male students not only use game systems more than female students but MP3 players as well which is contrary to the research.

About half (49.3%) of teen cellular phone owners reported having no responsibility for their cellular phone ownership, while others indicated some level of responsibility for owning a cellular phone. Responsibility levels were self-reported as: paying for a portion of the bill (4.7%), doing chores (24.5%), and keeping their grades up in school (15.3%). In comparing gender and ethnicity, there were no major differences between males. However for females of colour, only 37.7% of the parents paid for their cell phone with no other requirement, whereas white females self-reported parents paying at 62.3%. Evidently, a technology gap exists for females with white girls having more access than their counterparts of colour. In table 6 an analysis of teen cellular phone responsibility levels is provided.

Table 6. Who pays for teen cellular phones?

	Male		Female	
	White	Student of colour	White	Student of colour
Parents Pay (*N*)	225	35	256	29
(%)	56.1	41.2	62.3	37.7
Parents and Self (*N*)	20	5	22	2
(%)	5.9	5.0	5.4	2.6
Exchange Chores (*N*)	21	7	29	6
(%)	75	8.2	7.1	7.8

Maintain Grades (*N*)	53	13	59	14
(%)	13.2	15.3	14.4	18.2
NA (*N*)	82	25	45	26
(%)	20.4	29.4	10.9	33.8

Table 7 displays data gathered from question 10 of the TCS to evaluate differences between gender and ethnicity levels regarding the shared or individual ownership of a cellular phone or lack of owning a cellular phone. Differences were observed between ethnic groups for both males and females. White males and females considerably out-weigh students of colour for individual cellular phone ownership. It was evident students of colour are less likely to own a cellular phone or share one with their family than white students.

Table 7. Which option describes teen cellular phone ownership?

	Male		Female	
	White	Student of colour	White	Student of colour
Share Cell w/ Family (*N*)	8	8	10	10
(%)	2.0	9.4	2.5	13.0
Individually Own (*N*)	328	55	362	43
(%)	82.0	64.7	88.7	55.8
Do Not Own (*N*)	64	22	36	24
(%)	16.0	25.9	8.8	31.2

In examining the age at which teens reported first owning a cellular phone (question 11, TCS), no significant findings accrued among females and males or across ethnic groups (whites vs. students of colour). Ages 11 and 12 appeared to be the most common ages at which males and females of both ethnic groups acquired their first cellular phone. No significant differences occurred as to *why* teens obtained cellular phones (by self-request or parent purchase) by gender or ethnic group.

Question 15 of the TCS was the final question related to ownership. Teens were restricted to one choice. In table 8, a clear majority of student respondents identified texting as their primary reason for owning a cell in this forced-choice item. This was followed by calling (24.6%) and safety (15.8% of those reportedly owning a phone). Very few respondents selected games/music, video, or Internet access (all less than 1%). Similar results are found by [21] about the use of cellular phones among teens. Mostly teens use their cell phone for texting as it gives them liberty to communicate with more than one person at the same time.

Table 8. Primary reasons for owning a cellular phone

Primary Reason*	Frequency	Valid percent
Texting friends and family	429	58.4
Calling to stay in touch with family or call for a ride	181	24.6
Safety: It makes me feel safe	116	15.8
Games and music: Entertainment when I am bored	4	.5
Video and photos: Video and phone camera	3	.4
Internet access: To access Web tools	2	.3
Total	735	100.0
Missing System	240*	
Total	975	

* Items are in descending order by valid percentage.

** Reasons for large missing N unknown, except that teens may not have understood what was asked of them—data were analyzed only for students owning a phone.

Table 8 depicts the multiple ways that teens use their cell phones.

C. Research Question 2

What are teen perceptions of how they use cellular phones to communicate in their everyday lives? To evaluate teen texting and calling usage levels, Questions 13 and 14 of the *Teen Cellular Phones Survey* (TCS) were analyzed. However, using the original survey question intervals for the number of texts and calls made per day generated a large gap between the first choice and the remaining choices. To create a statistical measureable analysis for this data, the researcher recreated the variable as bivariate, with a broader range of intervals to eliminate counts < 5 in order to validate interpretation.

In table 9 the range of cellular phone calls made per day as self-reported by teens. Chi square indicated the existence of a significant difference between female students of colour vs. white (m^2, 1 df, = 4.10, p = .055). No significant differences across racial/ethnic categories among male respondents were found.

Table 9. Number of calls per day. Bivariate * ethnic: White vs. other. Self-reported gender cross-tabulation

Category	Male		Female	
	White	Student of colour	White	Student of colour
1–10 calls per day (N)	271	39	283	29
(%)	86.4	12.6	90.7	9.3
> 10 calls per day (N)	35	10	44	10
(%)	77.8	22.2	81.5	18.5

In table 10 the range of text messages sent per day as self-reported by teens are shown. There were no significant differences when completing a bivariate analysis of race with gender. However, a cross-tabulation comparison across genders indicates proportionately more females (50.5%) report texting more frequently than males (40.3%) over fifty-one text messages per day, which is in line with other researchers' findings [21].

Table 10. Text messages per day. Self-reported gender cross-tabulation

Category	Male	Female
0–10 text messages per day (N)	65	37
(%)	18.3	10.1
11–30 (N)	91	77
(%)	25.6	21.0
31–50 (N)	56	67
(%)	15.8	18.3
>51 (N)	143	185
(%)	40.3	50.5

Question 21 of the Teen Cellular Phone Survey (TCS) was generated to examine the overall method teens preferred to use in communicating with their friends. Table 11 compared the most common modes of communication by gender and ethnic groups, with frequency and valid percent displayed. Consistently, all groups reported face-to-face conversation to be their most popular mode of communication, ranging from 48.7% to 53.5%, followed by texting, calling, and social networking. E-mail is almost obsolete with this generation. When comparing communication modes, White females rank the highest for fact-to-face (53.5%), White males for texting (40.2%), male students of colour for calling (10.2%), and female students of colour for social networking (10.3%).

Table 11. Teen preferred modes of communication. Self-reported

Category*	Male		Female	
	White	Student of colour	White	Student of colour
Face-to-Face (N)	149	23	175	20
(%)	48.7	46.9	53.5	51.3
Texting (N)	123	19	129	12
(%)	40.2	31.7	39.4	30.8
Calling (N)	20	5	15	2
(%)	6.5	10.2	4.6	5.1
Social Networking (N)	12	2	8	4
(%)	3.9	4.1	2.4	10.3
E-mail (N)	2	0	0	1
(%)	.7	0	0	2.6

*Items are in descending order by valid percentage.

When teens were asked in question 22 of the TCS what their main purpose was for using their cellular phone to communicate with friends, overwhelmingly, responses were just to talk with friends. In table 12 frequencies for each category by all respondents can be found. Data indicates 103 of the 824 (12.5%) cellular phone owners did not respond to this item. Although this was a forced answer question, an Internet disruption occurred during the survey collection process which may have contributed to this inconsistency.

Table 12. Primary reason teens communicate with friends (self-reported)

Primary Reason**	Frequency	Valid percent
Just to talk with friends	539	74.8
To make plans	142	19.7
About assignments at school	22	3.1
To talk about concerns with friends	13	1.8
To talk about family problems	5	.7
Missing System*	103*	12.5
Total	824	-----

* Missing number largely represents those respondents that did not answer the question.

** Items are in descending order by valid percentage.

Displayed in table 13 is evidence of how students self-report using cellular phones to call or text while not in school, but in public. Overall, females self-reported behaviours which would constitute better etiquette than males when using their cellular phones in public. Both genders are more likely to text than call in front of others. While males are more likely to call anytime anywhere, females are most likely to text anytime anywhere. Public cellular phone use can be linked to research on cellular phone etiquette.

Table 13. Cellular phone use in public (self-reported by gender)

Category	Male	Female
Answer phone calls anytime, anywhere (N)	87	57
(%)	24.5	15.6
Put cell phone on vibrate or move to a private location (N)	88	79
(%)	24.8	47.3

Text messages anytime, anywhere (N) (%)	126 35.5	161 44.0
Text messages when I am not talking to friends or family (N) (%)	54 15.2	69 18.9

Conclusion

The evolution of technology continues with the growing popularity of the cellular phone and the increasing market of teen cellular phone owners increasing across the nation and the globe. It is evident from the study that, this trend holds true with this Midwest suburban community as middle level teen ownership surpasses national statistics. Parents influence teen cellular phone use as the percentage of cell phone only homes are quickly increasing, while landline-only homes are almost nonexistent. Parents also determine the age at which teens may own their first cellular phone; as in a majority of instances, trends indicated parents were paying for the device. The primary reason teens requested a cellular phone was for texting rather than calling. Cellular phones have been helping to reduce the digital divide as ethnic groups report higher percentages of cellular phone ownership and use. However, there are still gaps between ethnic groups regarding the most commonly used technology devices such as computers, laptops, Internet, and new technologies. The most significant gap in technology use occurs with female students of colour reporting less access to technology than white females or males in all ethnic groups.

Cellular phones were recognized as friend, foe, toy, and/or tool by Central Middle School teens. As a friend, cellular phones were recognized as a major safety tool to connect with parents, get help, communicate with friends, and as way to learn. As a foe, the anytime, anywhere, 24/7 access contributed to late night use, social pressures, etiquette concerns, and cyberbullying problems. As a toy, teens have been using cellular phones to take and share photos with family and friends, along with entertainment features to play music and games; yet differences were noted between gender and ethnic groups. As a tool, teens envision using cellular phones to access the Internet, use features for organisation, and to use applications for school.

A majority of Central Middle School teens reported sending fifty or more text messages per day and making ten or more calls. Although cellular phones appear to be popular with teens, over half would still prefer to communicate with their friends face-to-face. Despite school policy restrictions regarding cellular phone use during the school day, one-third of teens reported carrying or using their cellular phone during the school day to communicate with family and friends. Males were more likely to carry and use their cellular phone during the school. Parents contributed to teen cellular phone use during the school day as they would send text messages and/or call to leave messages for their children. A majority of school personnel assisted in reinforcing the school's no cellular phone use policy as violators were reported to the office. However, some inconsistencies did exist as to how the policy was being implemented at the time of this study which may have contributed to its ineffectiveness. Overall, students indicated an interest in using cellular phones beyond personal reasons towards learning in school.

Recommendations

There are multiple levels of recommendations for this study as it confirms or denies information in the literature in respect to trends in teen cellular phone use, and the natural human state to resist or adopt new technologies.

A. Recommendations for Schools

Identifying the current level of teen cellular phone ownership and how teens use them to communicate is the first step in understanding the impact cellular phones have on a school's environment. The key is to use data to recognize and develop an action plan for dealing with this new technology as it relates to the school environment. Schools may want to take the position of zero-tolerance for use of cellular phones in the school environment to avoid distractions. However, research in respect to the evolution of technology proves advancement in technology and society's reliance on these tools eventually transitions users and organisations in the direction of adoption. Thus, in learning from the past, the researcher recommends the Midwest Suburban School District and Central Middle

School take a proactive position in redefining school policy to embrace cellular phones in the school environment, is to empower teachers to identify strategies for its implementation for learning telephone etiquette which may be similar to the "Telephone Manners" found in the language arts textbooks of the 1950s. The school will need to establish standards and disseminate knowledge about the proper usage of cellular telephones to parents and students. If the school doesn't take the responsibility, the standards for cellular telephone usage will in all probability be established by outside forces. The school should establish disseminate the knowledge and developed standards to parents and students by such methods as seminars and eBlasts. The school needs to be engaged to encourage students to become responsible users of cellar telephone technology as twenty-first-century learners.

B.Recommendations for Parents

With assistance from the school in providing recommendations and guidance of cellular telephone usage, parents must assume the role of modeling, monitoring, and supervising their child's cellular phone use. As parents themselves increase their use of cellular phones for calling and texting, they must consciously take responsibility to teach and model cellular phone etiquette and appropriate use through their actions. This can be accomplished by defining the who, what, where, when, and how for their own cellular phone use and sharing it with their children.

It is recommended that parents communicate with their children to define rules and expectations for cellular phone use. As teens reported sending and/or receiving harassing or intimidating messages, parents should work with their children to define expectations and parameters prior to their cellular phone ownership and use. This can be done verbally and/or through a formal contract including topics such as defining a shared responsibility for cost, establishing hours for use, setting limits based on the calling plans, discussing prohibited or illegal behaviour (texting while driving, sexting, etc.), discussing cyberbullying and how to report cyberbullying, developing strategies for using the cellular phone as a tool, communicating parent expectations for monitoring cell phone use, limiting the number of people who have access to the child's number, completing random checks, and establishing natural consequences for inappropriate use.

It is recommended that parents support their child's safety by pursuing their own education on cyberbullying and initiating actions to deal with cyberbullying. This can be accomplished by parents owning their own cellular phone, becoming free-agent learners in respect to cellular phone use, attending workshops on cyberbullying, organizing a community campaign by parents to stop cyberbullying, getting to know their child's friends and parents, and working with their child's schools to understand and learn how the schools deal with cyberbullying.

C. Recommendations for Students

Teens must take responsibility for their use of technology, the latest being cellular phones. Students must recognize personal decision-making is part of the responsibility associated with being a cellular phone owner, and understand, "just because you can, doesn't mean you should." Today's cellular phones are powerful minicomputers, with multiple features including cameras, Internet access, and communication features. Teens must learn how to use cellular phones appropriately and to realize acceptable use now extends beyond the computer and Internet to today's handheld cellular devices. They must also apply moral and ethical behaviours to use cellular phones for good (organisation, communication, and learning) and not to be malicious (harassment, disruptions, sexting, and cheating).

Teens must develop cellular phone etiquette guided by the school and reinforced by parents so as to demonstrate respectful use of their cellular phones. Through the school curriculum and parental modeling teens must learn and identify pubic locations in which cellular phone use should be minimized or eliminated. They must understand their actions impact those around them when using, talking, or texting with a cellular phone. This can be done through a conscious effort to understand when it is the best time to use a cellular phone and apply self-controlling behaviours. Reinforced by the school and parents, teens must understand laws that mandate no texting while driving for safety reasons.

Teens should accept responsibility to learn and understand behaviours considered harmful: harassment, bullying, or cyberbullying. They must take responsibility to avoid initiating such behaviours and to report incidents when they occur. Teens should use cameras to take and transfer appropriate photos, and eliminate the sharing of photos classified as indecent or exposed. Students should report when they receive messages with indecent or exposed photos.

Teens should continue to utilise technology and their natural skills to learn and grow as free-agent learners to prepare for the twenty-first-century world in which they will learn and work. Teens with limited access to technology should advocate for themselves to school leaders to secure and integrate technologies within the school's curriculum to assist in preparing them for the future.

References

Andersen, E. (2009, August 31). *Cell phones in schools: Toys or tools?* Retrieved from Lincoln Journal Star: http://journalstar.com/lifestyles/article_b851282e-941c-11de-b6d1-001cc4c002e0.html.

Carr, V. H., Jr. (2011). Technology adoption and diffusion. Retrieved from http://www.au.af.mil/au/awc/awcgate/innovation/adoptiondiffusion.htm.

Commonsense Media. (2009). *Hi-Tech cheating: Cell phones and cheating in schools a national poll.* Retrieved from Commonsense Media Web site: http://www.commonsensemedia.org/sites/default/files/Hi-Tech%20Cheating%20-%20Summary%20NO%20EMBARGO%20TAGS.pdf.

Cotten, S. R., Shank, D. B., and Anderson, W. A. (2014). Gender, technology use and ownership, and media-based multitasking among middle school students. *Computers in Human Behavior, 35,* 99–106.

Cyberbullyalert.com.(2008). *Statistics thay may shock you!* Retrieved from http://www.cyberbullyalert.com/blog/tag/cell-phone/

eSchool News Staff. (2006). *Youth use cell phones as mini-PCs.* Retrieved from eSchool News Web site: http://www.eschoolnews. com/2006/04/05/youths-use-cell-phones-as-mini-pcs/

HarrisInterative. (2008). *CTIA advocacy.* Retrieved from CTIA Wireless Association Web site: http://www.files.ctia.org/pdf/ HI_TeenMobileStudy_ResearchReport.pdf

Kolb, L. (2010). *Toys to tools: Connecting student cell phones to education.* Eugene, OR: International Society for Technology in Education.

KVOA.com. (2009) *New 4 Tucson AZ.* Retrieved from KVOA. Com Web site: http://www.kvoa.com/Global/story. asp?S=8419930&nav=menu216_5_2.

Lenhart, A. (2009). *Teens and sexting.* Retrieved from Pew Internet Research Center Web site: http://www.pewinternet.org/Press-Releases/2009/Teens-and-Sexting.aspx.

Lenhart, A. (2010). *Cellphones and American adults.* Retrieved Pew Internet Web site: http://pewinternet.org/~/media//Files/ Reports/2010/PIP_Adults_Cellphones_Report_2010.pdf.

Madden, M., Lenhart, A., Duggan, M., Cortesi, S., and Gasser, U. (2013, March 13). Teens and technology 2013. Retrieved from http://www.pewinternet.org/2013/03/13/ teens-and-technology-2013/#.

OMNI Inc. (2004). *Toolkit for conducting focus groups.* Retrieved from OMNI Inc. Web site: http://www.omni.org/Default.aspx.

Parker, D. (2007). *Cell phones are allowing citizens to monitor/ record behavior of law enforcement personnel.* Retrieved from WG Wireless Galaxy Web site: http://www.wirelessgalaxy.com/

researchcenter/camera-cell-phones-are-allowing-citizens-to-monitor-record-behavior-of-law-enforcement-personnel.asp.

Pew Research Center for the People and the Press. (2010). *Asesssing the cell phone challenge to survey researach in 2010.* Retrieved from http://www.pewinternet.org.

Project Tomorrow. (2010). *Creating our future: Students speak up about their vision of 21ˢᵗ century learning.* Retrieved from Project Tomorrow Web site: http://www.tomorrow.org/speakup/pdfs/SUNationalFindings2009.pdf.

Project Tomorrow. (2011). *The new 3 e's of education enabled, engaged, empowerd: How today's students are leveraging emerging technologies for learning.* Retrieved from Project Tomorrow Web site: http://www.tomorrow.org/speakup/pdfs/SU10_3EofEducation(Students).pdf .

Saettler, P. (2004). *The evoluation of American educational technology.* Sacramento: Information Age Publishing.

Tulane, S. and Beckert, T. E. (2013). Perceptions of texting: A comparison of female high school and college students. *North American Journal Of Psychology,*15(2). Retrieved from https://www.questia.com/library/journal/1G1-331348554/perceptions-of-texting-a-comparison-of-female-high.

Chapter 8

Strategies for Effective Integration and Adoption of Educational Technology Tools into Teaching and Learning Process in Nigerian Schools

Jennifer N. L. Ughelu
&
Sylvester Akpan

Introduction

The process of technology integration is one of continuous change in teaching-learning and the success is hopefully improvement in educational processes. Technology is a catalyst for teaching and learning in the classroom. Developing a culture that embraces technology is also important to its successful integration. Therefore, "technologies" used in the classrooms today are often for the wrong reasons. Such incorrect reasons are convenience, pressure from school administrators, and the belief that students need to be entertained, and so on. Technology contributes to global development and diversity in classroom. For technology to be usefully in educational system, teachers and students must have access to technology in a contextual matter that is relevant, responsive, and meaningful to the educational practice which can promote quality teaching and learning of active students. Herrington and Kervin (2007) argue that technology presents powerful cognitive tools that can be used by students to solve complex and authentic problems. For such problem-solving to take place, however, technology needs to be used in all aspects. The students need it more than teachers' do to interact, socialize, and solve their daily exercises. Technology opens up many doors for students at all academic levels. It should be an integral part of how the classroom functions and be accessible. It should prepare the students for lifelong learning in a rapidly changing technological society through basic understanding of technology usage, processes, and systems. Successful technology integration goes hand in hand with the changes in teacher training curriculum and assessment practices. Integration of technology cannot be possible except if technology is available and accessible. Educational technology in school today is not like what it is used to be in the '90s. The past was dominated with tools such as charts, molding, and nothing like simulation and modeling. In yesteryears, teachers were long seeing as instructional experts due to the fact that there are few technologies out there in their possession, but today globally computers and other technological tools are there to complement, substitute, and augment teacher's efforts in the teaching and learning processes. In typical schools today in the United States of America, one computer is for three or four students, but in Nigeria, one computer is for a whole class of thirty to forty students, making it difficult to teach with computers.

Sometimes, it is difficult to describe how technology can be integrated into the teaching and learning process because the term "technology integration" is a broad umbrella that covers so many varied tools and practices.

Technology integration is the way technology resources such as mobile devices, social media platforms and networks, software applications, the Internet, etc., are used in our daily practices. Therefore, "technology integration" means harmonisation, that is, the combination of objectives, conditions for achieving the objectives, methods, resources, evaluation, and feedback in carrying out any educational activity in the processes of teaching and learning. It also means using many technologies to enhance teaching and learning experiences by providing "a range of pathways for students at varying levels" (Ficklen and Muscara, 2004). Integration depends on the kinds of technology available, how much access is available, how the technology can be used, and who is using the available technology and how to use it in the teaching and learning process.

For example, in a classroom with one only an interactive whiteboard and one computer, learning is likely to remain teacher-centred, and integration is likely to revolve around the teacher's needs, not necessarily around students' needs. This is not how things should be; the desired outcome is interaction among the students, the teacher, and the tools available in the classroom.

Willingness to embrace change is a major requirement for successful technology integration. Technology evolves continuously and rapidly in our everyday lives. This development is an ongoing process, and it demands continual learning and usage.

Integrating technology in the classroom practice can be a great way to strengthen engagement by working and making students a global audience, turning them into creators of digital media, and helping them practice collaboration skills that will prepare them for the future. For example, project-based activities can be in the form of Cyber-Hunt. Cyber-Hunt is used by the students to gain experience when exploring and browsing the Internet. It may ask the students to interact with the site, that is, to play a game or watch a video, record short answers to teacher's questions, read and write out a topic in an in-depth way. Another example is Virtual Field Trip, which is a Web site that allows the students to experience places, ideas, or objects beyond the constraints of the classroom. It gives the students an opportunity to experience new information. It is practical in nature and that is why the students liked it. A teacher can incorporate the use of hands-on material to further their knowledge about a new idea presented in a virtual field trip. Dede (2000) argues that teachers should be able to use technology in specific content areas, as well as to enhance personal productivity. The use of technology will also be based on the learner's age, level of understanding, type of methodology to be adopted by the teacher, and other things such as finance, available equipment, power, and conducive environment. The question is, How will a well-equipped classroom look like in the twenty-first century in Nigeria? How might the use of these tools reshape teaching and learning practices in Nigeria?

Questions about technology integration persist, even after more than half a century of documented research on the usage of technologies such as television and the benefits of using computers in a digital world. Even moderate goals have not been reached in the classroom. Students are already active users of technology outside of school, but in a school context, what is happening? Can technology be used to teach and learn in a school setting? The challenges are leveraging on the opportunities that technology creates so as to prepare learners for a global connection in the information-saturated world.

For effective technology integration in the classroom to be achieved, students and teachers should be able to select technology tools that will help them to obtain information in a timely manner, analyze and synthesize that information, and then present it professionally to one another. The technology should become an integral part of how the classroom should function and be accessible to students, along with other classroom tools.

Where Does Integration Start?

Integration started in the 1960s, during the collaboration of Swiss psychologists Seymons Papert and Jean Piaget. After the collaboration, Seymons came out with logo programming language. He began introducing the logo programming language to his students, and this logo software enabled his students to use the computer to control their learning environment. With a small amount of instruction, they were able to write and debug programmes that controlled the movements of a turtle robot. Not only did they gain deeper programming expertise, but they also showed an engagement in learning that is rarely in the mere traditional drill-and-practice classroom activities. Papert said, "With computers, there is a substantially bigger chance that one can lead a child with less effort into something he really likes doing."

The integration with a set of fun activities and with a set of educational learning activities is sufficiently big enough which should be able to keep every student internally motivated even without leaving the physical connection that technology brings about.

In 2000, SRI International identified four ways that technology enhances how children learn:

1. Technology offers active engagement, an opportunity to participate in groups, frequent interaction and feedback, and connections to real-world contexts.

2. Technology expands what concerns students by providing them with access to an ever-expanding store of information.

3. Technology integration is only one element in "what must be a coordinated approach to improving curriculum, pedagogy, assessment, teacher development, and other respects of school structure."

4. The increasingly interactive nature of technology, exemplified by Web 2.0 tools, creates new opportunities for students to learn, by allowing them to do a task, receive feedback on the task, and then build new knowledge from the task.

Technology can be used to achieve learning in the following ways:

1. Bringing exciting curricula based on real-world problems into the classroom.

2. Providing scaffolds and tools to enhance learning, such as modeling programs and visualisation tools.

3. Giving students and teachers more opportunities for feedback, reflection, and revision.

4. Building local and global communities that include teachers, administrators, and other interested people.

5. Examining opportunities for teacher learning.

Social media can be used to integrate technology into the classroom. Such media open new possibilities for connecting learners and taking education into a new direction or dimension.

Integration of technology can be achieved through the following processes:

1. Routine and transparency.

2. Accessibility and ready availability of the technology for the task at hand.

3. Support for curricular goals and help for students to effectively reach their goals.

There are other many ways that technology can also be an integral part of teaching and learning process. Such ways include:

- Online learning and blended classrooms.
- Project-based activities incorporating technology.
- Game-based learning and assessment.
- Learning with mobile and other hand-held devices.
- Instructional tools such as interactive whiteboards and student-response systems.
- Web-based projects, exploration, and research.
- Student-created media such as podcasts, videos, or slide shows.
- Collaborative online tools.
- Social media (WhatsApp, wikis, blogs, Twitter, etc.) to engage students.

Factors Necessary for Successful Integration of Technology in an Instruction Process

- Actively encourage and train faculty on the use of educational technology tools in teaching and learning.
- Use technology to assist in overall student productivity and to help support a student's own individual learning activities.
- Plan curricular activities that will be accomplished with technology.

Elements for Effective Integration of Educational Technology Tools

1. Some equipment must be available for both faculty and student use. That is, a computer on each faculty member's desk is the first step towards encouraging the faculty to use technology.

2. Providing appropriate software, an adequate network, and a good connection is necessary. This shows that learning does not happen by chance! There are important decisions that teachers must make to ensure learning when integrating technology into a lesson.

3. Ongoing faculty training and coaching. Such a program will be most successful if faculty training is designed to build a general awareness of the practical uses of technology followed by hands-on practical tasks where faculty members can develop personal applications of the technology. Well-planned instructional strategies incorporated with technology will promote learning regardless of the subject matter, the learner, or the learning environment.

4. The components of all instructional strategies include both content and learning outcomes, the instructional experiences, technology and media, a learning environment, and the people.

Kuhn and Udell (2001) state that teachers must plan and manage their learning setting so as to ensure that their students are both challenged and the outcome will be successful. The teacher is responsible for ensuring that the approach they used is to help students learn is effectively in their intended learning outcomes.

Importance of Technology Integration in Teaching-Learning Process

When there is an effective integration of technology tools in the teaching-learning processes, the teachers and students stand to gain:

1. Access to up-to-date primary source material for everyone.

2. Easier collecting and recording of data.

3. Collaboration with other students, teachers, and experts around the world.

4. Opportunities for expressing understanding via multimedia.

5. Learning that is relevant and assessment that is authentic.

6. Training for publishing and presenting new student knowledge.

7. The technology supports student performance of complex tasks that are similar to those performed by adult professionals and/or fill a genuine need of the student.

8. The technology is integrated into activities that are a core part of the classroom curriculum.

9. Technology is treated as a tool to help accomplish complex tasks (rather than as a subject of study for its own sake) that engage students in extended and cooperative learning experiences that involve multiple disciplines.

Methods of Incorporating Technology into Teaching and Learning Process

Technology can be integrated into different methods of teaching and learning processes. These methods range from presentation to demonstration, drill-and-practice, tutorials, discussions, cooperative learning games, discovery, simulation, problem solving, etc.

1. Presentation: This is a process of disseminating information to learners through the use of drama, storytelling, etc. During the presentation process, the teacher can interject questions students volunteer to answer or are assigned to answer. Or students can ask questions as the lesson is going on. Examples of information sources under presentation are available through Internet sites, audiotapes, videos, and book reading.

Integration of Technology into the Presentation Method

There are quite number of technologies that can be used to enhance presentation of information to the learners. These techniques can range from the use of a whiteboard for notes to video, audio, computer-based presentation, and overhead transparency, which the teacher might have prepared in the past, or a set of PowerPoint slides that include images downloaded from the Internet. Presentation does not necessitate that a teacher stands in front of the classroom lecturing, but he can decide to use of some of the aforementioned technologies to present his lesson without necessarily talking. For instance, the video can be played for the students to view and ask questions as necessary. But to use these technologies, the teacher must know the age and experience of his students before introducing them. Clickers can be used to present information to a large number of students. A clicker looks like a television remote control. It allows students to respond to questions during presentation and lets the teacher poll student opinions at key points. It comes with software for the teacher, who can later provide summary charts or graphs showing how the students selected each answer.

2. Demonstration Method: Demonstration shows the students how to do a task as well as why, when, and where it can be done. At this point, the learners view a real-life or lifelike examples of a skill or procedure to be learned then practiced. Demonstrations may be recorded and played back by means of media such as video and audio cassette. When there is two-way communication, a live instructor is needed in demonstration. The object for the learner to imitate can be on software programs on the computer. For instance, in some cases, the point is simply to illustrate how something works, such as a flip animated demonstration by the teacher, or teaching students how to copy and paste within a word-processing program.

Integration of Technology into Demonstration Method

Technology can be used by the teacher to demonstrate a lesson during class activity. The teacher can prepare a video of the demonstration before class to show to the class (along with talking points) when the students are viewing the video, that is, guiding the observation while the demonstration is going on. Students can provide demonstrations to their classmates on new skills or procedures they have learned. For instance, a good student who already knows how to move photos from a digital camera onto a computer can be asked to show others or the whole class the techniques behind the operation. Second, in chemistry lab, the titration interaction can be demonstrated to the students with a computer; students can observe and then go and practice what they have seen.

3. Drill-and-Practice: In this method, learners are led through a series of practice exercises designed to refresh or increase fluency in knowledge of specific content or a new skill. For instance, the use of software such as Mavis Beacon Teaches Typing to learn how to type and spell words on the computer is an example of the drill-and-practice method. Such learning can be treated as individualized learning. Drill-and-practice goes with feedback and reinforcement to correct responses. It goes a long way towards remedying errors that students encounter while practicing. The main goal of drill-and-practice is that students will master or learn the information without error. Technologies can be integrated into drill-and-practice of a lesson. For example, drill-and-practice can be used in the teaching and learning of some subjects such as mathematics facts, learning of foreign languages, building vocabulary, and computer-assisted instruction (CAI) etc. Some delivery systems and media formats can be used in drill-and-practice exercises. These tools include audio cassettes, flash cards, and worksheets that can be integrated into the teaching and learning process. They can also be used in the aforementioned subjects. Students can be placed in pairs to work through drill-and-practice exercises. Teachers can integrate these tools into the teaching and learning process. Student homework can be done using drill-and-practice because "practice makes perfect." The teacher can begin the class by using formulas that show how things work out. After his demonstration to the whole class on how these formulas work, students are asked to explain to others how to apply these formulas in one or two situations. There is software available for this situation, and students can at the same time develop their own software.

4. Tutorial Method: The tutorial method is a self-paced learning exercise or a lesson prepared so that a student can learn at his or her own speed or convenience. A tutor can be a person, computer software, or printed materials presented in the content, which poses a question or problem that requests a learner's response, analyzes the response, and supplies appropriate feedback to the learner. It can be used on a one-on-one basis or in a group, in subjects such as reading and arithmetic. The largest number in such an exercise cannot exceed 10 people at a go.

Technology Integration in the Tutorial Method

Tutors can be instructor-to-learner, learner-to-learner, computer-to-learner, or print-to-learner. Technologies can replace a teacher or tutor in the tutorial class. There is computer software designed to deliver lessons to students. Such a tool is an integrated learning system (ILS). Students are required to log on, entering a user name and password, to begin the tutorial exercise. For instance, a teacher can use short, animated clips at the beginning of the class to introduce the lesson of the day to his students. He does this by introducing basic concepts and using animation to illustrate, and after this, students work independently on the remaining lesson activities. The teacher or student can import preexisting data and can provide instruction on how to apply the target concept, or plot a chart.

5. Discussion Process: Discussing involves the exchange of ideas and opinions among students or among the students and their teacher. It can be used at any stage of instruction, with both large and small groups. It is a very good way of assessing the knowledge, skills, and attitudes of a group of students before finalizing instructional goals, especially when introducing a new topic. A major benefit of this method is the amount of interaction that occurs and the learning that results from that interaction. It provides the teacher with immediate feedback on students' understanding of course material. Discussion teaches content as well as processes, such as group dynamics, interpersonal skills, and oral communication.

Technologies can be integrated into discussion, which makes learning more effective, especially when introducing a new topic. Technology such as video can provide a common experience, and this experience gives the students opportunities to discuss and something to discuss. When the video is presented, the students observe critically before they can join in discussion on a topic or ongoing lesson.

6. Cooperative Learning Method: This is a grouping strategy in which students work together to augment each other's learning potentials. Johnson and Johnson (1999) outlined the components of cooperative learning as follows:

1.   Members who view that role as part of a whole team.

2.   Interactive engagement among the members of the group.

3.   Both individual and group accountability.

4.   Members who have interpersonal and leadership skills.

5.   The ability to reflect on personal learning and group function.

Integration of Technologies into Cooperative Learning

Introducing technology into the teaching and learning process is not the only way forward. Other useful methods include making the students designers and producers so that student-generated graphics become part of technology integration in the teaching and learning process. When a teacher carries out such integration, he or she should be part of the working team with the students. When learning is done in this way, the students become expert on a portion of the total content.

7. Games as a Method of Teaching and Learning: When a teacher uses games, an activity is performed, either alone or with others, for the primary purpose of entertainment. Game-playing provides a competitive environment in which learners follow set rules, as they strive to attain a highly motivating technique, especially for tedious and repetitive content. Playing games for learning can involve one learner or a group of learners. Games always go hand in hand with problem-solving skills and the ability to give a solution or to demonstrate mastery of specific content that demands a high degree of accuracy and efficiency. Students learn patterns that exist within a particular type of situation when games are used in the teaching and learning process (Moursund, 2006). For instance, games can be used in teaching mathematics calculation, multiplication, and addition and subtraction sets. As the students repeatedly play a game, they begin to develop better understanding of the rules of the game and ways to become better at achieving the desired outcome, which is winning. Games can be challenging and fun to play, and they can also provide variety to learning experiences.

Computer drill-and-practice software programs are examples of using games in the teaching and learning process. Students play the games a lot, and they thereby benefit from the play by extending their learning experience.

Games that can be included in the teaching and learning process include puzzles, crossword puzzles, cartoons, and the chess teaching and learning process: using such tools, students can, for example, learn their spelling words. Sudoku is a special kind of puzzle that can be used in solving mathematics issues. For example, a teacher can allow his students to practice mathematics facts using a bingo game. He provides cards to the students and uses flash cards, a whiteboard, or an overhead projector to create math problems that need to be solved. The students cover the appropriate number response on their own bingo cards.

Another example is creating a board game for learning of vocabulary words. At this point, students can work in pairs or small groups to practice the words, while enjoying the game.

8. Simulation Method: Simulation involves learners confronting a scaled-down version of a real-life situation. It allows realistic practice without the expense or risks of real life. Simulation could include dialog, manipulation of materials and equipment, or interaction with a computer. Simulation can be used for small groups or a whole class. Experiences that students might otherwise not have come in contact with can be gotten through simulation.

For instance, instead of bringing an entire automobile engine into the classroom, which is something that would be difficult to do, the teacher can bring in a model of and automobile engine to show students how the internal combustion engine works. Using the model, students will see all of the components of the engine and the function of each.

Integration of Simulation Tools into the Teaching and Learning Process

Computer software can provide a simulated learning experience. For example, the whole frog project engages students in a complex study of the frog, using types of information such as MRI imaging to detect internal sickness of the body. Mechanical students can use this technology in their learning process in the classroom. Students can use the computer to gather the data they need, and this process will help them to apply the information to advance their knowledge and skills.

Another example of technology integration is using computer simulation to learn interpersonal skills and watch laboratory experiments in the physical sciences. Role playing is yet another common example of simulation in the teaching and learning process. Software such as Tom Snyder's Decision, Decisions can provide a role for each member of the group in the real-life situation that needs to be resolved. This program (Decisions, Decisions) provides a process whereby one computer in the classroom can be used with a whole class.

9. Discovery: Discovery is an inductive or inquiry approach to learning. It presents problems to be solved through trial and error. The objective of the discovery strategy is to foster a deeper understanding of learning content through involvement with it. Learners can discover things through previous experience, based on information in reference boxes, or by learning from information stored in a computer database. This method is widely used by scientists who are involved in creating a hypothesis or question, trying out possible solutions, and analyzing the information at hand to determine whether a given approach works. When using this strategy with students, Marzano, Packering, and Pollock (2001) are of the opinion that it is not always better to let students find their own way through a skill or approach, rather than providing them with direct instruction. For instance, it is better to show students some skills, such as in mathematics computation, and let them practice with manipulative, than to have students try to learn the procedures on their own. The time required to learn such skills makes discovery less efficient than direct instruction, and it is possible that with discovery alone, students might not actually learn the information or skills they need to move on in their studies.

Integration Technology into Discovery Method

Technology in the instructional process can help promote discovery. For example, video may be used to discover a lot in the teaching and learning of the physical sciences. The teacher shows the video to the students who then view or observe the relationship represented in the video clips. For example, when students view a simple experiment about air using a balloon, the balloon is weighed before and after filling it with air. This procedure shows students that air has weight. From the discovery method, students can discover, or find, information that they wish to know about a specific topic of interest. A computer connected to the Internet can be used by the students to learn new ideas and access particularly difficult information.

10. The Global Classroom

Today's world is connected to an invisible digital network through the use of a complex satellite system: this fact makes today's classroom truly global. Students now get information from a multitude of resources that range from textbooks to live video conferences with people geographically separated by thousands of miles.

Teachers can use resources such as e-pals Global Network (www.epals. com/community) to connect with others outside the local area. This technology links together over 118,000 classrooms from about 200 countries, and students in classrooms in these countries participate in cross-cultural learning activities. Teachers can plan their lessons with one or two teachers and also engage their students in one of the main large interactive research projects involving children from around the globe.

Through live-streaming video, the world is also opened to students. This streaming video begins playing before the entire file is downloaded from the Web. This technology enables the students to see live shots from the South Pole, the streets of Vienna, Kenyan game reserves, the Eiffel Tower, the Bavarian forest, Mt. Fuji, a city market in Hong Kong, or the Olivetti Research Laboratory at Cambridge, England. If this type of technology is used in Nigeria, Nigerian students will utilize it to join their contemporaries in the outside world to discover themselves and creative in nature.

These sites have user friendly controls on the cameras so that students can freely explore the distant site from multiple viewpoints. This increases students' awareness about different countries and their differences in time.

11. Podcasting

Podcasts are like Web-based radio that is used to broadcast to the students. The students listen to computers, much like using the radio. Information and files can be sent to them over the Web subscriber's computer. Handsets can be used for this type of learning if they are well utilized. Students can use the technology to communicate with one another by creating their own podcasts as a learning activity. They can find information and share it among their members; doing so will help them to learn from one another.

12. Internet Radio

This is another technology that can be integrated into the classroom. Internet radio uses the Internet to offer online radio stations, which carry varieties of programming, such as music, sports, and science, and local, national, and world news to the users. Teachers in various fields can utilize these opportunities to teach their students.

13. Games

A classroom without elements of games and fun would be like a dry, barren landscape for students to traverse. The games appeal centres on the students' desire to compete and play. Games provide the teachers the opportunities for taking advantages of students' innate desire to get them to focus on the subject matter. When games are introduced into the teaching and learning process, their use encourages a very positive and healthy relationship in the classroom. The use of games helps the students to retain memories of the things they learn for a long period of time.

Issues in Integrating Technology into Teaching and Learning Processes

1. Increased costs of keeping up with technology: There is a need to cut some of the costs of technology usage so as to help everyone. Educators and policy-makers need a solid rationale for why funds used for instructional technology are well spent (Ringstaff and Kelley, 2002). There is need for substantial investment in technology infrastructure and teacher training. Hardware is expensive, too, in terms of maintenance, training and services, and utility. Cost alone can keep technology from reaching a particular target percentage of computer users.

2. Attacks by technology critics: Justifying technology expenditures by confirming technology's benefits is increasingly important in light of recent volleys of criticism from noneducators (Healey, 1998; Oppenheimer, 2003). Problems of implementing instructional technology include the high costs of updating resources, implementation difficulties, and technology's potential dangers to students. People also criticize the use of technology as making it easy to not challenge the user to think extensively to face the problems technology might let them bypass.

3. Low Teacher Usage: Some research indicates that even teachers who have sufficient training and access to resources are not using technology as much as had been expected (Cuban, Kirkpatrick and Peck, 2001); Norris, Sullivan, Peirot, and Soloway (2003) clearly stated that teachers are not nearing a convincing case for technology's benefit. In Nigeria precisely, the issues of light (power outages), low current, poor power grid infrastructure in rural areas, and so forth, contribute to teachers' low usage of technology. Because teachers often have no access to technology, given the above hindrances, then, how will those teachers be able to use such technologies in the teaching and learning process? Also, sometimes teachers do not regularly make use of technology due to personal lack of experience on how to use them.

4. The Influence of the Accountability Movement and the No Child Left Behind (NCLB). Act that was passed 2001. This act is predicted to dominate policy and drive educational funding for some time to come. One of its most controversial requirements is that funding for proposed expenditures must be tied to "scientifically based research" on effectiveness. At times there is no funding support to run these programs.

5. Teachers' might lack understanding on how to capitalize on the learning potential of technology. There can also be inadequate training of teachers to handle these tools properly in teaching and learning.

6. Students often have the handicap of not having their computer connected to Internet facilities and other technological devices.

7. There is a tremendous lack of Web site hosting: the cost of hosting a Web site can be exorcist. Also, the cost of maintaining such hosting can be an impediment to the integration of technology in teaching and learning.

8. Time consumption and restrictions issue: Because technology takes time to get around, i.e., bringing everybody along in

technological advances at the same time is time-consuming. Many people might not have the time to use technology in their classrooms, especially here in Nigeria, where a class period is forty minutes long. Expertise in using technology for teaching and learning needs time to develop the kind of technology that will match what a teacher is teaching. Time is needed to analyze what is needed by the learner; and to plan, reflect, and imagine how the technology can be used with the learner to produce a potential learning experience. It takes time to learn the technology's capabilities and what it has to offer, to develop the needed skills to use the technology appropriately, to prepare the lessons, implement the lessons, evaluate the integration, and try the technology out to see whether it works in the classroom.

How can these issues raised be solved in Nigeria educational system?

1. Schools and curriculum experts should have technology specialists to help the teachers to overcome these issues by employing technology coordinators and resource persons who can handle computers and its related technologies for school district.

2. Teachers can also be sent to be trained through in-service education to update their technological know-how. Constant training will go a long way to eliminate phobia about technology taking their job or not instead it makes them strong and grounded in their area of specialisation.

3. The issue of support should be resolve through government and individual support such support can come in terms grant in-aid, scholarship and setting up human capacity at different areas to tackle the issue.

4. Funding should also to be made available by the Government.

5. Electricity or power generation should be up graded so that everybody will have access to technology usage.

Conclusion

Technology integration can be used in different subjects in a school setting. The effects of technology integration cut across all disciplines and can enhance learning. Technology can be used by different number of learners. But without strong teacher's knowledge of the ways to use educational technology, a lot of precious instruction time can be wasted. For this to be accomplished, providing adequate technology access is paramount, equalizing technology access no matter the area, involving a majority of teachers in the use of technology in teaching-learning process and providing technical support for technology use and maintenance for proper caring.

References

Cuban, L. Kirkpatrick, H. and Peck, C. (2001). *High access and low use of Technologies in high school classroom. Explaining an approach paradox* (www.cited.org).

Dede, C. (2000). *Emerging influence of technology on school curriculum studies.*

Ficklen and Muscara (2003). *Fostering Meaningful Student earning through Constructivist pedagogy and Technology integration.*

Herrington, J. and Kervin, J. (2007). *Authentic learning supported by technology: 10 suggestion and cases of integration in classroom. Educational media international.*

Kuhn, D and Udell, W (2001). *The path to wisdom.* Educational Psychologist, 36(4), 261–264.

Marzano, R. J., Pickering, D. J., and Pollock, J. E. (2001). *Classroom instruction that works: Research-based strategies for increasing students' achievement. Alexandra, VA: association for supervision and curriculum development.*

Newby, T. J., Stepich, D.A., Lehman, J. D. and Russell, J. D. (2006). *Educational Technology for eaching and learning. Pearson Prentice Hall*3rd edition. www.prenhall.com/smaldino

Newby, T. J, Stepich, D. A., Lehman, J. D., Russell, J. D. and Leftwich, O. (2011). *Educational Technology for teaching and* Learning 4th edition, Pearson. www.pearsonhighered.com

Norris, C., Sullivan, T., Poirot, J., and Soloway, E. (2003). *No Access, No Use, No impact: Snapshot Surveys of Educational Technology in K-12," Journal of Research on Technology in Edu*cation. *ISTE, Vol. 36, number 1.*

Oppenheimer, T. (2003). *The Flickering Mind: The false promise of Technology in the classroom and how learning can be saved. New York, NY: Random House.*

Smaldino, S. E., Lowther, D. L., and Russell, J. D. (2008). *Instructional Technology and Media for Learning.* Pearson Merrill. Prentice Hall. New Jersey. www.prenhall.com/smaldino.

Rao, V. K. (2008). *Educational Technology.* S. B. Nangia, A. P. H Publishing Corporation. New Delhi-110002.

Ringstaff, C. and Kelley, L. (2002). The learning return on our educational technology investment. Retrieved April 8, 2015. http://www.wested.org/online_pubs/learning_return.pdf.

Roblyer, M. D. and Doering, A. H. (2010). *Integrating Educational Technology into Teaching.* 5th edition. www.pearsonhighered.com.www.edutopia.org/ technology-integration-guide-description. Retrieved April 23, 2004.

Chapter 9

Gender Differences in Achievement of Students Exposed to Concepts in Motor Vehicle Mechanics Work through Computer Simulation and Tutorial in Oyo State, Nigeria

J. A. Jimoh
S. A. Adebayo
&
I. O. Oguche

Introduction

One of the most topical issues in the current debate all over the world has been that of gender differences in achievement among students in schools. Gender refers to the socially constructed roles, behaviours, activities and attributes that a society considers appropriate for men and women. Uwameiye and Osunde (2005) refer to gender as a psychological term which describes behaviours and attributes expected of individual on the basis of being a male or a female. Gender equality has been the subject of much debate in technical and vocational education especially with emphasis placed on increasing manpower for technological development as well as increasing the population of females in science and technology fields (Ogunkola and Bilesanmi-Awoderu, 2000). In the whole of Africa and in Nigeria particularly, gender bias is still very prevalent (Arigbabu and Mji, 2004). Culturally, sex-stereotyped occupations of the male over the female have reinforced the notion of women into believing that it is taboo to venture into an occupation that is preserved of the males (Egun and Tibi, 2010). This notion has resulted in more male students than female students prefer to study technology courses (Oriahi, Uhumavbi, and Aguele, 2010). Consequently, there is a wide gender gap in the enrolments of male and female students offering trade subjects that involve workshop practice in technical colleges in Nigeria (Federal Ministry of Education, 2005; Fakorede, 1999; Erinosho, 1998). The poor access and limitation of vocational and technical education to the female students means low empowerment of a group that is needed for national development.

Many studies looking at gender differences in learning have indicated that gender influences students' academic achievement. For example, researchers have found significant differences between male and female students' achievement in science, mathematics, engineering, and technology disciplines. Students' achievement connotes performance in school subject as symbolized by a score or mark on an achievement test. It is quantified by a measure of the student's academic standing in relation to those of other students of his age (Anene, 2005). In an analysis of grade point average scores of high school students in the United States, Jabor, Machtmes, Kungu, Buntat, and Nordin (2011) found statistically significant gender effects on mathematics students' grade point average scores favouring female students. Levin, Sabar, and Libman (1991) reported that the achievement of male students in all subject area of their study (earth science, biology, chemistry and physics) was significantly better than the achievement of the female. Similarly, in a study of Australian secondary school students, Young and Fraser (1994) concluded that male students' achievement in biology, physics and chemistry was significantly better than female students. Jegede and Inyang (1990) found a significant gender effect favouring male students in integrated science achievement among junior secondary schools students in Nigeria. In their own study, Owoduni and Ogundola (2013) studied gender differences and academic achievement of students taught electronic works trades in the technical colleges in Nigeria and revealed that male students performed significantly better than the female students. They affirmed that male students possess greater vocational technical skills than female students. Male advantage in vocational and technical skills was supported by Blosser (1990) who concluded that male students were more likely than female students to report having attempted to fix electrical or mechanical devices.

However, some studies have indicated that gender differences in learning generally are small or nonexistent among students taught engineering trades in the technical colleges in Nigeria. Ogundola, Popoola, and Oke (2010) found no significant difference in the mean scores of male and female students taught mechanical related trades with constructivism instructional approach in the western Nigeria technical colleges. In the same vein, Owoso (2012) found no statistically significant difference in the mean scores of male and female students taught motor vehicle mechanics work with constructivist instructional approach in the technical colleges in Lagos state, Nigeria. In most of the studies though, male students performed better than the female students in the achievement test scores.

Motor vehicle mechanics work is one of the engineering trades in the technical colleges in Nigeria designed to produce competent auto mechanics craftsmen. The craftsmen have three options. These options according to the Federal Republic of Nigeria (FRN, 2004) is to either secure employment in the industries, pursue further education in advance craft in a higher technical institutions or set up their own business and become self-employed. Unfortunately, despite all efforts by the government to ensure qualitative education at the technical colleges and bring about high quality products both in academic and employability, National Business and Technical Examination Board (NABTEB) results of students indicate high failure rate in the main trades. Federal Ministry of Education (FME, 2000) reported that, in the NABTEB certificate examinations conducted in May 2000 the average failure rate (F9) were: electrical 25%; construction trade 41% and engineering trades, which include motor vehicle mechanic work 49%. In the same vein, the chief examiner's report for NABTEB examination conducted in May/June 2012 revealed that candidates recorded poor performance in motor vehicle mechanic works (NABTEB, 2012). In Oyo state specifically, NABTEB results showed that average failure rate in Motor vehicle Mechanic Work in the year 2006, 2007, 2008, 2009, 2010, and 2011 were 32.25%, 16.32%, 71.42%, 56.32%, 0.02%, and 34.92% respectively (NABTEB, 2006;

2007; 2008; 2009, 2010 and 2011). These results show that there is a consistent high failure rate in motor vehicle mechanics work in the NABTEB examinations conducted in the technical colleges in Oyo State. The high failure rate among technical colleges' graduates in Oyo State has resulted into increase in dropout rate and unemployment among the youth. The high rate of unemployment defeats the very fundamental objective of education for self-reliance emphasised in the National Policy of Education, Revised 2004. Besides, the cost of youth unemployment to economic and social development is extremely high because, it perpetuates intergenerational cycle of poverty and is associated with high levels of crimes, violence, substance abuse and the rise of political extremism (International Labour Organization or ILO, 2003).

High failure rate among technical colleges' graduates in technical trades which involve workshop practice have been attributed to gender (FME, 2005; Umunadi, 2009). This is based on the premise that by virtue of gender, male students tend to be more interested in learning technical trades and therefore have better achievement in the trades than the female students. In addition, FME (2000) noted that poor teaching accounts for high failure rate among technical colleges' graduates in Nigeria. Teaching methods such as demonstration and lecture methods which are teacher-centred are the main teaching methods employed by technical teachers for implementing the curriculum in the technical colleges (Owoso, 2012; Oranu, 2003). However, one of the challenges facing technical and vocational education in the twenty-first century demands learner-centred innovation and flexible approaches. In this context, United Nations Educational, Scientific and Cultural Organization (UNESCO, 2002) has recommended that full use should be made of contemporary educational technology, particularly, the Internet interactive multimedia materials, audiovisual aids and mass media, to enhance the reach, cost-effectiveness, quality and richness of programmes.

Connel and Gunzelmann (2004) also proposed creation of a supportive environment by including technology that provides equal opportunities in the classroom to help bridge the gap between male and female students' achievements. Moreover, motor vehicles are coming out with systems and devices which are controlled by microprocessor called the Electronic Control Unit (ECU). According to Nice (2001) automobiles seem to get more and more complicated because automobiles today might have as many as 50 microprocessors on them. If it is taken into account that technical college's students have difficulty in learning concepts in motor vehicle mechanics works due to introduction of new contents into the curriculum, it is important to make the concepts concrete and real using computer technology.

Computer technology can be utilized in many ways to improve teaching and learning. These include among others; computer simulation and computer tutorial. Computer tutorial provides information, to the students in much the same manner as a human teacher or tutor might. Schibeci (1997) described computer tutorial as similar to that of a teacher or textbook in explaining information or concepts to learners. It carries the full burden of instruction. As a result, computer tutorials typically use text, graphics and sounds to present contents. Embedded questions and review activities are used to assess the ability of the students in acquiring the contents. Computer tutorial usage helps students develop problem-solving skills, motivate students, and provide interactive feedback (Vockell and Schwartz, 2005). According to Mann (1995), educators who adopted the tutor mode of computer operation recommended that computer tutorial be designed to teach all manner of knowledge and skills to a wide range of students. Computer simulation, on the other hand, creates models of dynamic systems by combining words with animation (Nerdel and Prechtl, 2004; Schnotz and Bannert, 2003). Interactive learning by using computer simulations for abstract topics, where students become active in their learning provides opportunities for students to construct and understand difficult concepts (Orora, Keraro, and Wachanga, 2014).

Studies have examined effects of computer simulation and tutorial on students learning. Douglas, Millar, Kwanza and Cummings (2004) investigated the effects of computer simulation in science classroom. Results revealed that computer simulation was an effective mode for teaching science. Cole, Banks, and Tooker (1999) found computer visualisation and simulation effective for teaching electronics than the use of traditional method. Huang (2009) in his own study found computer simulation more effective than traditional method. While many educators have put and are putting tremendous effort into devising new ways of using computer technology in the classroom, with the clear expectation that such technology will dramatically increase students' academic achievement, the result of this study provide classroom teachers with cumulative bank of research-based evidence for the positive effect of computer simulation on students' learning. In another study, Fongsrisin (2004) determined effectiveness of computer tutorial on students' learning in photography. Result revealed that students taught using computer tutorial performed better than students taught with conventional methods in the achievement test but the high mean score was not found to be significant. However, Muhamma and Aktaruzzaman (2011) found students taught educational research with computer tutorial achieved better than students taught with traditional method. Onasanya, Daramola, and Asuquo (2006) also found secondary school students taught introductory technology with CAI in the form of tutorial performed better than those taught with conventional method.

As computer technology is introduced into the classrooms with a focus on improving students' achievements, gender difference in learning with computer technology has been of interest to researchers. Many of the studies on the gender differences in the achievement of students taught with computer technology have contradictory results. Spence (2004) found no significant influence of gender on the achievement of college students in mathematics when they were exposed to mathematics courseware in online and traditional learning environment. However, female online learners were significantly less likely to complete the course compared to their traditional female counterpart or male online counterparts. Kirkpatrick and Cuban (1998) found that when female and male students at all levels of education had the same amount and types of experiences on computers, female achievement scores and attitudes were similar in computer classes and classes using computer. Kumar and Helgeson (2000) reported that the use of HyperCard Program for teaching how to solve stoichometric chemical equations related problems produced no significant gender difference in academic achievement. Bernea and Dori (1999) found no significant difference between male and female students when exposed to building molecules in chemistry concepts using computerized molecular modeling software. Given these mixed results, this study determined gender differences in achievement of students exposed to concepts in motor vehicle mechanics work through computer simulation and tutorial in Oyo State, Nigeria

Purpose of the Study

The purpose of this study was to determine gender differences in achievement of students exposed to concepts in motor vehicle mechanics work through computer simulation and tutorial in Oyo State, Nigeria. Specifically, the study sought to determine gender differences in achievement of students exposed to concepts in motor vehicle mechanics work through three methods of teaching, namely; computer simulation, computer tutorial, and traditional teaching method.

Research Questions

The following research questions were posed to guide the study;

1. What is the effect of gender on the pretest mean score of motor vehicle mechanics work of students before treatment?

2. What are the gender differences in the achievement of students taught motor vehicle mechanic work with computer simulation and computer tutorial?

3. What are the effects of the three methods of teaching (computer simulation, computer tutorial and traditional teaching method) on the achievement of students taught motor vehicle mechanics work?

Hypotheses

The following null hypotheses tested at .05 level of significance guided this study:

HO$_1$: There is no significant effect of gender on the pretest mean scores of motor vehicle mechanics work of students prior to treatment

HO$_2$: There is no significant effect of gender on the pretest mean scores of motor vehicle mechanics work of students taught with computer simulation and computer tutorial

HO$_3$: There is no significant difference in the pretest mean score of the three treatment groups

HO$_4$: There is no significant effect of treatment on the achievement of students taught motor vehicle mechanic work with the three methods

HO$_5$: There is no significant interaction effect of treatment and gender on students' achievement in motor vehicle mechanics work.

Methods

Design and Area of the Study

The study adopted the quasi-experimental research design. Specifically, the study was a multifactorial pretest, posttest experimental (3x2) design comprising three independent variables, namely teaching methods at three levels (computer simulation, computer tutorial, traditional method as control) and gender at two levels (male and female). The study was conducted in NBTE accredited technical colleges offering motor vehicle mechanics work in Oyo state.

Population and Sample for the Study

The population for the study comprised all seventy of the second-year students studying motor vehicle mechanics work in all the four technical colleges offering motor vehicle mechanics work in Oyo state in 2012/2013 academic session. The sample size was fifty-four students. Simple random sampling technique was used to select three technical colleges offering motor vehicle mechanics work in Oyo State. Intact classes of the three technical colleges were randomly assigned to the three treatment conditions (computer simulation, computer tutorial, and traditional method as control). In all, three groups were formed namely; two experimental groups and one control group. The first experimental group (nine males and three females) taught with computer simulation, the second experimental group (eighteen males and three females) taught with computer tutorial and the control group (seventeen males and four females) taught with traditional teaching method.

Instrument, Validation and Reliability

The instrument used for data collection was motor vehicle mechanics work achievement test (MVMAT). A test blueprint was used to construct the MVMAT items in order to ensure content validity of the test. One hundred multiple choice items were drawn for the MVMAT, after which the MVMAT, the lesson guide for the teachers were subjected to face validation by three experts. A trial test was conducted on the MVMAT for the purpose of determining the psychometric indices of the test. The answer sheet were marked and used for computing the psychometric indices of the test items. A total of forty-five items of the MVMAT had good difficulty, discrimination and distractor indices. The coefficient of stability of the MVMAT was carried out using test-re-test reliability technique. The reliability coefficient of the MVMAT was found to be .85 using Pearson product-moment correlation coefficient. The data collected were analysed using mean to answer the research questions while t-test statistics, analysis of variance (ANOVA), and analysis of covariance (ANCOVA) were used to test the hypotheses.

Computer Simulation Package and Computer Tutorial Package

The computer simulation package was developed by the researchers and with the assistance of a computer programmer. The package covered the following topics: four-stroke spark ignition engine, two-stroke spark ignition engine, fuel supply system of petrol engine, working principle of injection system, working principle of carburettor system, ignition system, water-cooling system and air-cooling system, Computer simulation package was developed to display the contents to students using animation, words and sounds while computer tutorial package was developed to display the contents using words, sound, graphics, and pictures.

Control of Extraneous Variables

Teachers' Variability:

- To control the effect of variable such as teacher's variability, which could result to experimental bias, the regular technical teachers in the participating technical colleges, taught their own students. Hence, the researchers were not directly involved in administration of the research instruments and the treatments.

Teaching Guide (Lesson Plan) Preparation:

- To control variability in the development of the teaching guide (lesson plans) and to ensure uniform standard in the conduct of the research, the researchers prepared the teaching guide and organized training for participating lecturers. Three types of Lesson plan were developed, namely: simulation lesson plan, tutor lesson plan and traditional lesson plan.

Experimental Procedure

Administration of Pretest

The MVMAT was administered to all the three groups as pretest. Technical teachers administered the MVMAT to the groups in their respective schools. The researchers collected and marked the answer sheets of the MVMAT to obtain the students' Pretest before the treatment

The Treatment

The teacher who taught the group assigned to computer simulation used the simulation lesson plan as a teaching guide while the teacher who taught the group assigned to computer tutorial used the tutor lesson plan as a teaching guide. The teacher who taught the control group used the traditional lesson plan as a teaching guide. The same topics were taught to all the three groups. The lesson contents covered the following topics; four stroke spark ignition engine, two stroke spark ignition engine, fuel supply system of petrol engine, working principle of injection system, working principle of carburettor system, ignition system, water cooling system and air cooling system. Each lesson lasted for ninety minutes.

The computer tutorial package displayed contents of the lesson with text, graphics and sound. As the students clicked on the lesson contents, the computer tutorial package responded by displaying the contents. Students read the content in text and graphics as well listened to the sound.

The computer simulation package displayed contents of the lesson with animation, text, sound. As the students clicked on the lesson contents, the computer simulation package responded by displaying the contents. Students viewed the animation, read the text and as well listened to the sound.

The teacher who taught the control group used the traditional teaching method such as demonstration, lecture etc. as applicable to teach the content and as well wrote notes on the chalkboard for students to copy.

Administration of the Posttest

At the end of the treatment, the MVMAT was rearranged and administered to all the three groups as posttest. Technical teachers administered the MVMAT to the groups in their respective schools. The researchers collected and marked the answer sheets of the MVMAT to obtain the students' posttest scores.

Findings

Effect of Gender on the Students' Achievement before Treatment

Differences attributed to gender factor on students' achievement were investigated prior to the treatment. The mean scores of the pretest for male (n=44) and female (n=10) were 9.86 and 7.70 respectively, summarized in table 1.

Table 1. Pretest mean scores and standard deviation by gender and t-value

Gender	n	Mean $\overline{X}$	SD	df	t-cal	ρ value
Male	44	9.86	1.40	52	4.425	0.000
Female	10	7.70	1.33			

t-cal Significant at ρ< .05

An independent sample t-test was performed to detect if the mean difference was significant. The result in table 1, shows that the difference between the mean scores of the male and female was statistically significant (t=4.425, ρ>.05). Hence, HO_1 was rejected since the result shows that there was an effect of gender on students' achievement favouring male students prior to the treatment.

Gender Differences in the Achievement of Students in the Two Experimental Groups Descriptive statistics and analysis of covariance (ANCOVA) were used to determine if there were gender differences in the achievement of students taught with computer simulation and computer tutorial. Table 2 summarized the pretest, posttest and mean gain of male and female students in the two experimental groups.

Table 2. Mean achievement scores of experimental groups after treatment, by gender

	Computer Simulation				Computer tutorial			
Gender	n	Pretest	Posttest	Mean Gain	n	Pretest	Posttest	Mean Gain
Male	9	9.11	31.77	22.66	18	9.72	29.55	19.83
Female	3	7.53	25.33	17.80	3	8.00	20.33	12.33
Total	12	8.75	30.16	21.41	21	9.47	28.23	18.76

Table 2 shows the mean gain of male students in the two groups were higher than the mean gain of female students. The result shows that there was an effect attributable to gender on the achievement of students taught motor vehicle mechanics work with computer simulation and computer tutorial.

An analysis of covariance (ANCOVA) was used to determine if the differences in the male and female students' achievement scores among in the two experimental groups were statistically significant. Results of the ANCOVA revealed a statistical significant main effect of gender (F=43.540, p<.05), as shown in table 3.

Summary of analysis of covariance (ANCOVA) for test of significance of two-effects gender on Achievement of students taught motor vehicle mechanics work with computer simulation and computer tutorial.

Table 3. Summary of analysis of covariance (ANCOVA) for test of significance of two effects gender on achievement of students

Source	Sum of squares	df	Mean square	F	Sig.
Corrected Model	292.435[a]	2	146.217	22.444	.000
Intercept	813.941	1	813.941	124.937	.000
Pretest	19.019	1	19.019	2.919	.098
Gender	283.653	1	283.653	43.540	.000
Error	195.444	30	6.515		
Total	28125.000	33			
Corrected Total	487.879	32			

With this result, HO_2 was rejected as there was a statistical significant main effect of gender on the achievement of motor vehicle mechanics work students taught with computer simulation and computer tutorial favouring male students.

Differences in Achievement of the three Groups before Treatment

Descriptive statistics and analysis of variance (ANOVA) were used to determine if there were significant differences among the three groups pretest mean scores before the treatment was administered. Table 4 summarizes the pretest mean scores and standard deviation of the three groups.

Table 4. Comparison of pretest mean scores of the three groups in the MVMAT

	Teaching methods							
Computer simulation			Computer tutorial			Control		
n	Pretest Mean $\overline{X}$	SD	n	Pretest Mean $\overline{X}$	SD	n	Pretest Mean $\overline{X}$	SD
12	8.75	1.60	21	9.47	1.53	21	9.85	1.65

In the pretest mean scores, the twelve participants in the computer simulation group had a pretest mean score of 8.75 (SD=1.60); the twenty-one participants in the computer tutorial group had a pretest mean score of 9.47 (SD= 1.53); and the twenty-one participants in the control group had a pretest mean of 9.85 (SD= 1.65). Table 5 summarizes the result of the analysis of variance for test of significance difference in the pretest mean scores of the three groups.

Table 5. Summary of ANOVA for pretest mean achievement scores of the three groups

	Sum of Squares	df	Mean Square	F	ρ
Between Groups	9.366	2	4.683	1.836	.170
Within Groups	130.060	51	2.550		
Total	139.426	53			

There were no significant differences between the three groups pretest mean scores prior to the administration of the treatment (F=1.836, ρ>.05). Therefore, HO$_3$ was accepted. In other words, all the three groups, on the onset of this study, were statistically equivalent in their achievement in motor vehicle mechanics work.

Effects of the three Methods of Teaching on the Achievement of Students after Treatment

Descriptive statistics and analysis of covariance (ANCOVA) were used to determine if there were significant differences in the mean achievement scores among the three groups after the treatment was administered. Achievement of the Students in the three groups after the treatment was measured in terms of mean gain. Mean gain is the difference between the pretest mean score and posttest mean score of the same group. The mean gain of the two experimental groups and control are summarized in table 6.

Table 6. Comparison of mean gains of the three groups after treatment

Group	N	Pretest $\overline{X}$	SD	Posttest $\overline{X}$	SD	Mean Gain
Computer Simulation	12	8.75	1.60	30.16	3.78	21.41
Computer Tutorial	21	9.47	1.53	28.23	3.88	18.76
Control Group	21	9.85	1.65	18.52	3.68	8.67

Results in table 6 revealed that group taught with computer simulation had a mean gain of 21.41 and group taught with computer tutorial had a mean gain of 18.76, while the control group had a mean gain of 8.67. Hence, the group taught with computer simulation outperformed others and group taught with computer tutorial was better than the control group.

An analysis of covariance (ANCOVA) was used to determine if the differences in the students' achievement scores among the three groups were statistically significant. Results of the ANCOVA revealed a statistical significant main effect of treatments ($F=94.632$, $p<.05$), as shown in table 7.

Summary of analysis of covariance (ANCOVA) for test of significance of two effects: treatments and interaction effect of treatment and gender on students' achievement in motor vehicle mechanics work.

Table 7. Summary of analysis of covariance (ANCOVA) for test of significance of two effects: treatments and interaction effect

Source	Sum of squares	Df	Mean square	F	ρ
Corrected Model	1927.671[a]	6	321.279	67.513	.000
Intercept	646.232	1	646.232	135.798	.000
Pretest	.421	1	.421	.089	.767
Treatment	900.666	2	450.333	94.632*	.000
Gender	366.095	1	366.095	76.931	.000
Treatment * Gender	9.395	2	4.698	.987	.380
Error	223.662	47	4.759		
Total	35602.000	54			
Corrected Total	2151.333	53			

*Significant at $\rho < .05$

HO$_4$ was therefore rejected since there was a significant main effect of treatment on the achievement of students taught motor vehicle mechanic work. To determine where the differences among the three groups were, a post hoc test was conducted using Scheffe test. The Scheffe test for treatment condition was tested at .05 level of significance and summarized in **Table 8. Scheffe test results by treatment condition**

(I) Treatment	(J) Treatment	Mean difference (I-J)	Std. error	Sig.	95% Confidence interval Lower bound	Upper bound
Computer Simulation	Computer Tutorial	1.92857	1.36976	.378	-1.5252	5.3823
	Control	11.64286*	1.36976	.000	8.1891	15.0966
Computer Tutorial	Computer Simulation	-1.92857	1.36976	.378	-5.3823	1.5252
	Control	9.71429*	1.16814	.000	6.7689	12.6597
Control	Computer Simulation	-11.64286*	1.36976	.000	-15.0966	-8.1891
	Computer Tutorial	-9.71429*	1.16814	.000	-12.6597	-6.7689

*. The mean difference is significant at the 0.05 level.

As shown in table 8, the computer simulation compared to the computer tutorial revealed a mean difference of 1.928 and a significance of .378 which indicated that there was no significant difference between the two treatment conditions. This simply means that the mean difference between the group taught with computer simulation and those taught with computer tutorial was not statistically significant. The comparison of the computer simulation condition with the control condition revealed a mean difference (11.642) indicating a significance of .000. Also a mean difference of 9.714 with significance of .000 was found when comparing the computer tutorial condition and control, indicating a significant difference between the achievements of students in the two groups. Clearly, the post hoc test showed there was no significant difference between the achievement of students taught with computer tutorial and those taught with computer simulation, however, there were significant differences in the achievement of students taught with traditional methods and the other two experimental groups.

Table 7 also revealed that the interaction between treatment and gender on achievement of students in motor vehicle mechanic work was not statistically significant (F= .987, ρ> .05). Therefore, HO_5 was accepted, indicating that the effectiveness of the treatment on achievement does not vary depending on the students' gender.

Discussion

Prior to treatment this study found significant effect of gender on students' achievement in motor vehicle mechanics work favouring male students. This finding was not unexpected given that by virtue of gender, male students tend to be more interested in learning technical trades and therefore have better achievement in the trades than the female students. Additionally, Valentine (1998) noted that female students think and learn differently as well as interact with equipment differently in mathematics, science and technology classroom. Supporting this view, Klein (2007) explained that male students dominate class equipment and machines in engineering and technology classes, female students on the other hand fear handling equipment and machines during class project due to the feelings that they are incapable. This challenge faced by female students could probably result in lowered performance in engineering and technology trades (Elijah, Kimani, and Wango, 2014). Besides, some studies looking at gender differences in learning technical trades in the technical colleges in Nigeria have indicated that gender influences students' academic achievement favouring male students. Male superiority in achievement has been established in studies by Ogbuanya and Owoduni (2013) and Owoduni and Ogundola (2013).

After the treatment, this study found significant effect of gender on the achievement of students taught with computer simulation and those taught with computer tutorial. This implies that the use of computer simulation and computer tutorial do not overcome the gender disparity in achievement of male and female students taught motor vehicle mechanic work. This finding contradicts the findings of Yusuf and Afolabi (2010), Orora, Keraro, and Wachanga (2014), Kumar and Helgeson (2000) and Bernea and Dori (1999), who in their separate studies found no effect of gender on students' achievement when they were exposed to computer technology for learning. However, this study found no significant interaction effect of treatment and gender on students' achievement in motor vehicle mechanics work. This means that there were no differential effects of treatment over levels of gender.

Prior to the treatment, the two experimental groups and the control group were equivalent. After the treatment, students taught motor vehicle mechanics work through computer simulation outperformed others and group taught with computer tutorial performed better than the control group ensuring that resultant differential outcomes were attributed to the effect of teaching method. Post hoc test using Scheffe indicated that there was no significant difference between the mean achievement scores of students taught with computer tutorial and those taught with computer simulation, however, there were significant differences in the mean achievement scores of students taught with traditional methods and the other two experimental groups. This finding is consistent with Birkenholtz, Stewart, McCaskey, and Ogle (1999) who studied effects of using computer tutorial, drill and practice, and simulation strategies in vocational agric education. Birkenholtz et al. found student achievement is essentially equal when taught using any of the three microcomputer-enhanced teaching strategies. Moreover, the findings of this study agree with the findings of Bakaç, Taşoğlu, and Tkbay (2011) on electric current activities; Efe and Efe (2011) in biology; Aktaruzzaman and Muhammad (2011) in educational research; Douglas, Millar, Kwanza, and Cummings (2004) in science education; Cole, Banks, and Tooker (1999) and Onasanya, Daramola, and Asuquo (2006) which confirmed that computer simulation and computer tutorial have been effective in enhancing students' performance in other subjects than the traditional method of instruction.

Conclusions and Recommendations

Much of literature concerning the relative performance of male and female in technical trades indicates that male students outperform female students in most areas of technical education curriculum. The results of this investigation into gender differences in achievement of students exposed to concepts in motor vehicle mechanics work through computer simulation and tutorial in Oyo State, Nigeria, has provided useful information on male and female performance when taught using computer technology. The study has also provided optimistic support for the use of computer simulation and tutorial for teaching motor vehicle mechanics work. This study revealed significant effect of gender on the achievement of students exposed to concepts in motor vehicle mechanics work through computer simulation and tutorial in Oyo State. However, this study has found no significant interaction effect of treatment and gender on students' achievement in motor vehicle mechanics work. This simply means that the effectiveness of computer simulation and tutorial on students' achievement in motor vehicle mechanics work does not depend on the levels of gender. To this end, irrespective of nature of gender, male and female students would record improved performance in their achievement in motor vehicle mechanics work when computer simulation and tutorial are employed for teaching in the technical colleges in Oyo State.

References

Anene, G. U. (2005). Home economics and the academic performance of a child. *Journals of Home Economics Research*, 6 (1), 99–103.

Arigbabu, A. A. and Mji, A. (2004). Is gender a factor in mathematics performance among Nigerian preservice teachers? *Sex Role*, 51 (11 and 12). 749–755.

Bakac, G. B. Tasoglu, N. M. and Tkbay, D. F. (2011). Effect of computer assisted instruction with simulation on students' success in electric current activities. Retrieved January 4, 2011, from *http:// www.eurasianjournal.com/index.php.ejpce/pdf.*

Bernea, N. and Dori, Y. J. (1999). High-school chemistry students' performance and gender differences in a computerized molecular modeling learning. *Journal of Science Education and Technology*, 8(4), 257–271.

Birkenholtz, R. J. Stewart, B. R., McCaskey M. J., and Ogle, T. D. (1999). Effects of using microcomputers in education: Assessment of Three teaching strategist. Retrieved January 4, 2011, from *http://www.landersu.ngv/pdf.*

Blosser, P. (1990). Procedures to increase the entry of women in science-related career. Columbus, OH; ERIC Clearinghouse for Science, Mathematics and Environmental Education Cole, R. Banks, D. and Tooker, S. (1999). Improving interpretative skill with visualisation and simulation exercises in electromagnetic radiation, Retrieved January 10, 2012 from *http://www.udcphy.ucdavis.edu.*

Cole, R. Banks, D. and Tooker, S. (1999). Improving interpretative skill with visualisation and simulation exercises in electromagnetic radiation, Retrieved January 10, 2012 from *http://www.udcphy.ucdavis.edu.*

Connell, D. and Gunzelmann, B. (2004). The new gender gap. Instructor, 113(6), 14–17

Douglas, A., Millar, B., Jwanza, F. and Cummings, P. (2004). Usefulness of a virtual simulation in post-secondary education: students' perception. *Simulation and Gaming* 34(1), 23–28.

Douglas, A., Millar, B., Jwanza, F. and Cummings, P. (2004). Usefulness of a virtual simulation in post-secondary education: students' perception. *Simulation and Gaming* 34(1), 23–28.

Efe, H. A. and Efe R. (2011). Effect of Computer simulations on secondary biology instruction, *Scientific Research and Essay.* 6(10) 2137–2146.

Egun, A. C. and Tibi, E. U. (2010). The gap in vocational education: increasing girls access in the 21st century in the Midwestern states of Nigeria. Retrieved on December 7, 2014, from *http://www.academicjournals.org/IJVTE.*

Elijah, D. M., Kimani, E. N. and Wango, G. M. (2014). Gender-related challenges faced by students in learning technical courses in Machakos technical training institute, Machakos county-Kenya retrieved June 25, 2014, from *http://www.primejournal.or/PJSS*.

Erinosho, S. Y. (1998). Gender discrimination in science education in Nigeria. In Erionso S. Y (Ed) *Science Education for All in Nigeria, Which Way Forward*. Proceedings of seminar supported by FAWE Nairobi, Kenya 28–32.

Fakorede, A. D. 1(999). Survey into gender difference and students achievement in secondary school biology. A case study of Oyo State. *An unpublished M. Ed thesis*, University of Ibadan.

Federal Ministry of Education (2000). *Technical and vocational education development in Nigeria in the 21st century with the blue-print for the Decade 2001–2010*. Abuja: Federal Ministry of Education.

Federal Ministry of Education (2005). *Nigeria education sector diagnosis: A frame work for re-engineering education sector.* Abuja: Federal Ministry of Education.

Federal Republic of Nigeria (2004), National *Policy on Education*. Lagos; NERDC.

Fongrisin, N. (2004). Development of computer assisted instruction for introduction to photographic course. Retrieved on July 18, 2005 from *http://www.ihtit.co.org/pdf*.

Huang, H. J. (2009). Computer daylight simulation system. An experimental evaluation.

Retrieved January 10, 2012 from *http://www.aol.com*

International Labour Organization (2003). Working out of poverty. Retrieved, June 20, 2004. from *http//:www.ilo.org/document/1235452/pdf*.

Jabor, M. K., Machtmes, K., Kungu, K., Buntat, Y. and Nordin, M. (2011). The influence of age and gender on the students' achievement in mathematics. Retrieved on July, 14, 2014 from *http//:www.ipedr.com/vol5/no2/67-H10178.pdf.*

Jegede, O. J. and Inyang, N. (1990). Gender differences and achievement in integrated science among junior secondary school students: A Nigerian study. *International Review of Education,* 36(3), 364–386.

Kirkpatrick, H. and Cuban, L. (1998). Should we be worried? What the research says about gender differences in access, use, attitude and achievement with computer. *Educational Technology,* 38(4), 56–60.

Klein, S. (2007). *Achieving gender equity in technical education through education.* New York: Sage publication.

Kumar, D. D. and Helgeson, S. L. (2000). Effects of gender on computer-based chemistry problem solving: Early findings. *Electronic Journal of Science Education,* 4(4), 1–3

Levin, T., Sabar, N., and Libman, Z. (1991). Achievement and attitudinal patterns of boys and girls in science. *Journal of Research in Science Teaching,* 28(4), 315–328.

Mann B. L. (1995). Computer aided instruction. New York; Academic Press.

Muhammad, K., and Aktaruzzaman, M. D. (2011). A comparison of traditional method and computer aided instruction on students achievement in educational research. *Academic Research International,* 1(3), 246–253

NABTEB (2012). *NABTEB Chief Examiner's report on the 2012 May/ June NBC/NTC examinations.* Benin: NABTEB

National Business and Technical Examination Board (2006, 2007, 2008, 2009, 2010, and 2011).
May/June 2006, 2007, 2008, 2009, 2010 and 2011 NTC/NBC Examination Results in Oyo State technical colleges. Benin: NABTEB

Nerdel C, and Prechtl, H. (2004). Learning Complex Systems with simulations in Science Education. Retrieved June 25, 2007, from *http://Iwm-kmrc.de/workshops/SIM2004/pdf files/ Nerdel et al.pdf.*

Nice, K. (2001). How car computers work. Retrieved on January 18, 2003, from *http://www.howstuffswork.com.*

Ogbuanya, T. C. and Owoduni, S. A. (2013). Effects of reflective inquiry instructional technique on students' achievement and interest in radio, television and electronics works in technical colleges. *IOSR Journal of Engineering,* 2, 1–11.

Ogundola, I. P., Popoola, A. A. and Oke, J. (2010). Effects of constructivism instructional approach on teaching practical skills to mechanical related trade students in western Nigeria technical colleges. Retrieved on December 7, 2014, from *http://www. academicjournals.org/INGOJ.*

Ogunkola, J. B. and Bilesanmi-Awoderu, J. B. (2000). Effects of laboratory and lecture methods on students' achievement in biology, *African Journal of Education,* 5(2), 247–260.

Onasanya, S. A. Daramola, F. O. and Asuquo, E. N. (2006). Effects of Computer assisted instructional package on secondary school student's performance in introductory technology in Ilorin, Nigeria. *The Nigeria Journal of Education Media and Technology;* 12(1).

Oranu, R. N (2003). Vocational and technical education in Nigeria. Retrieved on July 18, 2005 from *http://www.ibe.co.org/curriculum/ Africapdf/lago2ora.pdf.*

Oriahi, C. I., Uhumavbi, P. O. and Aguele, L. I. (2010). Choice of science and technology subjects among secondary school students. Retrieved on December 7, 2014, from *http://www.krepublishers. com/....jss-22-3-191-10-874-Oriahi-C-ITt.pdf.*

Orora, W., Keraro, F. N. and Wachanga, S. W. (2014). Effects of cooperative e-learning teaching strategy on students' achievement in secondary school biology in Nakuru County, Kenya. Retrieved November 10, 2014, from http://*www.skyjournals.org/SJER.*

Owoduni, A. S. and Ogundola, I. P. (2013). Gender differences in the achievement and retention of Nigerian students exposed to concept in electronic works trade through reflective inquiry instructional technique. Retrieved on December 7, 2014, from *http://www. sciencedomain.org*

Owoduni, S. A. and Ogundola, I. P. (2013). Gender differences in the achievement and retention of Nigeria students exposed to concept in electronics works trades through reflective inquiry instructional technique. Retrieved July 13, 2014, from *http://www. sciencedomain.org.*

Owoso, J. O. (2012). Effects of collaborative learning and framing on psychomotor achievement and interest of automechanics students in the technical colleges in Lagos State, Nigeria. Proceeding of *Proceedings of Institute of Science, Technology and Education (ISTE) International Conference on Mathematics, Science and Technology, University of South Africa, 427–443.*

Schibeci, R. A. (1997). Tutorial simulation and drill and practice. *National Issues in Nigher Education, 18* (1985): 369–377.

Schnotz, W., and Bannert, M. (2003). Construction and Interference in Learning from multiple representation. *Learning and Instruct, 13;* *141–156.*

Spence, D. J. (2004). Engagement with mathematics courseware in traditional and online learning. Retrieved June 14, 2014, from *http://www.des.emory.edu.mfp/spenceDissertation2004.pdf.*

Umunadi, K. E. (2009). A relational study of students' academic achievement of television technology in technical colleges in Delta State of Nigeria. Retrieved January 1, 2013, from *http://www. scholar.lib.vt.edu/ejournals/JITE/v46n3/umunadi.html*

United Nations, Educational, Scientific and Cultural Organization and International Labour Organization (2002). *Technical and vocational education and training for the twenty-first century.* Paris: UNESCO.

Uwameiye, R. and Osunde, A. U. (2005). Analysis of enrolment pattern in Nigerian polytechnics' academic programmes and gender imbalance. *Journal of Home Economics research* 6 (1), 150–155.

Valentine, E. F. (1998). Gender differences in learning and achievement in mathematics, science and technology and strategies for equity: A literature review. Retrieved on October 5, 2014 from *http://www. chre.vt.edu.*

Vockell, E. L. and Schwartz, E. (2005). "Using Microcomputers to Teach Freshman English Composition." *National Issues in Higher Education, 18,* 369–377

Young, D. J. and Fraser, B. J. (1994). Gender differences in Science achievement: Do school effects make difference? *Journal of Research in Science Teaching,* 31(8), 857–871.

Yusuf, M. O., and Afolabi, A. O. (2010) Effects of computer assisted instruction (CAI) on secondary school students' performance in biology. *The Turkish Online Journal of educational Technology,* 9(1), 62–69.

Chapter 10

Restructuring Igbo Language Classroom for Modern Pedagogy: The Use of Information and Communication Technology

Okudo, Afoma Rosefelicia

Introduction

The people and countries of the world are being linked together economically and culturally, through trade, information technology, travel, cultural exchanges, the mass media and mass entertainment. Nowadays, technology has been revolutionizing every aspect of a people's life and has become an important part of most organisations and businesses (Zhang and Aimang, 2007). The impacts of these have been so rapid that they are being felt in every aspect of human endeavour. The world is also moving with a terrific speed in the use of ICT that only nations and countries that are committed in the use of ICT will compete in today's global market (Obiefuna and Enwereuzo, 2012). Obiefuna and Enwereuzo reiterated that the twenty-first-century learners have been described as the next generation learners or digital natives, because of their formal access to computer enabled technology. These twenty-first-century learners play video games, watch television, send text messages and e- mail and pay little or no attention to reading their books. Oblinger (2004) confirmed the assertion that the digital natives (Igbo language learners inclusive) by the age of twenty years, would have spent ten thousand hours playing video games, sent two hundred thousand e-mails, spent twenty thousand hours watching the television, spent ten thousand hours on the cell phone, but less than five thousand hours in reading. This would to a great extent affect their performances in school unless their new interests were adequately channeled and integrated in their academics work.

In education, technology is a compelling force that creates an engaged learning atmosphere which allows learners to become skilled technicians with an endless passion for learning. With the use of technology, students will become engaged in learning through problem solving, reflecting, synthesizing, evaluating and continuously applying their own skills. According to Brannigan (2002), when one combines enquiry based learning and technology integration there is a synergy created that really boosts students learning. The teacher controlled learning where deconstructed and reconstructed information were presented in a highly formal and standardized classroom setting have become very obsolete and may not be meeting the twenty-first-century learners' challenges (Obiefuna and Enwereuzo op. cit.). Teachers need to be fully aware of the technology that engages learners on daily basis and use that technology as a stepping stone to build upon in the learning environment. Educators must ensure that the prior learning experiences are appropriate and related to the concept to be taught that calls for the importance of laying the ground work for future gained knowledge. Rodgers, Rungon, Starreft and Holze (2006) were equally of the view that complaints were no solution to the problems but addressing the problems by finding adequate solution. Rodgers, Rungon, Starreft and Holze noted that children born from 1982 have different relationship with information and learning than children of previous years and as such some pedagogical skills/approaches for nondigital natives may no longer be relevant to the digital natives.

According to Dike (2000), teachers should make use of modern instructional technologies, such as the computers, the Internet, audiovisual equipment, video conferencing tools, projectors, and traditional software in their teaching and learning for maximum achievement. Dike (2000) reiterated that connecting the classrooms to the Internet was not enough, and that it is crystal clear that proper use of technology affects the way teachers teach and students learn. This to a large extent affects the teaching and learning of the Igbo language.

The pedagogical skills according to Davis, Preston and Cox (2009) entail integrating ICT tools into subject teaching rather than as a discrete school subject. This means that the school based subjects, such as Igbo language should be interactive, participatory with group collaboration rather than transmission, rote memorisation and passive based pedagogy. This chapter therefore specifically addressed the problems and prospects of integrating ICT tools into teaching and learning of Igbo language for qualitative education. Okafor (2013) added that:

> If the Igbo teacher is left with traditional instructional materials made up of chalk or whiteboard, textbooks, improvised instructional materials such as charts, shapes etc. while teaching and learning of other languages such as English continues to be computer assisted, one doubts whether the Igbo language would survive the 'ravages' of globalization ... (p. 154)

Information Communication Technology is just one of the tools that have played a major role in globalisation. While the impact of Information Communication Technology has been overly positive, it has led to certain challenges as well. Increasing global interdependence has profound influence on education at all levels, such as how to deal with a world with more permeable boundaries in which people are on the move more frequently (migration) than ever before in human history, and in which urbanisation is increasing at an unprecedented rate.

Adapting ICT Tools into Classroom Teaching and Learning for Qualitative Education

In this chapter, the term ICT is used to indicate the whole range of technologies involved in information processing and electronic communications, including the computers, Internet, e-mail, World Wide Web, word processors, Satellite, Global System Mobile Communication (GSM), and other allied electronic ICT needed for more efficient curriculum delivery in schools.

Computer technology has heralded in the contemporary world, especially in the area of computer technology and implementation of new and innovative curriculum delivery strategies particularly with the Internet revolution (Halat, 2008; Güzeller and Akin, 2012). In the world at large, it is acknowledged that ICT is growing at a rapid pace with emerging technologies continuing to develop. Information and Communication Technology (ICT) plays a vital role in the development of any nation. It has been an instrument for achieving social, economic, educational, scientific and technological development (Adesoji, 2010). The application of Information Communication Technology (ICT) is not only emphasised in the corporate world and the industrial sector, but it is an essential part of education at all levels. ICT has greatly influenced the educational sector especially the curriculum delivery, learning and research. As Kuthlau (2011) observed, global interconnectedness enabled by information technology calls for new skills, knowledge and ways of learning to prepare students for living and working in the twenty-first century.

According to United Educational Scientific and Cultural Organisation (UNESCO, 2002) educational systems around the world are under increasing pressure to use the new ICT tools to teach students the knowledge and skills they need in the twenty-first century. ICT tools have fundamentally changed the way people communicate and do their business and have the potential to transform the nature of education; where and how learning takes place and the roles of students and teachers in the learning process (Achukwu and Nnajiofor, 2012). The challenges confronting Nigeria educational systems is how to transform the curriculum and teaching and learning to provide students with the skills to function effectively in the dynamic, information-rich, and continuously changing environment. To meet these challenges, educational institutions must embrace the new technologies and appropriate ICT tools for learning. They must also move towards the goal of transforming the traditional paradigm of learning. They should integrate ICT in classroom instruction to facilitate effective teaching and learning. This would help in bringing up the twenty-one language learners who are efficient and effective ICT users. The acquired and developed ICT skills and knowledge might help them be able to be successful in the continuously evolving and competitive economy of today.

In Nigeria, the awareness of technology and the use of ICT tools are recent, but they are spreading drastically. Children and youths including those learning Igbo and other Nigerian languages have access to computer located in all nooks and crannies especially in the urban setting. The computer centres provide training and assistance on the use of technology to their learners such as how to read and send e- mails, word processing, social networking, etc. At homes, children are provided televisions with video games which also use the new technology; as such children are not only conversant with the new technology but spent many hours watching television at the detriment of their studies. All these pose challenges to the teachers especially as the informal ICT training programmes have no approved school curriculum guiding them. Also there was no supervision for the video games and television programmes watched at home. The use of ICT tools in Igbo language classroom becomes an avenue through which the Igbo learners will be properly guided by their teachers. Okafor (2013) echoed that the efforts of the Igbo language teacher certainly need to be supported by ICT if positive results are to be achieved and if the Igbo language will continue to be relevant in this digital era.

Therefore, the impact of integrating ICT tools in teaching and learning of Igbo Language should be a big and better asset in the educational system since the FRN (2004) maintained that Government shall provide facilities and infrastructures for the promotion of ICT at all levels. According to UNESCO (2004), reports show that ICT tools have helped to improve greater autonomy in learning, boost self-confidence and facilitate the learning of abstract ideas and theory. Jung (2005) supported that the use of computers in education can be more efficient, it can provide better learning results and it can be made adaptive to the individual learners. It can be fun to learn with computer through collaboration, critical evaluation, receiving feedback, planning and organisation. Abimbade, Aremu, and Adedoja (2003) also found that the use of computer software have actively helped the children's reading process, encouraged them to talk to each other and make decisions, argue and think through. Okudo (2008) also found that ICT programmes (The use of Igbo radio programmes) helped to improve the knowledge of Igbo language and culture among secondary school students. According to Olojo, Adewumni, Ajisolam (2012), the introduction of Information Technology to the educational systems is aimed at improving educational delivery and preparing children for a role in an information economy. Also, Onusanya (2002) found out that when attribute of computer animation products were considered, students exposed to computer instruction performed better in 3D animation than their manual group counterpart. Thus, integrating ICT tools in the classroom teaching and learning of Igbo language would yield an invaluable result.

Other benefits to be achieved if ICT tools are properly integrated into teaching and learning of Igbo language are presented by Friedman (1994); Gibbs and Gosper (2006); and Halat (2008) below:

1. *Meeting up with varying learners' needs*

Educational systems need to display more flexibility and variation in curriculum design and implementation to meet the increasingly varied and changing needs of learners due to migration, urbanisation and globalisation. Moreover, as a result from the dynamic developments in most academic disciplines and the requirement to remain well informed, it becomes imperative that many people including Igbo language learners would engage in life-long learning.

2. *Acculturation to improve harmonisation*

To play a role in a globalizing world, education structures, programmes, procedures, and agreements need to be harmonized across countries so that students, teachers, and researchers can move freely and choose the organisations, networks, and communities of their liking. Examples of harmonisation are the creation of international student exchange programmes, the adaptation to a unified course-credit system, and the conformation to globalisation implications for higher education international quality assessments. Such acts of harmonisation, bringing an international dimension to Igbo language classroom, can be seen as first steps towards harmonisation and internationalizing education programmes, structures and procedures for twenty-first-century Igbo language learners to be able to meet up with the challenges of the dynamic era. This will also help them to take a stand on globalisation issues.

Instructional Strategies for Teaching and Learning of Igbo Language in the Twenty-first Century

The implementation of Igbo language teaching and learning in the classroom for the 'digital natives' to be able to blend with this technologically developed era requires the use of learner-centred and problem-based instructional approaches. In this regard, UNESCO's (2010) teaching and learning instructional strategies should be adhered to by the Igbo language teachers. The strategies among others include:

1. *Future problem solving*: This strategy enables Igbo language learners to develop skills for analyzing and solving problems. They learn the steps and strategies in identifying a particular problem, its causes and effects as well as possible solutions.

2. *Experiential learning*: Igbo language teachers should engage their students in activities that have relevance to activities in their lives. This instructional strategy will help students to develop critical thinking, problem solving and decision making in contexts that are personally relevant to them" (UNESCO, 2010:8).

3. *Storytelling:* This involves appropriately integrating storytelling into the teaching units. This strategy is aimed at developing appreciation and skills in the instructional use of storytelling.

4. *Inculcating values and morals*: The development of values and attitudes is related to the affective domain of educational objectives. The aim of this instructional strategy is to help develop an understanding and required skills for morals and values education.

5. *Engaging them in enquiry learning*: This instructional strategy is used to develop the skills of thinking and problem solving

which are important skills for quality education in this twenty-first century era. Igbo language teachers need to possess skills in the planning, implementation and assessment of enquiry-based instruction.

6. *Assessing them appropriately*: Quality education requires evaluation strategies that integrate assessment with the process of teaching and learning. Also, Igbo teachers need to develop skills in assessing knowledge, skills, attitude and values in Information and Communications Technology for quality education.

7. *Learning outside the classroom and outdoor activities*: This strategy involves the use of learning opportunities outside the classroom to promote the use of Information and Communications Technology tools for quality education. These opportunities include different forms of in-school activities, excursions to relevant places such as factories, gardens, black smith's workshop, lakes etc.

8. *Community problem solving*: This strategy enables Igbo language learners to understand how problems are solved in the community. Participation in the solution of community-based problems helps to equip the Igbo language learners with the skills and abilities to solve problems in their local environment.

Effective Igbo language teaching and learning requires integrating information and communications technology tools for restructuring in order to achieve quality education, and it demands that Igbo language teachers should be knowledgeable and skilled in the above listed instructional strategies. The use of these strategies would promote the development of knowledge, skills and values dimensions of the use of information and communications technology for quality educational objectives in the Igbo language teaching classrooms.

Challenges Facing the Integration of ICT Tools in Igbo Language Classroom

Many obstacles seem to be facing the successful applications of ICT tools in teaching and learning of different subjects which Igbo language is included. Some of them according to (Okafor, 2013; Adesoji, 2012; Obiefuna and Enwereuzo, 2012; Iyamu and Ogiegbaen, 2006) are:

1. *Policy Problem*

The National Policy on Education (NPE) on computer education policy is not backed up with a political will. Consequently, ICT integration for Nigerian language teaching and in Igbo language to be precise is not tailored towards attaining the objective of the policy on language and computer education policy.

2. *Inadequate supply or lack of ICT literate Igbo language teachers*

Okafor (2013) posits that teachers are insufficiently trained and are in short supply. Poor linguistic abilities of the teachers themselves are transmitting to learners' inability and non-knowledge of modern technological tools to be used as teaching tools. Further, ICTs are still a novelty to many Igbo language teachers; they are not being exploited for solving educational problems. Most of them are not computer literates and may lack motivation to acquire computer literacy or surf the Internet for teaching materials and innovative ideas. Developing Igbo language teachers' ICT skills is then very imperative. According to Pelgram and Law (2003), competencies that need to be developed include: the training of teachers in the use of common office application programme, sending of e-mail, making use of the Internet, use of ICT in subject based teaching and classroom practice, production of multimedia course materials, data analysis and so on. Indeed, most Igbo language teachers could be said to be in the world of their own, isolated from developments in teaching indigenous languages elsewhere with integration of ICT tools.

3. *Costs*

There is lack of adequate funding; poor infrastructure, materials and equipment. ICT materials seem to be expensive to many Igbo language teachers. Some schools may also not be able to provide ICT materials in their libraries. In case of maintenance of existing tools, the repair are often done by many people through trial and error as most technicians, especially roadside ones are not likely to have enough technical knowledge of the ICT tools.

4. *Impact on learning Achievement*

There appears to be a mismatch between methods to measure effects and type of learning promoted. Standardized tests tend to measure the results of traditional teaching strategy in Igbo rather than new knowledge and skills relating to the use of ICTs. It is therefore important that researches need to be conducted to understand the complex links between ICTs' learning achievements as it hinders the development of the teaching and learning of Igbo language which it supposed to benefit.

5. *Curriculum Content*

The perceived quality of the content and its value to instruction are also one of the challenges of integrating ICTs in Igbo language classroom. Notwithstanding how innovative the school may be, and regardless of what concerns they bring, all the educational/ instructional developments must begin and end with emphasis upon the enhancement of teaching effectiveness. Professional development programmes must offer the schools tangible benefits before they embrace new approaches. Curriculum has to be restructured in a way as to portray the intent of achieving the aims and objectives of the twenty-first-century Igbo language learners.

Conclusion

The importance of ICT tools in teaching and learning of Igbo language cannot be over emphasised due to the fact that this is a digital age. Information Communication Technology tools have potentials for far reaching instructional changes towards qualitative education especially in Igbo language classroom. It may offer Igbo language learners the opportunity of ever benefitting from world level transformation strategies; therefore catching up to the level where they can use and participate in world development. This implies that if the Igbo language teachers fail to adjust in the new epoch and integrate ICT tools into their classroom teaching and learning they cannot sustain the Igbo language learners' relevancy in the global village.

Recommendations

In view of the potential of integrating ICT tools for qualitative education and that of Igbo language in particular, it is recommended that:

Government should increase funding for the entire education sector with particular emphasis on ICT usage in the teaching and learning of Igbo language and other Nigerian languages.

Educational Planners and policy makers must see Information Technology a priority in the teaching and learning of Igbo and other Nigerian languages. Those concerned would always help them to invent new words and vocabularies that will blend with the dynamic technology. This will help to enrich the vocabularies of Igbo and other Nigerian languages as well as their standardisation. For example in Igbo language, there is no standardized word to describe some modern numeracy like trillion, quadrillion; these words were substituted with indigenous words.

Schools should be equipped with enabling technological tools such as computers, audiovisual equipment, projectors, Internet access, etc. Teachers of Igbo language should be trained on the use of ICT tools for quality and effective teaching and learning. There is need for training of both staff and students so as to be aware that there are some instructional programmes, standardized Igbo language practical exercises, Internet and CD-ROM available for Igbo language curriculum delivery. The Igbo language teachers should be trained also to be aware that they could use e-mail to ask teachers in other locations or countries questions on effective instructional strategies and even download or record of lectures and books from the Internet, which could be used for drills and memorisation. With such improved level of awareness, Igbo language instruction in this digital era might be ICT-driven and then restructured.

Students should take advantage of opportunities offered by technologies to enhance learning and even to streamline their learning processes. This is very important because innovations will not happen unless Igbo teachers are given adequate time, resources and support to make the changes, to integrate technologies into their curricula (Gibbs and Gosper, 2006). Strategies for encouraging and supporting staff in exploring technologies to enhance their teaching, should include raising the profile of digital literacy as an attribute that may encourage Igbo language teachers and indeed other Nigerian languages to design curricula accordingly.

Suggestion is therefore being made that appropriate integration of ICT tools in Igbo language classroom be promoted so as to create a major change in both the learners' and teachers' professional development. This would help to promote and improve students' understanding, motivations, perceptions, and beliefs about learning the Igbo language and the use of technology.

References

Abimbade, A., Aremu, A., and Adedoja, G. O. (2003). *Providing information communication technology (ICT) environment for teaching and learning in the Nigerian education system.* Lagos-Nigeria. Macmillan Publishers.

Adesoji, F. F. (2012). Undergraduate students' perception of the effectiveness of ICT use in improving teaching and learning in Ekiti State University, Ado-Ekiti, Nigeria. *International Journal of Library and Information Science 4*(7), 121–130.

Friedman, J. (1994). *Cultural identity and global Process*, Sage, London.

Gibbs, D., and Gosper, M. (2006). The upside-down-world of e-learning. *Journal of learning design, 1*(2), 46–54.

Güzeller, C. O. and Akin, A. A. (2012). The effect of web-based mathematics instruction on mathematics achievement, attitudes, anxiety and self-efficacy of 6th grade students. *International Journal of Academic Research in Progressive Education and Development (1)* 242–254.

Halat, E. (2008). A good teaching technique: WebQuests. *The Clearing House: A Journal of Educational Strategies, Issues and Ideas, 81*(3), 109–112.

Iyamu, E. O. S., and Ogiegbaen, S. E. A. (2006). Assessment of the use of educational technology by social studies teachers in higher education in Western Nigeria. *E-Journal of Instructional Science and Technology,* 8(1). Retrieved October 14, 2006, from http:// www.usq.edu.au/electpub/e- jist/html/commentary.htm.

Kuhlthau, C. C. (2010). Guided inquiry: School libraries in the 21st century. *School Libraries Worldwide,* 16 (1), 17–28.

Obiefuna, C. A. and Enwereuzo, N. M. (2012). An appraisal of retraining programmes for in-service teachers. The need for a paradigm shift from 20th to 21st century pedagogical skills, *UNIZIK Orient Journal* 6(1).

Oblinger, D. (2004). Education the next generation: Keynote address delivered at the 2004 Educase Conference Denver, Colorado, USA.

Okafor, M. N. (2013). Re-tooling the teacher of the Igbo language for optimum productivity in a world in transition. *Igbo Studies Review.* 1,151–161.

Okudo, A. R. (2008). The impact of Igbo radio programmes on the knowledge of Igbo language and culture among upper basic II and III student Igbo native speakers in Lagos State. Unpublished M. Ed Thesis. University of Lagos, Nigeria.

Olojo, O., Adewumni, M., and Ajisolam, K. (2012). E-learning and its effects on teaching and learning in a global age. *International Journal of Academic Research in Business and Social Sciences 2* (1), 203–210.

Oye, N. D., Iahad, N, Madar, M. J. and Rahim, A. B. (2012). The impact of e-learning on students' performance in tertiary institutions. *International Journal of Computer Networks and Wireless Communications 2, 2.*

Rodgers, M. Rayon, D. Starrett, D, and Holzen R.V. (2006). Teaching the 21st century learner: The 22nd annual conference on distance and learning. University of Wisconsin. Teachers ICT base Skills (n.d). Retrieved 13/01/2011) from http//.www.eapoedu.org/inservice/skillsall htm.

UNESCO (2003). Mapping globalization: Selected cross-cutting issues: role of ICT in bridging the digital divide in selected area. Economic and Social Commission for Asia and the Pacific.

UNESCO, ILO (2002). *Revised recommendation concerning technical and vocational education (2001).* Paris: UNESCO: Geneva.

Chapter 11

E-management and Secondary School Effectiveness: Implications for Policy and Practice in Nigeria

S. A. Oladipo
&
A. A. Adekunle

Introduction

Education is the greatest force that brings about developmental changes. It is the greatest investment that a nation can embark upon for rapid sociopolitical, economic, and human resources development. The avenue through which this could be achieved is the school, which is an institution established to carry out special functions so that certain public goals and objectives could be achieved. Specifically, the Federal Republic of Nigeria (2004) described secondary education in Nigeria as the form of education children receive after primary education and before the tertiary stage, with the broad aims of preparing the students for useful living within the society and for higher education.

There is no gainsaying the fact that the achievement of the goals of secondary education in Nigeria is, to a large extent, dependent on how well it is managed, hence, the need for school management effectiveness. School effectiveness involves the degree to which schools are successful in accomplishing their educational objectives or fulfilling their administrative, instructional or service functions (Ogunu, 2000). While educational management is 'an executive function for carrying out agreed policy' (Bush, 2010). Babalola

(2006) sees educational management as a concept that goes along with the quest to put the formal education system under control, regulation or supervision in its attempt to use carefully (that is, to manage or economize) available scarce resources through co-operative efforts when establishing institutions of learning, enrolling learners, attracting best staff, conducting teaching, learning and research, as well as graduating learners at all levels of education in an effective and efficient manner.

Obviously, the classroom is a curious, demanding, crowded and chaotic place where groups, individual strangers, are packed and are expected to work together in harmony. Anumnu (2008) observed that the classroom is the main component and compartment of the teaching-learning process that entails the use of human and material resources, which should be efficiently managed to accomplish educational goals and objectives.

Be that as it may, it has been observed that the management of schools is becoming more complex, and the use of information and communication technology (ICT) in education is receiving more attention globally. In most Nigeria secondary schools, the electronic application in the day-to-day administration has not been given due consideration as clearly evident in the use of the traditional method of manual record keeping. Most schools also do not have the ICT devices and where available, they are not properly utilized (Salam, 2001). In addition, the human resources that are required to operate the devices are ill-prepared, as they do not possess the technical knowhow as to enhance their effective usage to bring about the desired results.

Since information technology (IT) has turned the world into a global village, it has shown that geographic distance, national borders and language are no longer regarded as obstacles to the movement of ideas and intellectual capital. There is, therefore, the need for urgent change in orientation from the manual record keeping and management in schools to computerized system, so as to link with the ongoing rapid development in the global village.

It is against this background that this chapter presents the importance of and constraints to application of ICT in school management in Nigeria.

Conceptual Clarifications

Management

Management has been defined severally by scholars. Olagboye (2004) sees it as the organisation and coordination of the activities of an enterprise in order to achieve desired objectives. Management is often involved as a factor of production along with machines, materials and money. According to the management guru Peter Drucker (1909–2005), the basic task of management includes both marketing and innovation. Management therefore can be said to consist of the interlocking functions of creating corporate policy and organizing, planning, controlling, and dissecting an organisation's resources in order to achieve the objectives of that policy.

According to Stoner and Wavell (1988) as cited in Adeyemo, Folajin, Sotannde, and Adekunle (2012), since organisations can be viewed as systems, management can also be defined as human actions, including design, to facilitate the production of useful outcomes from a system. Management in all organisational activities is the act of getting people together to accomplish desired goals and objectives using available resources efficiently and effectively. Management, therefore, comprises planning, organizing, staffing, leading or directing, and controlling an organisation or effort for the purpose of accomplishing a goal.

E-management

E-management, otherwise referred to as automated management system (AMS), has to do with electronic application in management. It is therefore the application of electronic devices or information and communication technology in carrying out the management functions in an organisation. According to Farlex (2013), e-management was coined by Francis Ohanyido as part of the new evolving concepts around e-Governance. It is about the policy of getting people together to accomplish desired goals. E-management has to do with planning, organizing, staffing, leading or directing, and controlling an organisation (a group of one or more people or entities) or efforts for the purpose of accomplishing a goal through the deployment of ICT and manipulation of human, financial, and natural resources.

In educational management, like every other organisation, the automated devices are essential tools for running the day-to-day operations, enhancing productivity and communicating with the school functionaries, students and the public. Educational managers use these devices for a variety of reasons, including keeping their team on track, budgeting and planning, monitoring inventory and preparing documents and presentation. Managers therefore need to understand not only the basic functions of the corporate software tools used in the office but also the Internet and other external computing tools that can improve the way they manage their schools affairs.

Electronic Application in Secondary School Management

Information and communication technology is increasingly becoming a powerful means of enhancing human man's ability to think, to learn, to communicate and to use one's brain creatively and logically. According to Imison and Taylor (2001), ICT provides the means by which one can search out vast stores of up to date, relevant as well as archival information. It gives every school the opportunity of becoming instantly recognized worldwide. The electronic devices such as computer (hardware and software) networking, telephone, video, multi-media and the Internet are required in the management of school as they are very instrumental to effective service delivery. As observed by Anumnu (2008), the use of these devices in the teaching-learning process makes students and teachers to become proud of their work, increase their motivation and become actively involved in their own learning and progression. The application of electronic devices in school management can manifest itself in different ways. Some of these, according to Braide (2003), include:

Records Keeping

Record keeping is the detailed and meticulous account of any phenomenon. It can be set down permanently in writing, on a flat plate, film strip, computer diskette or plate or other permanent forms. A record serves as a memorial or authentic evidence of a fact or event. A record, if well preserved, lasts from one generation to another or handed down (Adeniyi, 2006). Obadara (2006) sees records as papers, written or printed books, documents or drawings, maps or plans, photographs, microfilms or any coping thereof of which has been made or is required by law to be received for filing, indexing or reproducing by an officer in connection with the transaction of public business. It refers to information created, received, and maintained as evidence by an organisation or person, in fulfillment of legal obligations or in the transaction of business (ISO, 2001).

School managers keep track of a lot great deal of information that is vital to the success and survival of the school system. Using electronic devices to store and manage documents reduces the amount of physical storage an institution needs and also allows managers to have easy access to their files using simple document search methods. Additionally, Braide (2003) further explained that by keeping records managers can easily share information about employers, students' job performance and history with other stakeholders in the school system. School records can therefore be seen as the documentation of events, programs and activities that go on in an educational institution which could be retrieved when the need arises.

Electronic Record Keeping (E-record Management)

This refers to the application of information and communication technology in collecting, storing, processing and distributing school records. Oginni (2005) sees electronic records as information of all kinds, including words, numbers and pictures acquired, analyzed, manipulated, stored and distributed using the system of ICT. In the same direction, Awonor (2010) defines electronic records as the pieces of information collected, stored, processed and delivered using a combination of technologies.

The adoption of e-record management in secondary schools will provide for safety and retrieval of a large amount of information, which could occupy large volumes of files, reports, book, etc. This could be stored in a flash drive, compact disk, etc., which are easier to keep and move about. A single hard disk can contain information previously occupying twenty rooms (Oginni, 2005). Also, some software are user-friendly and have templates of reports, memo, brochures, business letters, etc., which a user can choose from and apply in routine office work. Similarly, the use of passwords provides confidentiality for making it impossible to access top secret documents, thus, ensuring confidentiality of official documents.

As observed by Salam (2005), ICT is an important and useful tool in educational management, planning and administration, proper data management is very vital to the success of educational institutions. According to Salam (2001), effective management of school will be enhanced when the following areas are computerized:

Information on the employees, student administration, bursary, examinations, and records as well as welfare, as this allows for the wants and needs of staff and students to be easily accessed by the school authority and processed for immediate action.

Records management is a field of management responsible for the efficient and systematic control of the creation, receipt, maintenance, use and disposition of records, including the processes for capturing and maintaining evidence of and information about business activities and transactions in the form of records. The International Standards Organization (2001) states that the practice of records management involves:

- planning the information needs of an organization;
- identifying information requiring capture;
- creating, approving and enforcing policies and practices regarding records including their organization and disposal;
- developing a record storage plans which includes the short and long run housing of physical records and digital information;
- identifying, classifying and storing records;

- coordinating access to records internally and outside of the organization, balancing the requirements of business confidentiality, data privacy and public access; and
- executing a retention policy on disposal of records which are no longer required for operational reasons, according to organizational policies, statutory requirements, and other regulations.

In the same vein, Krishnaveni and Meenakumari (2010) identified three functional areas of e-management in the day-to-day administration of educational institutions as follows:

- student administration
- staff administration
- general administration

According to them, in student administration, it could be seen in the usage of electronic media by students to apply for admission, usage of computers for students' registration/enrollment, availability of time table/class schedule in electronic form, usage of computer for maintenance of attendance of students, communication of academic details of students to their parents/guardians through e-media, usage of e-media for notifications regarding hostel accommodation and usage of e-media for notifications regarding transportation.

In staff administration, it could manifest in the usage of computer for recruitment and work allotment of staff in the school, automation of attendance and leave management of staff members, usage of electronic media for performance appraisal, communication with staff using e-media, e-circulars regarding official matters, as well as making e-kiosks available in the institution.

In general administration, it could come in form of using e-media for scheduling/allocation of halls for examinations, usage of e-media for the display and processing students' results as well as making fees payment electronically.

Therefore, records management principles and automated management systems aid in the capture, classification, and ongoing management of records throughout their life cycle.

Electronic Mail

The use of communication software for e-mail and the Internet provides opportunity for school administrators to link any part of the world and obtain information that would be of benefits for the day-to-day running of the school system. Okure (2008) opines that it is now possible to send an electronic message to one or more individuals in an organization for storage in their electronic mail boxes on magnetic disk devices, whenever they are ready, they can read their e-mail by displaying them on the video screen on their terminals, personal computers or intelligent work stations. So, with only a few minutes of efforts (and a few micro seconds of transmission) a message to one or many individuals can be composed, sent and received. So in the school system notices of meetings, minutes of meetings and other information can be sent into the mail boxes of teachers.

Educational Planning

Educational planning can take up a lot of valuable time of the planner, but electronic application can make it easier. For instance, the use of e-mail programs like Outlook or Gmail (Google mail) to set appointments, tasks and deadlines. Other software such as Excel, Access and some others could also serve as financial tools to develop budgets and project proposals, as well as for using computer for instance to planning the day-to-day activities of the school. School managers can also use the Internet to conduct research, and to look for ideas to help them create plans to provide better services for their clients (the students), thus helping in the achievement of goals of the specific level of the school system.

Administrative Use

Schools often use electronic devices in administration. This could be in form of maintaining students' record and managing school information. Computers may be used to track students' attendance and grades.

Communication

The use of automated devices in education is also for communication. Their teaching-learning process has to do with the transmission of messages and ideas, and effective administration of the school is also hinged on effective communication. Communication is therefore essential among the stakeholders in the school system. The use of e-mail and instant messaging programmes allows the gathering of information among teachers. It also allows educational administrators to delegate tasks and make necessary follow up.

The Need for e- Management in School

The importance of e-management in schools in the twenty-first century cannot be overemphasised. These include among others, the following:

- It provides for safety, storage and retrieval of a large amount of information which could occupy large volumes of files, reports, books, etc. This could be stored in a flash drive, compact disks, diskette which is easier to keep and move about. A single hard disk can contain information previously occupying twenty rooms.
- The use of word processor like Microsoft Words, WordPerfect, Publishers, etc. with grammar, spelling checker and thesaurus, assists records managers to record perfect and accurate minutes of meetings devoid of spelling mistakes.
- Some software are user-friendly and have templates of reports; memo, brochures, business letters, etc. which a user can choose from and apply in routine office work.
- The use of passwords provides confidentiality for making it impossible for intruders to access top secret documents.
- The use of communication software for e-mail and the Internet provides opportunity for school administrators to link any part of the world and obtain information that would be of benefits for effective day-to-day running of the school system.

- The use of scanner makes it possible for images and specimen signatures to be scanned and transferred into the computer.
- In the area of personnel management, database software allows prompt retrieval of information, printing of records and updating of records quickly and easily. It involves small size workforce while a single computer performs accounting, database, word processing tasks and dispatch mails.
- The application of electronic devices in record management makes it possible to access information anywhere. This is because all data are registered once only and can be accessed by the administrator anywhere. For instance, members of the departments may access departmental subaccounts, students' files, lecture schedules, budgets, etc. without having to ask some persons in the central administration.
- The electronic application into school records keeping is useful in administrative activities in the school. The usefulness can be viewed from three broad interdependent categories. First, there is the use for the development and implementation of professional data processing systems to support the daily operations of individual offices. The second is the use for the management information system (MIS), providing information to assist decision making at all levels. The third and rapidly growing field is that of office automation covering word processing, spread sheets, electronic mailing and filing.

Electronic mails enable organizational members as well as other stakeholders in the school business to exchange messages among themselves. A single electronic mail can be sent to several recipients. Examples of messages that can be sent electronically include policies and directives, correspondence, work schedules, assignments, agenda and minutes of meetings, drafts of documents, final reports and recommendations.

Similarly, e-management enables school managers to apply their knowledge and skills of management and technology in the planning, analysis, and supervision of works; solve problems logically, creatively and analytically based on sound facts and ideas; communicate effectively across a range of contexts and audiences; respond with high integrity and adapt readily to changing situations.

E-management (electronic management) as a new concept is an invention which is Internet based with real-time update of records, accurate information and precise decision making at the press of buttons.

According to Adegbija (2012), technology can be described as a product in the sense that it is the end result of the systematic application of scientific knowledge in addressing educational problems, including examination related problems. The concept "new technology" is an indication that the technology is not stagnant but keeps bringing new ideas, knowledge, inventions and skills that should be applied. The new technology gave birth to the current use of e-School powered by the computers and other information technology (IT) products such as microcomputers, mainframe computers and the Internet. E-school is the end-to-end electronic assessment processes where IT is used for the presentation of assessment process from the perspective of learners, tutors, learning establishment, awarding bodies and regulators and the general public (e-management).

Therefore, e-management is therefore the aggregate of all electronic means (e-school, e-learning, e-library, e-fee etc.) adopted to aid management to accomplish educational goals at unit, district or national levels i.e. e-management is synonymous to ICT in school administration. Ejiogu (2010) describes the new age of information technology as information super highway designed to achieve global proximity in the face of rapid globalization of social, political and economic activities. Through a creative and maximum use of portable computers and computer networking it is now easier for a principal to interact and quickly share information (school records) not only with his teachers but also with his superiors (ministry officials), stakeholders and parents / students both locally and global via Internet facilities. Through the technology (e-management), the principals can easily gain access to important information such as revenue, detail of students and teachers from a remote location outside the school premises. In this way, information is decentralized, thereby loosening managers' grip over certain vital information.

From the foregoing, it can be seen that the application of electronic devices helps to transform the system significantly. The electronic software packages have become indispensable office delight; information can now be processed, stored and retrieved with ease, and with a general system improvement. It has also impacted positively on the school office routine and procedures. It has offered school administrators and record managers the unlimited opportunities and prospects for their personal and official effectiveness.

Information and Communication Technology Policy in Nigeria

The Federal Executive Council of Nigeria in March, 2001 approved a National Information and Communication Technology Policy, and this led to the setting up of the National Information Technology Development Agency (NITDA), which is saddled with the responsibility of implementing the policy. The policy sees the private sector as the driving engine of the ICT sector. NITDA is expected to enter in strategic alliance, collaboration and joint venture with the private sector for the actualization of the IT vision which is to make Nigeria an ICT capable country in addition to using ICT as an instrument for sustainable development and global competitiveness. It is also used for education, job creation, wealth creation, and poverty eradication. Emphasis is to be laid on development of National Information Infrastructure Backbone (NIIB) as well as the human resources development (Federal Republic of Nigeria, 2011).

The National Policy on Information and Communication Technology provides for:

1. ensuring that ICT resources are readily available to promote efficient national development;

2. guaranteeing that the country benefits maximally and contributes meaningfully by providing the global solution to the challenges of an information age;

3. empowering Nigerians to participate in software and IT development;

4. ensuring local production and manufacture of ICT components in competitive manner;

5. empowering the youth on ICT skills to prepare them for global competitiveness;

6. integrating ICT into the mainstream of education and training;

7. creating ICT awareness and ensuring universal access in order to promote ICT diffusion in all sectors of our national life;

8. stimulating the private sector to become the driving force for ICT creativity and enhance productivity and competiveness;

9. building a mass pool of IT expertise using the National Youth Service Corps (NYSC), National Directorate of Employment (NDE), and other platforms as "Train the Trainer" Scheme (TTT) for capacity building;

It is worthy of note that some of the above objectives have not been achieved, though an attempt has been made to integrate ICT into every sphere of the Nigeria's national life. Baro (2011) observes that it is sad to note that Nigeria has not made headway in terms of implementing the policy on ICT, and according to him, Nigeria still depend on foreign countries for importation of computer hardware, software packages and depending of foreign experts for the technical knowhow.

The Trend of ICT Adoption in Nigerian Secondary Schools

The adoption of ICT in Nigerian secondary school could be traced back to 1988, when the Nigerian government enacted a policy on computer education. According to Okebukola (1997) (as cited in Aduwa-Ogiegbean and Iyamu, 2005), the intention then was to establish pilot schools and diffuse computer education innovation first to all secondary schools, and then to primary schools. Unfortunately, the project did not really take off beyond the distribution and installation of personal computers.

The National Policy on Education (FRN, 2004) in recognition of the prominent role of ICT in the modern world has integrated ICTs into education in Nigeria. The policy document provides that at the junior secondary school level, computer education is a prevocational elective, and at the senior secondary school, a vocational elective. However, the new senior secondary school curriculum has placed computer studies/ICT as a core subject at the senior secondary School level in Nigeria. As observed by Okebukola (1997) (as cited in Aduwa-Ogiegbean and Iyamu, 2005), computer is not part of classroom technology in more than 90% of Nigerian public schools, as the chalkboard and textbook continue to dominate classroom activities in most Nigerian Secondary Schools. Though, this observation was made about eight years ago, the situation still prevails.

The Federal Ministry of Education launched an ICT driven project called School Net aimed at equipping all schools in Nigeria with computers and communication technologies. According Adomi and Kpangban (2010), at the African World Economic Forum held in Durban in 2003, the New Partnership for African Development (NEPAD) launched the e-school initiative, intended to equip all African high schools with ICT skills to young African in primary and secondary schools, and to harness ICT to improve, enrich and expand education in African countries.

Similarly, the federal government of Nigeria has commissioned a mobile Internet unit (MIU) operated by the Nigerian National Information Technology Development Agency (NITDA). The MIU is a locally made bus that has been converted into a mobile training and cyber centre. Its interior has ten workstations, all networked and connected to the Internet. The MIU is also equipped with printers, photocopies, and a number of multimedia facilities. Internet is provided via VSAT with a 1.2m dish mounted on the roof of the bus. It is also equipped with a small electric generator to ensure regular power supply. The MIU takes the Internet to some urban areas and various primary and high schools (Ajayi, 2003). It should be noted that the buses are so small, such that it covers a very few areas and most if not all schools in rural areas are not covered.

It has however been discovered that ICT components are not available in schools and this has hampered their effectiveness, and as observed by Adomi, Okiy, and Rutiyan (2003) this has led students to resort to cybercafés for Internet access. According to them, most of the cybercafé clients in Nigeria are students.

Constraints to e-Management in Schools

The automation of school records management in Nigeria currently faces some challenges. Some of these, according to Oginni (2005), include:

Incompatible Systems

A major constraint for many educational institutions in electronically filling their records computerizing its records keeping is the different functional areas that were developed at different point in time with different options, each choosing the best equipment and software available for their particular needs.

Internal Tradition

Internal traditions and the culture of institution also play important role in deciding the extent to which one can exploit the new information technologies, and what procedures to employ in planning, deciding, implementing and managing such activities. A conservative management team may not welcome the innovation for change.

Dwindling Financial Allocation

It is a common knowledge that educational institutions in Nigeria have been badly affected by dwindling financial allocation. Educational institutions at all levels in Nigeria are ill equipped in terms of physical facilities.

Epileptic Power Supply

It is a fact that for electronic devices to work there is the need to power them. It is a known fact that in Nigeria today, the supply of electricity has been erratic; hence, it affects the effective use of electronic appliances in school management.

High Cost of Operation and Maintenance

The application of electronic devices in record keeping requires huge amount of money to purchase, operate and maintain.

Similarly, it is more difficult to ensure that the content, context and structure of records are preserved and protected when the records do not have a physical existence.

There is also the concern about the ability to access and read electronic records over time since the rapid pace of change in technology can make the software used to create the records to be obsolete, thereby leaving the records unreadable and irretrievable.

Problem of Interconnectivity

Offices in most institutions are not connected to one another, and this makes it difficult for inter office on-line exchange of electronic information.

Implications for Policy and Practice

For effective electronic application in records management in educational institutions, the following measures are suggested:

- There is the need for constant and regular upgrading of the software used for record keeping.
- Educational managers and other key players in the education industry should adapt to change and innovation in technology, by embracing the ICT application in school management.

- Institutional managers should ensure that well equipped information and communication technology centres are provided in their schools.
- The government should subsidize the process of getting electronic devices, so as to enable institutions have a large number of these devices.
- There is the need for the provision of Internet connectivity and facilities in our schools, so as to provide opportunities for easy access to the Internet.
- Institutional managers should ensure that provision for back up for electronic records is made, in case of crash or corruption done to the documents.
- Acquisition of skills in Information and Communication Technology should be made mandatory for management, staff and students.
- Regular source of power should be provided to educational institutions.
- Staff members should be encouraged to go for seminars, workshops and conferences on the use of Information and Communication Technology.

Conclusion

Securing resources for any educational institution activity requires long-term planning, bringing together well considered plans, wide participation and strong factual argument. Experience seems, for example, to show that the application of electronic devices in general, reduce the need for additional staff, improve the quality of work, and facilitates improved methods of work and working conditions. To make a strong case for securing the necessary resources, systems development should be considered as part of the general improvement of administrative efficiency. Securing funding is a matter of priority and should be pursued by educational institutions management to facilitate complete computerization of the institutions administrative system. There is no doubt that when computers are introduced into an organization, because of the benefits they can provide, they usually affect enhance effective work performance of the employees of the organization. Though, some jobs may be lost, there will be a demand for training and retraining in order to meet the challenges of globalization.

References

Adegbija, M. V. (2012). New technologies and the conduct of e-examinations: A case study of the national open university. *Journal of research in education. An official journal of the collaboration of education faculties in West Africa (CEFWA), 1 (2)*, 104–113.

Adeniyi, I. A. (2006). Records and records keeping in educational organization. In Ajayi Kayode, Joshua Oni and Olasunkanmi Gbadamosi (Eds.). *Fundamentals of educational management.* Ago-Iwoye: Department of Educational Management and Business Education.

Adeyemo, B., Folajin, T., Sotannde, W., and Adekunle, A. (2012). Essentials of educational management. Abeokuta: Goad Publishers.

Adomi, E. E., Okiy, R. B., and Ruteyan, J. O. (2003). Survey of Cybercafés in Delta State, Nigeria. *The Electronic Library 21(5)*, 487–495.

Adomi, E. E., and Kpangban, E. (2001). Application of ICTS in Nigerian secondary schools. Retrieved from www.wabpages. uidaho.edu/mbolin/ado.

Aduwa-Ogbiegbean, S. E., and Iyamu, E. O. S. (2005) Using Information Communication Technologies in Secondary Schools in Nigeria. *Education Technology and Society 8(1)*, 104–112.

Ajayi, G. O. (2003). *NITDA and ICT in Nigeria.* Retrieved from: http:// ejds.org/meeting/2003/ictp/papers/ajayi.pdf.

Akinyele, T. A., and Olowookere, O. O. (2004). Internet services: Implications on the teaching and learning of Business Education. *Journal of Vocational Education, 5(1)*, 86–91.

Anumnu, S. I. (2008). Information and communication and technology for sustainable classroom management. In Babalola, J. B., Akpa, G. O., Imamn, H., and Ayeni, A. O. (Eds.). *Managing Education for Sustainable Development.* Ibadan: NAEAP.

Awonor, O. O. (2010). Imperatives of ICT for teachers, in L. O. Ocho (Ed.) *Reforms in Nigerian Education,* Enugu: New Generation Books.

Babalola, J. B., Ayeni, A. O., Adedeji, S. O., Suleiman, A. A., Arikewuyo, M. O. (2006). *Educational Management: Thoughts and Practice.* Ibadan: Codat Publications.

Baro, E. E.(2011). A critical examination of information and communication technology policies: effects on library services in Nigeria. *Library Philosophy and Practice (e-journal), 1(1).* Retrieved from http://digitalcommons.unl.edu/libphilprac/464.

Bello, S. A. (2005). The role of records in school management, In R. A. Alani (Ed.) *Managing the Educational System,* Ilesa: Triumph Providential Publishers.

Braide, E. (2003). *ICT and the evolving learning environment.* A paper presented at NAEN Annual National conference held at the University of Port Harcourt, 4th–7th November.

Bush, T. (2003). Theories of Educational Leadership and Management. London: SAGE publications Ltd.

Ejiogu, A. (2010). *Total involvement management: A 21ˢᵗ century imperative.* Lagos: Chartered Institute of Administration.

Forlex, P. C. (2013). *Comprehensive Approach to Computer Usage.* Lagos: Stepherson Publishers.

Freedman, T. (2000). *Using computer in classroom.* ICT in education site. www.ictineducationorg/conpit.htmgenial. Retrieved on June 25, 2011.

Imison, T. and Taylor, P. (2001). *Managing ICT in the Secondary School.* Jurden Hill: Oxford, Heineman Educational Publishers.

International Organization for Standardization (ISO). (2001). Information and documentation records management- Part 1: General. New York. Author

Krishnaveni, R., and Meenakumari J. (2010). Usage of ICT for information administration in higher education institutions-A study. *International Journal of Environmental Science and Development, 1(3),* 282–286.

ISO (2001). *Electronic records management.* Unpublished.

Obadarae, E. O. (2005). *Fundamentals of Educational Administration,* Lagos: Gafet Publications.

Oginni, O. (2005). *Computer literacy, University Records System and the challenge of ICT.* Paper presented at a workshop on quality management and positive changes in Higher Educational Institutions, organized by ICENS, Abeokuta, Ogun State.

Ogunnu, M. (2001). *Introduction to Educational Management.* Edo: Mabogun Publishers.

Okure, S. J. (2008). Assessing the Benefit from ICT Compliance for Sustainable Development in Nigeria: In Deference of the Open University. In Babalola, J. B., Akpa, G. O., Iman, H., and Ayeni, A. O. (Eds). *Managing Education for Sustainable Development.* Ibadan: NAEAP.

Olagboye, A. A. (2004). *Introduction to educational management in Nigeria.* Ibadan: Kemsio Educational Consultants.

Onuma, N. (2009). Promoting quality education in higher institution of learning in Nigeria through Information and Communication Technology In. B. G. Nworgu and E. I. Eke (Eds) *Access, quality and cost in Nigerian Education.* Abuja: Nigerian Education Research and Development Council.

Salam, C. O. (2001). Impact of Information and Communication Technology on Socio-Economic and Educational development of Africa. *Journal of Science and Technology, 3 (4),* 145–158.

Stoner, J. A. F. and Wankel, C. (1988). Management (3rd ed.) New Delhi: Prentice-Hall.

SECTION 4

The Cultural Factors Affecting Integration of Educational Technology Tools in Education

Chapter 12

Cultural Factors Affecting Integration of Educational Technology Tools in Education in Private Learning Institutions of Malawi

Joseph Boniface Mwaimwai Maere

Introduction

This chapter examines how culture influences the adoption, usage, and application of educational technology tools in three levels of education: primary, secondary, and tertiary education. Culture is one of the main influences of behaviour as it has been highlighted by many scholars (Roberts, 2006; Wood and Roberts, 2006; Matsomoto, 2007), and its impact on educational technology tools cannot be overemphasized.

Cultural factors in Malawi have some of the most intricate areas of consideration for most development initiatives, and these can also be singled out as factors behind the lag in most development initiatives to some extent. For instance, in the combat against the spread of HIV/ AIDS, cultural factors display a significant role in either promoting the prevention of or accelerating the spread of the virus across the nation as some people consider the use of contraceptives such as condoms a cultural taboo (Malawi Human Rights Commission, 2010; Bradley, Schwandt, and Khan, 2009), and the fight against gender-based violence is realistically under the plight of cultural reorientation as suggested by different scholars and policy analysts (Basu, 1992; Blayo and Blayo, 2003). In a country with almost 80% of its population located in the rural areas, culture remains the main player in all development initiatives, even in the urban centres of the country.

Malawi is one of the least developed countries in the world and according to the recent UNDP ranking, the country is in position 170 based on the recent 2014 Human Development Index (HDI). Malawi introduced free primary education (FPE) in 1994 immediately after the ushering in of multiparty democracy in 1993 and after its general parliamentary and presidential elections. According to Ng'ambi (2010: 3), the initiative increased access to primary and secondary education while creating an additional burden on the education system, which was already weak in terms of human and financial resources. The increased intake necessitated more qualified teachers, provision of teaching and learning materials, and even an increased use of educational technology tools. During the introduction of FPE in 1994, there were 49,138 teachers in government-run primary schools across the country, but in 2005, the number dropped to 45,705 and further down to 45,507 in 2009 (Ng'ambi, 2010: 8–9). All these reductions surfaced in the face of the increasing teacher-to-pupil ratio of 1 to 81 and that of qualified teacher-to-pupil ratio of 1 to 92, which are the highest teacher-pupil ratios in the world (Ng'ambi, 2010: 9). All these developments emphasize the point that there is no country that needs strengthening its structures to attain a sound educational system and, therefore, seriously need the promotion of educational technology tools more than Malawi.

Educational technology tools (ETT) have not been completely embraced in the country because of a lack of financial resources and training. However, the country has benefited from a number of educational technologies in advanced educational levels such as secondary and tertiary education. The current tools that have been observed to take ground include Internet education, e-learning facilities, use of LCD projectors in class, social media education, and others, especially in the country's renowned universities and colleges and private institutions. The concern of this chapter is to indicate the extent to which cultural orientations have made an impact on the usage, learning, and adoption of ETT in the Malawian educational system and vice versa. Focus of the chapter shall be on university educational system where ETT has been partially employed.

Understanding Educational Technology Tools

Educational technology is the study and the ethical practice of facilitating learning and improving performance by creating, using, and managing appropriate technological processes and resources (Richey, 2008). Educational technology tools, therefore, refer to all tools or technological applications, which could be hardware and software, that are applied to assist facilitation of teaching and learning. According to Malik and Agarwal (2012), these are an assortment of tools that might prove helpful in student-centred learning, problem-based learning, or care-based learning. Educational technology is a recent trend to knowledge impartation, which places the "teacher on the side" as Malik and Agarwal (2012) described it; it is like placing the "guide on the side" rather than the "sage on the stage."

Educational technology tools are placed in several categories, but the commonly known categories include the following: (1) tools of the trade, (2) online courses, programmes, and academies, and (3) applications also known as apps (My Teaching Degree, online, 2014). Tools for the trade include all gadgets that assist the facilitation of learning, which can be used by both the trainers (educationists) and the students. These technologies include, among others, laptops, interactive whiteboards, tablets, LCD projectors, video-conferencing technology, clickers, and Kindles. The online courses and programmes include all lessons, which are in forms of text, audio, and video files, which are uploaded on the Web, and students are offered access codes to open and use them. These educational technology tools` may also be available for free in open source Web sites and other social media Web sites such as YouTube. Lastly, the common category of ETT includes the usage of apps, which are tailor-made software applications that can be used for specific discipline or skills development in an interactive mode. These software apps are usually contained in tablets or smartphones, they are interactive and used by many students. Considering the approach of ETT, which to some extent places the educationist "on the side," we may not be fully surprised to realise their impact on the cultural orientations of the learners and educationist alike.

Cultural Orientations Dominating the Educational Sector

Cultural orientations are cultural inclinations having prevalence on a specific programme or development initiative. To better understand cultural orientations, it is imperative to define culture and its influences on other matters. According to Taylor (1958:1), "culture is that complex whole which includes knowledge, beliefs, art, morals, customs, and other capabilities acquired by man as a member of a society." Culture is the delicate inheritance from the society; it becomes part of us, and it can be "thrown away" through enculturation and transculturation. A French writer and politician, Herriot, defines culture as "what remains when all is forgotten." Culture is the way of living which is practiced consciously and unconsciously. The way of dressing, speaking, addressing, and general living speaks volume of us as cultural beings than as human beings.

In spite of the fact that we all share similar biological features, as human beings, we do not live homogeneously. Culture has an impact on our mental programming and social conduct. According to Hofstede (1997), there are three main aspects of "human programming," namely personality, culture, and human nature. He specifically calls culture as the "inherited mental software" which is learned and distinguishes an individual or group of people from the rest. Human nature is shared across proportionately, just as personality traits are. The diagram below illustrates this point.

Figure 1 : Three Levels of Uniqueness in Human Mental Programming

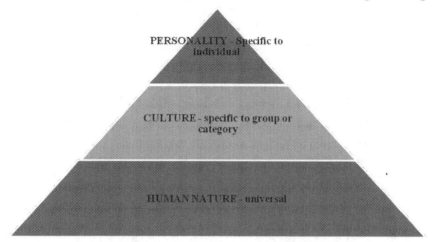

Source: Hofstede, 1997

Culture and education form an intricate connection, which can influence the acculturation or transculturation processes in students. Just as it has been illustrated above that culture is learned, in the same way, one cultural trait can be learned and displace another cultural trait. Educators are cautioned against rejecting one particular cultural orientation over another. Education should not entail an overthrow of cultural identities over other superior cultural identities. Third-world cultural identities are at the mercy of the dominant Western cultural identities, just as it had been illustrated by Samuel Huntington's *Theory of Clash of Civilization.* According to Huntington (1996), the end of the ideological cold war signalled the beginning of a cultural and religious conflict among nations. The cultural values underpinning different nationalities and ethnicities ought to be respected and reserved to minimize this conflict. Cultural barriers to development can be amended, but cultural values that do not conform to the path of development ought to be respected.

Elements of Culture

Sociologists and anthropologists notice that all culture constitutes both the outward and tangible elements and the inward and intangible elements. The most common elements of every culture that are tangible include clothing, dressing, food, housing, writing, and artefacts, while intangible elements include respect, values, attitudes, language, music, dances, and others. The complexity of a culture is easily noted through the complexity of its elements. Understanding a particular culture would also mean understanding the utility and frequency of use of its elements and cultural values. Giddens (2001) states the following elements of every culture, despite the fact that culture variably changes from society to society: beliefs, values, norms and sanctions, ideologies, symbols, material culture, language, and technologies.

This study examines how the introduction of technology tools would affect universal elements of culture and how educators can harness the use of educational technology in such a way that they do not impede the cultural dynamics of the nations for which they are meant to support.

Beliefs. These are the set beliefs, theories of the way natural things happen. These are expressed worldviews, or explanations of realities. They could be agreed or disagreed, but mostly they are illogical and unscientifically proven. According to Leung et al (2002), beliefs are explanations of a perceived relationship between two objects or concepts. African cultures have always believed in rain-making rituals, an exercise of offering sacrifice to the gods to seek a return reward of abundance of rain. Education has helped to provoke an attitude of questioning some of the embedded beliefs that are wholesomely unsubstantiated. For instance, the fight against HIV/AIDS has faced resistance in regions whose beliefs contravene medically proven HIV/ AIDS prevention measures.

Values. Values are an important element of every culture because how social behaviour is viewed is partly caused by dominant values and ideologies (Leung and Bond, 1989). Values define the direction of the morals in every society. In terms of their operationalisation, values are known through the preferences of the individuals in every society. Some African societies prefer traditional dances around the fire, and we cannot do anything to stop them because the practice gradually becomes their value. Japanese societies consider the taking off of shoes upon entry into every house as a value, not just tradition. In education, African values may include respect for the elders and their privacy, decency in dressing, and observance of rituals and rites of passage.

Norms and Sanctions. Norms and sanctions include the set rules and regulations that prescribe acceptable behavioural conduct in a specified culture. According to Hofstede (2001), Inglehart and Baker (2000), and Smith, Dugan, and Trompenaars (1996), norms are important cultural phenomenon that prescribe desirable values, behaviours, and conduct for an ethically consistent society. Norms form the basis of institutions, which are set rules and regulations. They can be confused with values and ideologies, but norms are prescriptions.

Sanctions, on the other hand, are punishments attracted to violations of specific norms. Some norm violations may attract severe punishments, whereas other norm violations may attract lenient punishments. It is not the nature of consequences that defines the severity of punishment as it is in legal systems, but it is the prescription and value system that determines the severity of punishment.

Ideologies. These are the grounded paradigms that reflect the direction of all the cultural values, doctrines, and beliefs. An ideology is the underpinning belief behind all beliefs, values, and behaviours, and it is an ideology that prescribes values, which later prescribe the norms. For instance, an ideology that women are subservient to men is said to be the most dominant obstacle to efforts for gender equality and female education.

Symbols. Symbols are objects or designs that have acquired special cultural meaning. This is the reason why the same objects may carry different meanings in different cultural settings. A symbol is a perceived outlook and its underlying meaning. For instance, children born with little offshoots of upper teeth were be killed, for they believed that upper teeth at birth are a symbol that the child would be a wizard or witch. Symbols could stand for what we call icons in the modern day. Traditionally, a small gesture would speak louder than volumes and volumes of words, and in education, symbols are more commonly used to enhance understanding than mere words. Students are supposed to be acquainted with the traditionally defined symbols being communicated to them.

Language. In a cultural setting, language includes not only written and spoken communication, but also nonverbal cues, such as the use of eyes, hands, and body and their interpretations (Zion and Kozleski, 2005). Misinterpretation of language can be a cause of conflict in many societies. Language is the most valuable but dynamic element of culture, as it binds the society together; it can be used as a cultural identity, and it undergoes a continuous linguistic mutation as it is passed on from one generation to the other. Introduction of advancement in ICT has come with a new brand of linguistic mutation in the form of introduction of new terms in the English language. For instance, in the nineteenth century, words like *surfing, browsing, clicking,* and others were either used to mean completely different actions or did not exist in the English language altogether. The study focuses on how language has been affected through the introduction of ETT.

The emergence of educational technology tools poses a serious threat of cultural diffusion and rejection of some of the cultural values considered unwanted with fashion. There are different cultural interests in the adoption of educational technology tools, mostly dominant ones, wondering whether the fully fledged adoption would not be at the expense of indigenous cultural heritage. The interest of the study would border on an analysis of how far the educational technology tools have made an impact on culture in Malawian context, considering the most common cultural elements mentioned above. The study focused on two private institutions with two different levels of education: one secondary school and the other a university.

Methodology

The study adopted both qualitative and quantitative research approaches to ensure balance and the ability to generalise the findings from this study. The study used questionnaires and key informant interviews targeting both students and their teachers or lecturers.

These findings were analysed using SPSS for quantitative data and discourse analysis for qualitative data. The exercise valued the consent of the participants, and no interview was carried out without the consent of the participants.

The study sampled two institutions in a comparative case study approach using convenience sampling technique, and these schools include one secondary school, St. Patrick's Academy, and a tertiary institution, HBI Institute of Communications and Management. Quota sampling technique was applied in the exercise, and the selection was based on availability of educational technology tools on campus and proximity. Data collection involved the use of questionnaires for students (learners) and key informant interviews for teachers and lecturers (educators). The approach considered all ethical requirements of voluntary consent and strict confidentiality. The questionnaires targeted 20 students, while key informant interviews targeted four candidates in both of these institutions.

Educational Technology Tools in the Sampled Institutions

According to both the questionnaire and the key informant interviews, the three learning institutions utilised different educational technology tools in the following categories: tools of the trade such as interactive whiteboards, PowerPoint presentations, LCD projectors, videoconferencing facilities, and learning management system (LMS), online courses, YouTube, Wikipedia, and lesson blogs. Few respondents acknowledged the usage of the tools under the Apps category. Figure 2 below indicates the availability of educational technology tools.

Figure 2. Availability of Educational Technology Tools—Tools of the Trade

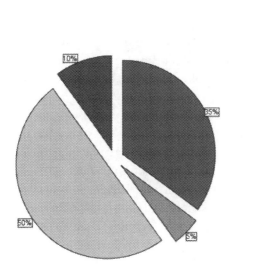

Source: Author, 2015

Figure 2 indicates the videoconferencing as the most frequently used tool with 50% of the respondents acknowledging it, especially those from HBI Institute for Communications and Management. In the other categories of online courses, programmes, and academies found in figure 3, a combined 66% of the respondents acknowledged the use of YouTube with 44% of the responses, and 22% indicated the use of Wikipedia and Wikibooks as available tools. These statistics indicate that online courses are not frequently used in this part of the region.

Figure 3. Availability of Educational Technology Tools—Online Courses Source: Author, 2015

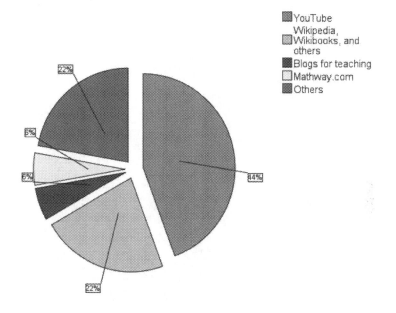

On usage of different apps, the responses did not affirm that these different educational apps are frequently used in the institutions of learning as illustrated in the figure below:

Figure 4. Educational Technology Tools—Usage of Apps

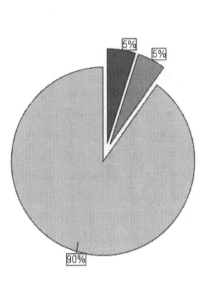

Source: Author, 2015

The highest percent of the respondents (90%) indicated that apps are not frequently used as educational aids in their institutions, and even the responses that affirmed their usage indicated personal usage of apps, not for academic purposes. It was noted that most students and teachers have not yet adopted the use of apps as a form of teaching and learning.

Cultural Aspects at Stake with the Integration of Educational Technology Tools

The discussion on how cultural orientations are affected by the use educational technology tools and how cultural orientations affect the usage of educational technology tools are discussed below.

Beliefs. Based on the discussions from the three educational institutions, all the key informant interviews confirmed the influence of cultural beliefs in the usage of educational technology tools. Specifically, the IT teacher from St. Patrick's Academy stated that their beliefs are being eroded through the use of educational Web sites on the Internet. Students are learning aboutthe beliefs alien to them through exposure to the Internet. The teachers noted that the beliefs that boys are superior to girls and that girls are less intelligent than boys are being challenged through the use of the Internet (personal communication, 18th March 2015). On the same vein, the interview with the HBI Institute of Communications Management IT lecturer asserted that beliefs in Malawian culture that are detrimental to educational progress are equally being challenged with the use of videoconferencing. The interviewer stated that the belief of the teacher-student (face-to-face) learning system as the perfect way of learning is no longer relevant. He commented that learning has turned flexible and has become convenient for very busy students who could be working. This flexibility has further created an advantage of productivity and lessened chances for students who would need to resign to study or get fired for failing to come in to the office due to studying (personal communication, 18th March 2015).

According to the data collected, most of the respondents indicated that ETT has had some impact on their cultural beliefs, especially making them consider the value or benefits of certain beliefs. The table below indicates the distribution of responses on the influence of ETT on cultural beliefs.

Table 1. Influence of ETT on beliefs

Influence of ETT on Beliefs	*Institution*		
	St. Patrick's Academy	*HBI Institute of Communications and Management*	*Total*
It has made me ignore certain beliefs	0	2	2
It has made me ignore some cultural practices	1	0	1
It has made me totally abandon some cultural practices	1	4	5
It has made me start a new way of living from other cultures	2	4	6
Others	6	0	6
Total	*10*	*10*	*20*

Source: Author, 2015

Values. When it comes to values, the study revealed that cultural values have been greatly affected by the educational technology tools. According to a key informant interview from St. Patrick's Academy, students who rely on Google lessons through YouTube are exposed to different cultural values that make them compromise their own values (personal communication, 18th March 2015). The students imitated the "way of living" from the West, which creates a cultural conflict with the "way of living" in Africa and Malawi in particular.

According to the IT teacher of St. Patrick's Academy, the fact that the institution is an international institution with students from different ethnic backgrounds makes the campus a multi-cultural environment for all students. However, the extent of divorce from the local cultural values to those of the West remain observable in the way students behave in class and dress and the nature of music they listen to.

On the contrary, at HBI Institute of Communications Management, cultural values have been enhanced by the introduction of educational technology tools. According to the director of HBI Institute of Communications and Management, the Malawian cultural value of making "women play a background, domestic role instead of being in the forefront" has been seriously challenged by their exposure to the way women behave in Asia. He stated that the television programme that the institute runs has seen the emergence of boldness in women who had been previously shy in front of the camera. Furthermore, based on the report from the IT lecturer from the same institution, the video conferences offer a great interface that challenges our way of learning and values. For instance, he stated that students are inspired to learn that the Asian communities continue to embrace cultural values of modesty in dressing, respect, and language even in the face of technological advancement at their disposal. Queried on the impact of ETT on values through the questionnaire, the following table displays their responses.

Table 2. Influence of ETT on cultural values

| *Influence of ETT on Values* | Institution | | |
	St. Patrick's Academy	*HBI Institute of Communications and Management*	*Total*
It has made me forsake some cultural values and ideologies	*0*	*2*	*2*
It has made me question the importance of some cultural values	*2*	*4*	*6*
It has made me ignore most of the cultural values	*1*	*0*	*1*
It has made me create new cultural values	*1*	*3*	*4*
Others	*6*	*1*	*7*
Total	**10**	**10**	**20**

Source: Author, 2015

From the responses documented in table 2, it shows that culture has been influenced by ETT, and ETT has also influenced culture.

Norms and Sanctions. This study revealed that the educational technology tools water down the local sanctions and norms, thereby creating a more liberal society that is not bound by specific cultural restrictions in the form of norms and sanctions. According to the IT teacher (personal communication, 18th March 2015), the students at St. Patrick's Academy have a liberal attitude towards education, and they do not tolerate cultural restrictions on what they can do with the technology. The students could access every Web site they find fit and could even access Web sites that are illicit and morally bankrupt. He noted that the question of censorship to online courses is dependent on the service provider and not the students. He appreciated that in line with norms and sanctions, educational technology tools truly places the "guide on the side."

Symbols. On the point of symbols, the study noted that educational technology tools do not interfere with cultural symbols in any way. Following the key informant interview at HBI Institute of Communications and Management, cultural symbols are not, strictly speaking, imparted from the Asian community through educational technology tools. According to the director, it was noted that the universally accepted symbols are the most commonly used symbols in most of the videoconferencing lessons the institute holds, such that no cultural conflicts arise on the usage of symbols and signs through the lessons. It was noted that universally accepted symbols are the symbols that share similar meanings across varying cultural settings.

Language. On language, all the institutions had consented to noting changes in language usage with the increasing use of educational technology tools. This was an assertion we could not test but simply agree with the respondents. According to the interview with the St. Patrick's Academy IT teacher, language on campus at St. Patrick's Academy had started to change with new words being introduced in Chichewa as a result of the use of educational technology tools. Words like *kugugula*, meaning "to use Google," and *kubulauza* meaning "to surf the Internet" were new versions of Chichewa. These words indicate a complete submersion of Chichewa semantics and creation of new terms that are simplified English terms spoken in a Chichewa fluency rate. The IT teacher continued to illustrate that language on the campus had been largely English, as the institution is an international academy of different people from different cultural and ethnical backgrounds, but the angle of 'sifting' vulgarity in the English language was something lacking. He stated that foreign language has come with its vulgar language, and the students barely distinguish the decency of the language in the usage.

According to the results from the HBI Institute of Communications and Management, language did not create a cultural conflict, on the context that the lessons were emerging from areas whose mother language is not English. The director stated that through videoconferencing classes, the lecturers on the other side of the world would speak English with local fluency. In fact, he stated that even if the lessons were held in English, it was visible that the fluency was Asian, and courtesy and politeness could be observed in the tone. The educational gadgets could not be blamed for the deterioration of local languages, and probably, lessons would be drawn from the conferences on the possibility of adopting a foreign language without a complete abandonment of one's language and culture (personal communication, 18th March 2015).

The Implications of a Tradeoff of Some Cultural Orientations against Technological Development in the Educational Sector

According to this study, the director of HBI Institute of Communications and Management commented that there are some impacts of technology. He indicated that technology has assisted in the effective and efficient delivery of educational services. He pointed out that students can, with the tools, learn on their own, write exams, and expect to receive results of the examinations instantly. The hassle of waiting months for results has been significantly removed. It was even further noted that the educational technology tools reduces the risks for teachers from getting infections from the traditional chalk-and-board system.

Analysis of the data collected from St. Patrick's Academy revealed that the educational tools have both positive and negative effects, but besides the fact that they have negative effects, with proper harnessing, they would yield a lot of benefits to the society. An accounting teacher of the institution stated that it is easier for him to demonstrate examples using educational technology tools and even for students to quickly grasp accounting examples through the use of educational technology tools (personal communication, 18th March 2015). The IT teacher stated that the tools help in creating a good platform for students to develop their cognitive skills if they seek not to abuse them. He further commented that with guidance and proper supervision, educational technology tools would fast-track learning and teaching for both teachers and learners.

Recommendations and Lessons Learned

From this study, important lessons can be drawn, and the following constitute some of them:

- Educational technology tools are needful for overall educational development, but they need to be custom-made to suit the cultural orientations of the users.
- Usage of elements of culture through educational technology tools needs to be reviewed in terms of its relevance and

applicability. Usage of some elements may lead to the dispensation of other valuable cultural elements of the users of the tools.

• The integration of ETT must accompany orientation of how these tools operate, how they can be used, abused, and more importantly, how we can harness them in line with our cultural values and orientations.

Conclusion

The secret to Africa's development is high-quality education that creates a platform for an industrious and learned society. The emergence of ETT in Africa is a positive development to the revitalization of the sinking standards of education in Africa. However, the pursuit for development needs to correlate with the institutionalisation of cultural values that harness it. Cultural values that compete or impede developmental progress become more irrelevant with progress. This study has clearly indicated how the integration of ETT has become a barrier to some of the most important elements of culture and how we can ably harness the tools with respect to the diverse cultural orientations of both the teachers and learners who use such tools.

References

Basu, A. M. (1992). *Culture, the status of women and demographic behavior: Illustrated with the case of India.* Oxford: Clarendon Press.

Blayo, C. and Y. Blayo. (2003). *The social pressure to abort. In The Sociocultural and political aspects of abortion: Global perspectives,* ed. A. M. Basu Westport, Connecticut: Praeger Publishers.

Bradley, S. and Schwandt, H. (2008). *Levels, trends, and reasons for contraceptive discontinuation.* Calverton, MD.: Macro International.

Giddens, A. (2001) *Sociology: Introductory readings,* Polity: Cambridge Hofstede, G. H. (1997). *Cultures and organizations: Software of the Mind,* London: McGraw Hill.

Huntington, S. (1996). *The clash of civilization and the remaking of world order,* New York: Simon and Schuster.

Inglehart, R. and Baker, W. (2000). Modernization, cultural change, and the persistence of traditional values. *American Sociological Review,* 19–19.

Leung, K. and Bond, M. H. (1989). On the empirical identification of dimensions for cross-cultural comparisons. *Journal of Cross-Cultural Psychology,* 20, 192–208.

Leung, K., Bond, M. H., Reimel de Carrasquel, S., Munoz, C., Hernandez, M., and Murakami, F. (2002). Social axioms: The search for universal dimensions of general beliefs about how the world functions. *Journal of Cross-Cultural Psychology,* 33, 286–302.

Malawi Human Rights Commission (2010). *Cultural practices and their enjoyment on human rights, particularly rights of women and children in Malawi.* Lilongwe: MHRC.

Malik, S. and Agarwal, A. (2012) 'Use of multimedia as new Educational Technology tool—A study, *International Journal of Information and Education Technology,* Vol. 2, No. 5, October 2012

Matsumoto, D. (2007) 'Culture, context and behaviour.' *Journal of Personality,* 75:6, December, 2007.

Ng'ambi, F. (2010) *Malawi: Effective delivery of public education services, A review of and open society initiative of Southern Africa,* OSISA: Johannesburg.

Richey, R. C. (2008), Reflections on the 2008 AECT definitions of the field. *TechTrends.* 52(1). pp. 24–25.

Roberts, B. W. (2006). Personality development and organizational behaviour,' in B. M. Staw (Ed.), *Research on Organizational Behaviour.* Elsevier Science, JAI Press.

Sarah, E. K. Bradley, Hilary M. Schwandt, Shane Khan (2009). Levels, trends and reasons for contraceptive discontinuation, *DHS Analytical Studies*, No. 20, Maryland, September, 2009.

Smith, P., Dugan, S., and Trompenaars, F. (1996). National culture and the values of organizational employees: A dimensional analysis across 43 Nations. *Journal of Cross-Cultural Psychology,* 231–264.

Wood, D., and Roberts, B. W. (2006). Cross-sectional and longitudinal tests of the personality and role identity structural model (PRISM). *Journal of Personality,* 74, 779–809.

Zion, S. and Kozleski, E. (2005). *Understanding culture.* National Institute for Urban School Improvement, October, 2005.

Chapter 13

Science Communication for Effective Teaching and Learning for Societal Impact

Caleb Ademola Omuwa Gbiri

Introduction

Science is a systematic enterprise that builds and organizes knowledge in testable and predictable manners (Harper, 2010). Communication is scientific way of transferring (conveying) information (knowledge) through the exchange of ideas, feelings, intentions, attitudes, expectations, perceptions, or commands. Communication is conveyed by speech, nonverbal gestures, writings, behavior, and possibly by other means such as electromagnetic, chemical, or physical phenomena and smells from one place to another or from one person to another involving two or more participants (Harper, 2010). However science communication refers to public communication presenting science-related topics to nonexperts and learners for proper understanding. Science communication uses educational technology tools to improve teaching, learning, and interactive learning environment for effective information dissemination.

Although science communication differs from communication science (schools of scientific research of human communication [Rogers, 2001]), science communication incorporates the use of communication science for appropriate information dissemination. In this era of growing information communication technology (ICT), its integration into the teaching and learning will make teaching and learning much easier and better delivered in both natural and applied sciences. The proficiency in communication of educational material in all aspects of science education (physical and applied, including medical sciences) depends on the proficiency of the science communicator in the utilisation of ICT for communication and teaching. Though applied and medical sciences' researches and teachings are conducted in professional-specific ideologies using dedicated terminologies and processes, the use of communication science improves its proper communications and understanding. Hence, integration of ICT makes the terminologies communicated to the learners possible through environmentally friendly and culturally acceptable manners with an intention for reproducibility.

Importance of Science Communication in Science Education

The ultimate goal of training or teaching in science is to improve the knowledge base of the learners towards providing quality, accessible, affordable, acceptable, applicable, and responsible scientific delivery for the improvement of the quality of life, functionality, and general well-being of human being and other creatures. A high scientific knowledge without understanding and acceptability is absolutely useless. Hence, scientists must create good training environments and venues for proper dissemination of information in simple and understandable language with specific consideration for the peculiarity of the population in terms of culture (norms and values systems), language, and belief systems.

Contemporary sciences, most especially applied and medical sciences, cannot be effectively communicated without the use of information technology. Scientists must improve their proficiency in science communication for effective training and practice for societal impact using environmentally friendly ICT applications. Hence, science communication must incorporate the appropriate ICT application to enhance information dissemination. Until a research result is understood by the second party through effective communication and is used by the larger public with proven efficacy, it still remains a mere hypothesis and bookshelf property. The productivity and impact of a research finding is assessed through its societal use and applicability and the value they placed on it. A research with no societal value can be said to be irrelevant and a waste of resources.

Science Education and the Utilisation of Information Technology in Teaching and Learning

Science education involves the acquisition of knowledge and effectiveness in communicating the knowledge for societal values. Although the acquisition of knowledge is good, knowledge is useless when it cannot be effectively communicated for utility. Utilisation of the content of communication is measured by the understanding, reproducibility, acceptability, and use of the information by the recipients. Science communication, most especially in medical science, should be made simple for applicability. This will improve the learning system and learning environment and training and practice, including confidence and proficiency. Therefore, nothing must be left for personal interpretation or discretional utilisation; information must be devoid of ambiguity and misuse.

The integration of ICT into science communication is becoming more important in medical sciences as many treatment options in the medical world are delivered using ICT applications. Information on medical treatment is available on the Internet, and learners must be introduced to how, where, and when to source for it to enhance their learning and subsequent practice. More so that science education, most especially applied and medical sciences, are so complex and dynamic that a teacher must be able to teach the learners on how to have interactive and complementary collateral information for effective practice. This should be integrated at the training stage of the science and medical educations.

Information Communication Technology and Science Communication

The advent of ICT has made it inevitable for medical scientists to get familiar with the appropriate use of information technology in clinical training, clinical research, clinical management, and communication of clinical findings to the learners and the general population. A trainer must have good proficiency in the use of ICT in teaching and integrate the learners to its use. The importance of ICT in training and learning includes but is not limited to:

- Improving the number of trainees a trainer can effectively reach, supervise, mentor, and assess within a specific time frame,
- Improving the communication of both the trainer and the trainees in a learning environment,
- Increasing the exposure of both the trainer and the trainees to new information and methods in the subject area,
- Enhancing the exposure, skill, and proficiency of both the trainer and trainees to new technology and methods of communication,
- Improving the globalisation of both the trainer and the trainees in a specific area of expertise,
- Improving trainer-trainee interaction and integration in the learning environment,
- Improving the maximisation of the available resources for efficient learning between the trainer and the trainees,
- Promoting internal content in the teaching and learning environment within globally acceptable norms,
- Exposing to both the trainer and the trainees to contemporary information and development in their subject area,
- Improving culturally acceptable scientific-based teaching and communication a subject area,
- Improving the approval and acceptability of the teaching by the learners and their ability to reproduce the same at the appropriate time,
- Creating a dynamic environment for teaching and learning within a globally acceptable concept

Science Communication and Educational Technology: The Trainers, the Trainees, and the End Users

The best idea of a teacher will be totally irrelevant and unpopular when it is not well communicated, understood, appreciated, embraced, participated in, and used by the targeted trainees. To make an idea relevant, it has to be well communicated to the learners, which can be facilitated by the integration of appropriate educational technology applications. Apart from the fact that the receivers (learners) must understand the message, they must be able to reproduce the same to the end users (general population) in a simple, understandable, and unadulterated language for effective use and reproducibility. During training, a trainer must avoid ambiguity in the style of communication and also make sure that the learners understand, conceptualise, internalise, and are able to interpret the same message. Teachers must use ICT to enhance the quality of teaching delivered to the learners. Integration of ICT will not only improve the quality of teaching, but it will improve learning interaction in a friendly environment. However, a teacher must appreciate that there is no single technological solution that applies for every teaching, every course, or every view of teaching (Mishra and Koehler, 2006). Hence, quality teaching requires developing a nuanced understanding of the complex relationships between technology, content, and pedagogy and using this understanding to develop appropriate context-specific strategies and representations (Mishra and Koehler, 2006). The ingenuity and dynamism of a teacher is highly imperative in creating an interactive and friendly learning environment and culture. The use of familiar examples within the environment should be used most often by the trainer to improve the communication strategies to trainees using contemporary ICT applications. Various forms of information technology applications are available to improve science communication, and a scientist must be able to effectively use these to promote teaching and learning.

In communication and, most importantly, in effective science communication, there are important factors that need to be considered for effective teaching and learning using appropriate ICT applications.

1. ***Knowing your audience.*** One important factor a teacher should first understand is the peculiarity and diversity of the learners. Understanding who your audiences are is very important in your effective teaching and their consequent understanding of the subject area. This understanding includes knowing the background of the trainers, understanding their strengths to leverage upon, and knowing their limitations to

improve upon. Some learners learn faster through pictures and diagrams while others require repetition and voice reinforcement. A teacher should have proficiency in blending audiovisual apparatuses to enhance learning. The use of annotated diagrams and figures will improve interaction and assimilation of the subject and improve learning interaction and integration. The personality of the learners can be well understood by the trainers using appropriate ICT application assessed anonymously. This will improve the knowledge of the trainers about the trainings and would make the trainer treat the class within its peculiarity. It will improve the dynamism, and the application of an ICT-enhancement programme will ensure adequate communication and assimilation. There are online collaboration hubs that provide customised classrooms for every teacher and learners with their peculiarities and make learning of medical sciences fun. This should be used most often by teachers to promote bonding among the learners.

2. ***Knowing who the information is meant for.*** Another factor for the trainer to consider is who the information is meant for. This requires the trainer to understand the culture and belief system of the learners and integrate it appropriately into teaching the learners. It is only information that is culturally acceptable and is integrated within the belief system of the people that can be effectively utilised. Many culturally friendly and acceptable ICT applications are available that can be used by the teacher to transmit the information to the learners without interfering with their belief system. Medical education has some peculiarities as some of the contents may be in contrast to a culture and belief system. A trainer must know how to use the appropriate ICT application to deliver the lecture in a friendly and interactive manner. The peculiarity of the end users (learners) should be accommodated in choosing appropriate ICT application that will improve communication and can also be used by the learners for their home and group learning.

3. ***Being a good negotiator.*** A trainer must be a good negotiator. A teacher must develop the skill of involving the trainees

or the audience in important decision making. This will involve the use of appropriate ICT application to ensure that the learners contribute to important decisions about the learning process. This does not mean that the power to make an important decision should be transferred to the trainees, but it requires that the trainee should be considered and be involved in decision making while the trainer holds the ace. This will improve trainer-trainee interaction and the learning environment. A trainer can get the view of the learners through specific information software. Even when a learner is not confident enough to face the trainer, he or she can provide his or her view through anonymous responses. This will definitely promote class interaction and decision making towards effective learning. A teacher should leverage on ICT platforms that offer a way for teachers to communicate directly with learners by connecting administrative software and giving real-time updates on class activities.

4. ***Integration of a feedback mechanism or loop***: A trainer should not just be a giver but also should be able to receive. A trainer should, at intervals, ask a leading question and assess the understanding and concentration of the trainees as the lecture proceeds. A trainer should also be able to call upon a trainee to demonstrate a skill and assesses his or her proficiency. If a trainee knows that he or she can be called upon to demonstrate a process or answer a question within a lecture hour, it will improve their concentration, participation, and consequently their understanding of the subject. A feedback mechanism does not necessarily mean a face-to-face feedback loop. There are online collaborative platforms that are designed to complement classroom instructions through interactive activities, where assignments can be submitted, and discussion can be generated for everyone's benefit. A teacher must know and initiate the learners into these programmes to improve learning. A classroom blog can be created that will improve communication among the learners and between the learners and the teachers. This will improve communication on the grey areas in the topic, and solutions may be provided without

necessarily having the whole class together in a face-to-face interaction.

5. ***Encouraging learning interaction***: A trainer should be able to access a platform through which questions can be accommodated at the appropriate time during a classroom learning without a trainee feeling intimidated or insulted. There are ICT tools that allow a teacher to collect and organise resources and connect the classroom with the other similar classes in order to collaborate with the outside world or learners on the same course in other areas of the world. This allows for exchange of ideas among learners in the same, similar, or related fields in other parts of the world. Questions are asked and answered by virtual contacts. This also improves the external validity and globalisation of the lecture and also improves the face value of the learners in the medical knowledge. This approach will improve the confidence of the trainees and improve interaction in the learning environment. Freedom should be created for improved learning interaction. However, this freedom must be conceptualised within the limit of the rules and regulations of the learning environment and must not infringe on the right or freedom of other members of the learning society. A trainer must also improve and encourage interpersonal interaction among the trainees. A teacher should get familiar with applications that will encourage the class to bind together and work for positive skill development. There are ICT applications that provide instantaneous updates on all activities in the learning environment, including attendance, seating arrangements, and contribution to class activities that can be employed by the teacher to improve class interaction.

6. ***The use of pre-recoded and take-home applications***: A teacher should have pre-recorded information, most especially the hands-on practical demonstration that the learners can play and watch in private or as a group to improve their skills and proficiency in a particular course. Pre-recorded audiovisual materials can also be shared with the learners to reinforce

the demonstration in the classroom. Learning should not only be based on face-to-face interaction. Information technology such as the Internet and other software should be employed to promote interaction between learners and trainers. Lectures and information should be communicated using IT and responses also received through the same medium either using a class-structured system (assignment) or an anonymous one depending on the aim of the assessment or the dissemination of the information. Virtual learning environments should also be created to improve dissemination of information to the learners. Both the learners and the trainers should improve their proficiency in the use of IT to provide dynamism in the learning environment. There should be development of environmentally friendly applications and software for use to enhance effective teaching and learning. Some learning information could be place on video, Internet applications, and some other friendly IT mode for easy communication. Lectures could be uploaded for easy accessibility by the learners at any time and anywhere without any hindrance. Lectures should not be monotonous; dynamism should be introduced using appropriate IT within a global concept. In a learning environment, a trainer should have easy access to IT to enhance his or her training. A teacher can use ICT to provide a collection of games and activities that are designed based on medical concepts to be used in the classroom, covering a variety of topics in the subject area. A high-quality multimedia content on medical education and medical sciences should be made available to the learners to improve their knowledge of the subject. Teachers should also introduce the learners to applications that demonstrates and explains a variety of topics using video and illustrations so that it will enables them to learn more about how specific procedures are performed.

Effective Science Communication in Science Educations

Effective communication occurs when a desired effect is the result of intentional or unintentional information sharing, which is interpreted between multiple entities and acted on in a desired way (Robbins,

Judge, Millett, and Boyle, 2001). This effect also ensures that messages are not distorted during the communication process (Robbins *et al*, 2001). Hence, for a communication to be effective, it must generate the desired effect (understanding and utilisation) and maintain the effect with the potential to increase the effect of the message. Therefore, effective communication serves the purpose for which it was planned or designed. Possible purposes might be to elicit change, generate action, create understanding, or inform or communicate a certain idea or point of view. When the desired effect is not achieved, factors such as barriers to communication are explored, with the intention being to discover how the communication has been ineffective. For communication to be effective, it requires the involvement of a sender and a receiver. This means communication is meant to change the understanding, perception, and orientation of the receiver about a specific thought. For information to be well understood and to change the perception of the receiver, it has to fulfill certain criteria.

Barriers to Effective Science Communication and the Solutions through Educational Technology Apparatus

Barriers to effective communication can retard or distort the message and intention of the message being conveyed, which may result in failure of the communication process or an effect that is undesirable. The major issue in science is the inappropriate understanding of information by the end users. This has retarded effective service delivery and service utilisation as well as appropriate responses by the public to the services. This often leads to abuse and misuse. A trainee who does not grab the concept of a communication will find it difficult to disseminate the same to the end users. Barriers to effective science communication often include filtering, selective perception, information overload, emotions, language, silence, communication apprehension, gender differences, and political correctness. It also includes a lack of expressive communication, which occurs when a person uses ambiguous or complex and professional jargon or descriptions of a situation or environment that is not understood by the recipient. However, the use of appropriate ICT platforms can solve many barriers in science communication to produce effective teaching and learning.

Physical Barriers

Physical barriers are often due to the nature of the environment. This barrier often exists in science communication, most especially in applied and medical sciences. This barrier usually presents in the nonavailability of contemporary and functional equipment in the field for teaching and learning. The failure of use or the lack of exposure of the trainer to new technology, may also pose problems in teaching and learning. Disproportionality in the learners-to-trainers ratio may also serve as a physical barrier towards effective teaching and learning. In this process, the trainer may not be able to properly demonstrate and monitor the progress of the information communicated to the learners. This often creates a situation in which some of the learners may not be adequately informed to make an informed decision towards utilisation and dissemination to the end users.

The solution to this barrier is that a trainer should be trained on the use of appropriate ICT for training and lecture delivery. There must be efficient trainer-trainee ratio for effective communication for training. Classes or training environment should be made conducive for learning, and the learners should be broken down to smaller, efficient, and accountable units. There should be close supervision and monitoring of the trainees to ensure adequate understanding. This will conversely promote understanding and effective communication. There are ICT applications that can be used to create this environment and be a barrier-blocker to physical barriers for effective learning.

Physiological Barriers

Physiological barriers are those barriers that are concerned with the normal functionality of an individual trainer and trainee. For example, a trainer with a low voice tone will not be able to deliver a lecture effectively to a large class. The use of a public address system and audiovisual lecture aids will improve the trainer's effective communication with the trainees. A trainee with visual or hearing impairments but do not have visual or hearing aids should be positioned appropriately in a learning environment to improve his or her participation. Appropriate ICT application should also be employed to improve the learning and participation of such a learner. A trainee with ambulatory challenges should be given a privilege position to improve accessibility and participation in the learning environment. A learning environment should have good lightning and be well-ventilated for comfort. Every physiological deficiency of either the trainer or trainees should be well-accommodated and overcome using appropriate ICT without promoting indulgence. A trainer must be able to control his or temperament and avoid emotional outbursts even in the face of extreme provocation. Although a trainee is expected to follow and obey certain rules and regulations, a trainer must not be easily influenced by external factors and should not be given to prejudices in order to promote an adequate learning environment and culture. A trainer who is stereotypical with his or her mode of teaching or assessment may become predictable and taken advantage of by the trainees. A trainer should use the appropriate ICT to improve class interaction and overcome physiological barriers.

Psycho-Social Barriers

Psycho-social barriers are those that result from individual exposure and experiences in life. Some people have never experienced good and favourable conditions before. They tend to be governed by previous but better experiences, which then influences their behaviour and interpersonal relationship. The use of ICT and exposing the individual to brain-stimulated environmentally friendly learning environments can improve the emotional disposition of the individual favourably. Either the trainer or the learner can be affected equally by psycho-social barriers. Because of past experiences, they compare every reaction to their past experiences and they express phobia, incompetence or becoming moody. A learner who has been abused throughout life has not seen a better life. This will definitely affect his or her learning and relationship with others. A trainer who was brutalised throughout his or her training period tends to react in that manner to the learners. Identification and treatment of this will definitely improve the learning environment and interaction.

System Design Barriers

System design barriers refer to problems with the structures or systems in learning or training environment. These problems often present in learning or training structure that is clumsy or cumbersome or a learning system that makes it confusing for a learner to know who to communicate with in respect to a certain concern. Sometimes, a learning environment is too hierarchical and makes responsibility dissemination difficult either to the trainer or the trainees. In another situation where there are many trainers for a topic, since each trainer has his or her own methods of teaching, it makes the understanding of the topic difficult for the learners. For example, in a situation in which a topic in a subject area is being taken by many trainers in parts, it is often difficult for the trainee to know to who a specific obligation is to be directed among the trainers. This will lead to inefficient or inappropriate information dissemination, inappropriate training, inadequate supervision, and lack of mentorship. Since the trainee lacks the appropriate information, the information given to the end users will be distorted and inappropriate, leading inevitably to misapplication and quackery.

The way out to overcome this is that a specific topic in a subject area should be handled by a specific and dedicated trainer who will be the line manager to the trainees and will be ready to solve any issue that may arise from the topic. This is not advocating that a subject should be taken by a single trainer, as each of the trainers has his or her own proficiency in an area of interest, but a topic must be taken by a trainer within the subject area. This will also promote confidence and improve the expectation of the trainee for accountability and assessment for effectiveness and proficiency. There should be the development of information technology platforms through which a learner can reach the trainers or appropriate authority and get a quick and satisfactory response without any delay.

Attitudinal Barriers

Attitudinal barriers present itself in the form of trainers-learners interaction and integration. When a trainer's style of training does not give room for feedback and accountability, the effectiveness of the communication may be inadequate. If the training is not issue-based and practice-based, it will be difficult for a learner to internalise the concept of a discussion and be able to appropriately transfer it to the end users. A situation in which a large class is taught as a whole rather than breaking them into smaller groups could also pose an attitudinal barrier both from the trainer and the learner. The trainer may be carried away by the response of the fast learners in the group while the slow learners may often be neglected. However, on the part of the learners, it promoted redundancy. Learners may not have the attitude of catching up when they lag behind if they observe that their deficiencies may not easily be noticed by the trainer. This creates lack of motivation both to the trainers and the learners. When there is inadequate avenue for interaction between the trainers and the learners, this may result in lack of consultations with trainers and between the learners and interpersonal conflict, most especially between the learners, and lack of motivation to learning. The way out of this is for a trainer to break a large class to smaller intractable and accountable groups. The strength and weakness of every individual in the group can be spotted and come adequate measure will be put in place to bring them up to speed with the rest of the class. This will also promote commitment and dedication on the side of the learners as they know that their attitude and contributions are being monitored and assessed.

Linguistic Barriers

Although every profession has its own terminologies and jargon, the use of jargon makes communication unattractive to the receiver and therefore creates apathy towards the information irrespective how lofty such information may be. The use of jargon is more profound in the field of applied and medical sciences and is sometimes unavoidable. However, effort should be employed by the trainer to use these terminologies sparingly, most especially when it is not mandatory. An ICT communicator can also be used to improve the linguistics of the learners in the subject area. Words sounding the same but having different meaning can convey a different meaning altogether. Hence, the teacher must ensure that the learners receive the same meaning. It is better if such words are avoided by using alternatives whenever possible. Although jargon is inevitable in the applied and medical sciences, each of those jargon should be explained to the trainees to improve understanding for onward utilisation. An ICT application specifically built for some professional terminologies is very essential for use here. However, if possible, the use of jargon should be avoided or limited to essentials used. The use of jargon or difficult or inappropriate words in communication can prevent the recipients from understanding the message (Scott, 1980). Poorly explained or misunderstood messages can also result in confusion. However, research in communication has shown that confusion can lend legitimacy to research when persuasion fails (Scott, 1980). These will complicate the learning process and distort the content of the same information to the end users. This often leads to misapplication of information for genuine intention.

Cultural and Belief Barriers

Culture plays a great role in communication and the mode of communication. Included in culture is the language and belief system. Some languages and modes of communication are more acceptable in some cultures than others. Cultural differences affect communication between people and the subsequent acceptance of the content of the communication. The perception of individuals about information in relation to his or her culture affects the acceptability and the readiness of the recipient to disseminate such information. The ability of a communicator to conceptualise his or her information content constructively within the acceptable cultural norms and values improves the in-depth penetration of the information among the people. The use of body gestures to communicate, most especially in the learning environment, should also be discouraged as different body movements may mean different things in different cultures and convey different messages to different receivers. Although some body movements such as the nodding of the head to indicate agreement or the shaking of the head to indicate disagreement seem to be interpreted the same way in most parts of the world, it should be used sparingly by trainers in science communication. Nothing should be left to cultural interpretation or belief insinuation in science communication. A science communicator should use culturally acceptable examples and connotations during training and dissemination of information in the public domain.

A trainer should be familiar with the culture of the people and use it often for training. This will automatically instill that culture in the trainee who will indigently use it to communicate with the end users. Although science cannot be reduced to the content of cultural values only, the use of cultural values to communicate science for effective acceptability and utilisation by the end users is inevitable. Most importantly, a trainer must not be clouded by his or her religion or belief inclinations in the execution of his or her duty. He or she should be seen as neutral as possible and use natural examples rather than belief systems as examples during training. Specific culturally acceptable information technology platforms should be used to promote learning and communication to the learners.

References

Harper, D. (2010). "Communication" *Online etymology dictionary*. Dictionary.com. available at www. http://dictionary.reference.com/browse/communication.Accessed 09/11/2014.

Mishra, P, and Koehler, M. J. (2006). Technological pedagogical content knowledge: A new framework for teacher knowledge. *Teachers College Record*. 108 (6), 1017–1054.

Robbins, S, Judge T, Millett, B, and Boyle, M (2001). Organisational Behaviour. 6th ed. Pearson, French's Forest, NSW. pp. 315–317.

Rogers, E. M. (2001). The Department of communication at Michigan State University as a seed institution for communication study. *Communication Studies*. 52 (3): 234–248

Scott, A. J. (1980). "Bafflegab pays." *Psychology Today*. p. 12.

SECTION 5

Emerging Trends in Educational Technology Tools in Education

Chapter 14

Using Social Media Tools to Enhance E-Learning

Michele T. Cole
Blessing F. Adeoye
Louis B. Swartz
&
Daniel J. Shelley

Social media has been defined as a "collection of Internet-based applications that build on the ideological and technological foundations of Web 2.0 and allow the creation and exchange of user-generated content" (Kaplan and Haenlein, 2010, p. 61). From the playground to the boardroom, social media seems to be everywhere. How are people using it? To communicate, surely, but to what end For our purposes, the answer hinges on the way that students and educators are employing the various applications to enhance learning. By enhance, we mean to improve comprehension, to facilitate understanding, to more actively engage participants in the process of learning, and to create a community of learning that bridges geographical boundaries. The last purpose is cultivated and advanced by virtue of the technology, which provides the foundation for online instruction.

Referring to the broad range of Web 2.0 and blended technologies, Wankel and Blessinger (2013) remind us that "the power of using these tools is in their ability to break down barriers (physical, geographical, political, economic, social, technological) and create more agile, inclusive, and democratic learning environments" (p. 5).

The following literature review provides the context for the discussion of the realities and implications of using social media tools in the classroom and online to enhance student learning.

Literature Review

Online Learning

In 2014, Allen and Seamen reported on the continuing importance of online education to institutions of all sizes in the United States, finding that 71% of academic leaders responding to their survey felt that online education was mission-critical. As institutions of higher education struggle to keep up with new technologies, needs of adult and distance learners, and the ever-increasing costs, academic leaders are looking to online instruction as part of the value proposition. Four years earlier, Picciano, Seamen, and Allen (2010) predicted that there would be an educational transformation through online learning. In their view, the foundation for transformation was already in place for higher education in the United States. However, whether widespread transformation was actually occurring yet was difficult to determine. "More simply put, adding technology without changing the pedagogy does not necessarily result in any major change to teaching and learning" (p. 28). The authors refer to the definitions of "transform" as meaning (one) to change completely or essentially in composition or structure and (two) to change the outward form or appearance" (pp. 17–18).

If online instruction leads to transforming higher education and, by so doing, mitigates obstacles to access, one could argue that online learning enhanced by technology can be the means to leveling the playing field, that is, eliminating external causes that affect the ability of students to access learning. However, without faculty buy-in, which remains at 28% (Allen and Seamen, 2014), it is difficult to imagine that the pedagogical changes required to make technologically enhanced online instruction work can be effected.

Adoption and Integration of Educational Technology Tools Online

Tunks (2012) proposed a framework for bringing instructors on board using Web 2.0 tools to facilitate behaviours that are critical for effective e-learning, such as personalization and presence. In this context, "personalization" refers to the effective use of social media to reinforce the instructor's presence in the course. "Presence" refers to the ability of students to interact without sharing the same physical space. Tunks developed a guide for instructors teaching online to help them understand which educational technology tools might best be integrated into their courses, including blogs, wikis, videos, and photo sharing, as well as social media, among others.

In the study that is the focus of this chapter, the authors were interested in exploring how technology was being used to enhance student learning. As online instruction continues to spread, using technology as a medium of instructional design seems logical, given the accessibility of Web-based resources.

Tung (2013) suggested that the integration of social networking tools with open and distance learning held "tremendous promise" (p. 236) for enhancing student learning. Advances in technology have given students and their instructors expanded opportunities for building collaboration, knowledge sharing, and critical-thinking skills development. Tung emphasized Siemens and Conole's assessment of the role that technology will play in transforming education, stating that the emergence of new Internet technologies is changing how people and societies interact and learn. "The implications for education are significant" (Siemens and Conole, p. i).

However, as Czerkawski (2013) noted, it is not technology alone that facilitates interaction and enhances learning; rather, it is the integration of technology into instructional strategies. What makes Web-based emerging technologies significant is their potential to extend access as well as to support instruction, thereby enhancing learning. "Emerging technologies" are defined here as "tools, concepts, innovations, and advancements, utilized in diverse educational settings to serve varied education-related purposes" (Veletsianos, 2010, p. 3). As higher education increasingly embraces online instruction, emerging technologies become an important part of online instruction. Although he felt that the research on the impact that new technologies had on teaching and learning was inconclusive, Czerkawski did conclude that for courses leading to a technology degree, the integration of technology into the curriculum was appropriate to enhance learning.

New technologies had major effects on teaching and student learning in the course that formed the basis of Varela and Westman's (2013) research. They found that when used to promote active learning, technology changed how they themselves taught. Varela and Westman's study was limited to one course in popular culture, but the suggestion is that, together with other instructional strategies, technology can make a significant positive impact on student learning. Facebook and Twitter in particular were found to be aids to active learning because they enabled students to increase their level of interaction with each other and with the instructor and to continue discussion beyond the boundaries of the instructional session.

Others have been more cautious in their endorsement of emerging technologies as the key to transforming education. In his consideration of educational and online technologies, Burke (2013) wrote that what is needed before committing to new technologies as the answer to enhancing learning in online environments is a better understanding of what and how technology is used. In his view, the indiscriminate use of educational technology leads to cognitive overload, not enhanced learning.

Grossek (2009) posed the question of whether "to use or not to use" Web-based educational technology in higher education. Her response was positive. She concluded that once educators, not just enthusiasts, become engaged with Web 2.0 tools, they would find that technology fostered collaboration among students, researchers, teachers, and the community worldwide. Concurring, Majid (2014) cited Grossek's assessment of the significance of Web 2.0 tools as supports to key learning trends, stating that "higher education is undergoing a major transformation enabled by Information Technology (IT)" (p. 88). Grossek described Web 2.0 as referring to "the social use of the Web which allow people to collaborate, to get actively involved in creating content, to generate knowledge and to share information online" (p. 478).

Yet integrating educational technology tools into course delivery does not provide a one-size-fits-all response to enhancing student learning. In searching for a balanced technology, Frantzen (2013) found that a technology-intensive curriculum alone did not result in enhanced student performance. He was comparing the application of technology in the instructional design of face-to-face, hybrid (blended), and online courses. His results were mixed. The researcher found that while there were positive effects from the use of technology, some students still faced barriers in fully benefiting from the integration of technology into the curriculum. With regard to adult learners, the study determined that a technology-intensive curriculum did not significantly affect that population, suggesting that additional measures would be needed before determining if technology would have any meaningful impact on their learning. Echoing a theme prominent in the literature on cultural influences on education, the author also noted the importance of recognising that the use of technology across course models and vehicles is diverse so that connecting tools and content with the different student populations is necessary for technology to fulfill its potential as an educational tool. The part that the instructor plays in integrating technology into course design and delivery is critical as well. In summing up, the author noted, "Understanding the limits and benefits to using technology . . . has important implications . . . as distance education courses become more popular and integrated into the culture of higher education" (p. 576).

Viewing the adoption of social media as educational technology from the student's perspective, Tuten and Marks (2012) urged instructors to use the tools that students, "digital natives," use constantly in order to realise what researchers have envisioned the role that social media can play in education. Prensky (2001) has defined "digital natives" as 'native speakers' of the digital language of computers, video games, and the Internet" (p. 2). In another description of Web 2.0, Tuten and Marks refer to the "read/write Web" or the social Web. "It is a broad term used comprehensively to refer to technologies that enable users to consume, contribute, share, and augment content online, often in the context of social media" (Tuten 2008). (p. 201). In support of enhanced student learning, the authors argue that the use of social media tools can address course management issues and provide opportunities to meet course objectives. Similarly, Hammerling (2012) noted that the use of social networking sites could be employed effectively to assist in student development, a position consistent with the value that her university places on technology as a means to educational quality, efficiency, and distribution. Her discussion centred on best practices in online education for an undergraduate clinical science course.

Yet despite the argument that students are immersed in social media and the use of Web 2.0 technologies, the connection to their adoption and use as tools for learning has not been automatic. Lowe, D'Alessandro, Winzar, Laffey, and Collier (2013) argued that in order for students to accept the use of Web 2.0 technologies in the course, they needed to perceive that the technology would be easy to use and helpful to them in learning course material. Crampton, Ragusa, and Cavanagh (2012) also argued that students needed to be made aware of the value of educational technology used in online instruction and how best to employ that technology in order for them to truly benefit from such tools. In their cross-discipline review of student performance in selected online courses, they found a positive correlation between access to low-cost and easy-to-operate online resources and student success.

Integrating Educational Technology in the Classroom

While most of the discussion has focused on the application of educational technologies, particularly social media in the online learning environment, the adoption of emerging technologies has not been limited to online instruction. Many of the issues with regard to how effectively these tools can contribute to student learning are the same. McCabe and Meuter (2011) studied the implementation of technology in the classroom from the perspective of the students. Did the students feel that the use of technology enhanced their learning? Researchers examined whether technology created a more effective learning environment within the context of the seven principles of good practice. That is, did the integration of technology into the classroom instruction result in: contact between students and faculty; reciprocity and cooperation among students; active learning; prompt feedback; time-on-task; high expectations; and, respect for diverse talents and ways of learning? Their results were consistent with other studies that found that some technological tools were used more by students than others. They also found that the content management system (CMS) used, Blackboard, was more suited to realizing some of the seven principles than it was to realizing others.

Buzzard, Crittenden, Crittenden, and McCarty (2011) investigated the use of educational technologies in the classroom, seeking to determine if and how instructional technology could be an effective pedagogical tool for teaching and learning. They found that, on the whole, students preferred more traditional instructional technology, such as Web sites, office suite, and social networking, while instructors showed a preference for incorporating course enhancements available from publishers and university support services. "Instructional technology" as used here has been defined as including "hardware and software, tools and techniques that are used directly or indirectly in facilitating, enhancing, and improving the effectiveness and efficiency of teaching, learning, and practicing" (p. 132). The authors looked at usage of instructional technology and the relationship between usage and student engagement and learning. In summing up, Buzzard, Crittenden, Crittenden, and McCarty reported that while both students and instructors were interested in employing technology, there were issues for instructors related to the differences among disciplines; and for students, there were issues regarding instructors' ability to assist them with using technology. The authors concluded that traditional instructional tools appeared to be sufficient for enhancing student learning without the "more sophisticated or advanced Web 2.0 digital tools" (p. 138).

In their study of how the integration of technology into an on ground class environment contributed to a student-centred classroom, Thiele, Mai, and Post (2014) found that technology in the classroom can enhance learning by making the classroom more student-centred and more active. They also found that while students were familiar with computer concepts, they were limited in their use of the Web. They also found that students liked using technology to learn, but, at the same time, they recognized its limitations. However, it should be noted that the applications that the researchers studied were chosen specifically for a doctoral program in physical therapy and included Raptivity, Camtasia, Jing, and Triptico. The learning management system (LMS) used was Moodle.

Cultural Factors Affecting Integration of Educational Technology into the Curriculum

Are research findings consistent across cultures? Are students immersed in social networking in all cultures? Are instructors' struggles with issues of how best to integrate which educational technology tools the same regardless of their location? Are issues of enhancing student learning universal?

Adeoye and Oni (2010) tell us that in order for students to develop the critical thinking skills required for learning, a certain degree of personal and social transformation must take place. That transformation, they maintain, can be enhanced with e-learning. Hence, knowing how culture interacts with and affects online education is particularly relevant. In their study of the sociocultural dimensions of e-learning systems, the authors looked at how students from different cultural backgrounds responded to a particular e-learning system. Relying on Hofstede's (1997) cultural dimensions (power distance, collectivism vs. individualism, femininity vs. masculinity, uncertainty avoidance) and Nielson's (1993) usability attributes (learnability, efficiency, memorability, accuracy, satisfaction), the researchers found that culture did in fact substantially impact students' ability to use online learning systems. Perhaps most importantly, their findings demonstrated the need to integrate cultural awareness in online instructional design to effect enhanced student learning.

In summary:

Not everyone in a society fits the same cultural dimensions precisely, but there is enough statistical regularity to identify trends and tendencies. These trends and tendencies should be recognized as different patterns of values and thought. In a multi-cultural world, it is necessary to cooperate to achieve practical goals without requiring everyone to think, act, and believe identically (Adeoye and Oni, 2010, p. 155). In an earlier study, Adeoye and Wentling (2007) compared how useful a particular e-learning system was to students from different cultures studying in the United States. Again, differences were found based on national culture as well as on gender.

In their case study of a professional development program in Ghana, Nijhuis, Pieters, and Voogt (2013) point to the failure of curriculum developers to take cultural characteristics into consideration when designing courses for educational change. Curriculum reform that has been initiated to enhance student learning and improve instructional practice, they say, often fails because of that. What the authors maintain is necessary for success is a culture-sensitive approach to curriculum development. Acknowledging the difficulty inherent in understanding another culture, the authors recommend three steps in developing curricula:

1. Conduct extensive context analysis early on

2. Interpret and analyze the needs analysis as part of the context analysis keeping the culture and the related educational settings at the forefront

3. Continue formative evaluation activities throughout the curriculum development processes

"Culture" as used here encompasses "the knowledge and ideas which give meaning to beliefs and actions of individuals and societies and can be used to describe and evaluate those actions (Stephens 2007)" (p. 246).

To better understand how learning style affects students' receptivity to instructional methods in an online environment in different cultures, Adeoye (2011) considered the case of Nigerian university students. He found that tribal affiliation and learning style did affect how these students responded to e-learning, but noted that instructional design that took ethnicity and cultural diversity into account and provided tailored instructional support could help to overcome these obstacles. The author's recommendations for instructional designers in a multi-cultural arena included:

- Consider cultural dimensions and learning styles.
- Recognize that in a global economy, there are variations in political and economic as well as sociocultural settings that impact the learner.
- Take into account differences in culture and group when selecting the quality and quantity of instructional materials.

In their study of how information literacy programs were developed and exported from one culture to another, Dorner and Gorman (2006) made similar observations of the role that culture plays in learning. To ignore culturally and socially determined differences, differences that reflect how people feel and communicate as well as learn is to forget how cultures differentiate and distinguish groups. While their focus was on students in East Asian countries, they, as did Adeoye, relied on Hofstede's (1997) cultural dimensions to understand how students approached learning. In the end, to be successful, instructional designers must ask how cultural dimensions affect how students learn. In particular, developing new curricula involves:

- Defining the subject in terms of the country, region, and culture
- Determining if students would learn better in groups or as individuals
- Whether students 'learn by doing' or by specific example
- How to encourage students to communicate what they have learned

In his discussion of cultural factors that influence instruction, Assie-Lumumba (2012) focused on the role that teachers have historically played in developing countries and the position they now hold rather than on instructional design and the use of educational technologies. Nonetheless, his consideration of cultural factors as determinants of the teaching profession reinforces the argument that cultural awareness is critical in any educational setting. Understanding the importance of a well-educated work force in a global economy, the author was somewhat pessimistic about the influence that teachers can exert to transform education systems and policies.

In a study of e-learning technologies as they were being used at the Open University of Tanzania, Nihuka and Voogt (2011) found that despite limited access, students and instructors alike had positive perceptions of the usefulness of technology for distance learning and were sufficiently competent in computer and Internet skills to take advantage of the educational technology available to them.

In their research on the role that distance education has played in overcoming social and cultural barriers to expand access to education, Wankel and Blessinger (2013) described the current generation of distance learning as drawing together technology and pedagogy. The authors refer to this iteration, the fifth generation, as the intelligent flexible learning model in which technology has been integrated and enhanced by Web 2.0 and Internet capabilities. In this framework, learners are knowledgeable, self-assured, and able to access information to enhance learning. They live in a "world of digital connections" (p. 118), familiar with social media. Some suggest that learners in this generation care about social presence and developing social capital. Teachers are viewed as enablers and "fellow voyagers."

Continuing Research

Research continues on the impact that technology has on education, particularly as to how the integration of emerging technologies will transform online instruction and enhance student learning. In their review of the most recent studies on effective online instructional methods for students, Tsai, Shen, and Chiang (2013) concluded not surprisingly, that studies of the integration of technology into instructional design increased significantly between 2003 and 2012. That trend can be expected to continue.

Problem Statement

Does the use of social media in course design enhance e-learning? Krentler and Willis-Flurry (2005) observed that while the use of technology, including social media, in curriculum design is widespread, there has been little empirical research to support the contention that students' use of technology actually does enhance learning.

To add to the research on whether the use of social media in course design really does enhance student learning, the authors conducted two surveys of students, both graduate and undergraduate, at two institutions, the University of Lagos in Akoka, Nigeria, and Robert Morris University in Pennsylvania (RMU), United States, to determine whether and how students were using social media as aids to learning in the online environment.

Study Purpose

One purpose of the study was to investigate the use of social media tools by students enrolled at two culturally distinct universities to determine whether the use of interactive, collaborative tools such as Facebook, YouTube, Twitter, Wikipedia, LinkedIn, blogs, and wikis were perceived by the students themselves as aids to student learning.

Research Questions

RQ. 1. Are students using social media to help them learn course material?

RQ. 2. Have students found the use of social media to be helpful in learning course material?

Methodology

Researchers used QuestionPro, an online survey software package, to develop the surveys. Results from the initial survey conducted in summer, 2013 raised a question about the relationship between students' use of social media outside of the formal learning environment and their use of the similar tools to master course material. The initial survey was modified to measure the relationship in the second survey.

The courses that formed the basis for the surveys were a mix of fully online, hybrid (blended), and traditional classroom courses that had an online component. Students from five undergraduate and six graduate courses were asked to participate via e-mail solicitations from the instructors. Participation was voluntary and responses were anonymous. Survey results were transferred from QuestionPro to SPSS for analysis of selected responses. Results are reported in the aggregate.

Sample

The combined sample was composed of 258 students, 119 from the University of Lagos and 139 from RMU. Graduate and undergraduate students at both institutions were included in the sample.

Of the 382 students who began the surveys in 2013, 258 completed them, for a completion rate of 67.5%. One hundred and nineteen were identified as responding to the University of Lagos solicitation, or 46% of the sample. One hundred and thirty-nine participants, 54% of the sample, were studying at Robert Morris University.

Procedure

In summer, 2013, researchers administered a survey to graduate and undergraduate students at the University of Lagos (Unilag) and Robert Morris University (RMU) to establish the baseline and to determine if additional survey questions should be included in the fall survey. Six questions were added to the second survey; two other questions were modified. With both surveys, the purpose was to determine the frequency with which students were using certain digital technologies, particularly social media tools, to aid their learning.

With regard to the questions on effectiveness of specific technologies for learning, a question on how instructors used technology was included. To determine if students' use of social media in daily activities might encourage their use of the same tools to aid learning, researchers asked how comfortable students were with using technology. In both surveys, researchers asked how soon after release students would adopt the new technology.

Research Instrument

The research instrument was a twenty-four question survey composed of Likert-style questions measuring students' familiarity with and use of technology, open-ended questions to allow participants to expand on their responses, and single choice questions for demographic and background information.

Results

RQ. 1: In the first survey, researchers found a statistically significant difference at the .01 level (.002, equal variances assumed) for students' use of wikis in coursework. The mean score was 2.93, with 56 students responding. The mean score for students at the University of Lagos was 3.08. For students at Robert Morris University, the mean score was 2.83.

In the second survey, researchers found statistically significant differences at the .01 level for students' use of social media to help learn course material based on university attended: Facebook (.000, equal variances not assumed), Twitter (.001, equal variances not assumed), for LinkedIn (.000, equal variances not assumed), Wikipedia (.000, equal variances assumed), Blogs (.000, equal variances not assumed), and Wikis (.000, equal variances not assumed). In all instances, students at the University of Lagos reported greater use of social media for learning.

RQ. 2: With regard to how useful students found selected social media tools or learning course material, in the first survey, researchers found statistically significant differences in the mean scores at the .01 level, equal variances assumed with regard to the ratings for Facebook (.000), Twitter (.000), LinkedIn (.017), and Blogs (.019). Students from the University of Lagos reported finding these social media tools to be more helpful to learning than did students at RMU.

In the second survey, there was a statistically significant difference at the .01 level for Facebook (.000, equal variances not assumed), and at the .05 level for Wikipedia (.013, equal variances not assumed). In both cases, students from Unilag reported greater usefulness for learning using Facebook and Wikipedia than did their counterparts at RMU.

Of particular interest were the responses on the final question asking when the respondent would most likely to explore or begin to use new technology. In both surveys, responses from students at the University of Lagos indicated that they would be more likely than those at RMU to explore or use new technology as soon as it became available. Table 1 presents those results.

Table 1. Comparison of study samples on adoption of new technology

Sample	University	n	M	T	Sig. (2-tailed
Pilot	RMU	42	2.60	3.079	.003
	U. Lagos	20	1.60		
Fall	RMU	94	2.40	4.592	.000
	U. Lagos	72	1.61		

Discussion

The study examined the use of social media tools by students and instructors at two universities comparing their use and perceptions of effectiveness as aids for learning course material by school, gender and age (Cole, Shelley, Swartz, and Adeoye, 2014a, 2014b). Results reported here focus on students' use of social media to facilitate learning. Responses were compared by university. With regard to this focus, the authors found that there were significant differences between the two student populations both in their choice of social media and in their assessment of the value of social media to learning. In both surveys, the reported use of social media to enhance learning was greater at the University of Lagos. In the second survey, there were significant differences between the two universities in their students' use of social media. For example, students in the University of Lagos sample reported greater use of selected social media tools (Facebook, Wikipedia, wikis, blogs, Twitter, and LinkedIn) for learning course material than did students in the RMU sample. Of these, Facebook and Wikipedia were rated significantly higher as helpful to learning by respondents from the University of Lagos than by those from RMU. In both surveys, students from the University of Lagos were more likely than their counterparts from RMU to adopt new technology "as soon as it becomes available."

In the second survey, students were asked about the use of social media in daily activities and about their comfort level with using technology. While both survey questions elicited positive responses, there was not enough data to establish a correlation between students' use of social media in daily activities, their comfort level with using technology and enhanced learning. Respondents were both comfortable with and positive about the use and application of technology to learn course material. There were differences of degree as to which technology tools were more favored by respondents at the University of Lagos and those at RMU. There was insufficient information to determine if those differences were based on or related to cultural differences.

Saeed, Yang, and Sinnappan (2009) made the connection between preferences in technology, and their impact on student learning to students' learning styles. In this study, researchers did investigate students' comfort level with technology, their technology preferences in learning course material, and their assessment of the effectiveness of using those technologies; however, the results were inconclusive with regard to establishing a causal relationship.

What might account for the differences in the adoption and integration of educational technology tools in the online learning environment found in these two surveys? What role might cultural factors play in the integration of educational technology into the curriculum? At a conference presentation of the study results in 2014, some members of the audience suggested that the differences in responses between the two student groups could be attributed to an urgency to advance felt by those in many non-Western countries that was not shared by students in the United States. One participant indicated that in his opinion, the difference in early adoption of new technologies could be traced to American students' "complacency."

Finding a correlation integrating educational technology into course design and enhanced student engagement and by extension, enhanced student learning, Hedberg (2011) argued that course design needed to focus on the technology-enabled student to achieve desired learning outcomes because today's students are increasingly "digital natives."

In their study of Internet connectivity and usage among university students in Southwestern Nigeria, Adeoye, Adebo, and Olakulehin (2011) found that despite a confidence level of 78% for Internet usage, student success was not assured, in part due to challenges presented by limited access to reliable Internet services. As a result, any discussion of how effective the integration of educational technology tools for enhancing e-learning must factor in environment as well as culture. Adeoye, Oluwole, and Blessing (2013) argued that because ICT is an indispensable part of today's world, "culture and society have to be adjusted to meet the challenges of the knowledge age" (p. 177). While recognizing that culture and environment are part of the equation in higher education, these authors appear to be shifting the emphasis from instructional design accommodating culture to culture needing to adapt to educational and work force demands.

Enhancing E-learning: Can Social Media Play a Role?

By studying two distinct student populations and asking how each viewed the use of selected educational technologies for learning and how each used the same technologies in daily activities, researchers hoped to begin an investigation into how best to integrate emerging technologies into e-learning platforms, taking into account differences in culture as well as the challenges presented by limited access to resources.

The authors' study looked at how students were using social media in course work and asked how effective their use was for learning.

They asked how students perceived the usefulness of selected educational technology. Future research would need to connect student perceptions with the more objective quantitative measures, including course grading. As noted earlier, there were significant differences in many instances between the two student groups' use of social media. One factor not accounted for in the studies was the difference in the structure of the two educational environments. Another was the mode of instruction—online, hybrid (blended), or on ground. One U.S. instructor used YouTube extensively to enhance course material. In the Nigerian example, the instructor included Facebook among his tools for course delivery. Future research would need to control for the variances. However, to determine if social media actually did enhance student learning, researchers would need to look at how students' learning (assessment of learning/course grades) changed with the integration of educational technology, that is, before social media was employed.

Not to be overlooked in the equation is the importance of instructors' willingness and ability to integrate technology into their instructional designs. A recent survey of faculty across higher education institutions in the United States (FTI, 2015) found that while technological innovations affecting access, instructional design, and course delivery have provided a foundation for change and improvement in American higher education, there remain obstacles to realization. Of particular note is the role that instructors play in creating an overall strategy for institutional change. Key to implementation is faculty buy-in as "potential creators and drivers of innovation, and as the direct, front-line drivers of student success" (preface).

One study indicated that while instructors know about new technology and high-tech instructional methods, few report using them (Fabris, 2015). The 2015 FTI study found that instructors were most likely to incorporate technology and adopt innovative teaching tools if they could see the relationship between adoption and student success. From those surveyed, 40% responded that they were interested in using different teaching techniques to enhance student learning. Of the 40%, half responded that they were using the educational tools listed. Among the strategies for increasing instructors' use of new technologies, the researchers cited "providing evidence base for student outcomes" (p. 5).

Another consideration is the role that education, particularly higher education can play in narrowing the divide between developing nations and the West. In 2012, Lwoga argued that it was higher education that would be critical in bridging the gap between students' skill with social media in daily activities and the incorporation of those skills in learning course material. Again, the integration of educational technology into educational programs is seen as an important step in this regard. However, as Adeoye, Adebo, and Olakulehin (2011) point out, until the disparities in access to the Internet can be resolved, its promise for transformational learning will remain unfulfilled.

As has been suggested in the majority of studies to date on the integration of educational technology tools online and the critical role that culture plays in the success of that integration, additional research is necessary to develop a comprehensive strategy to effect enhanced student learning with technology.

Study Limitations

Survey participants were drawn from students studying at one of two universities, one in Western Pennsylvania and one in Nigeria. Graduate and undergraduate courses were selected based on researchers' ease of access. The courses were taught by different instructors using different technologies.

Conclusion

"Social media" is more than a grab-bag of Web sites and social networking applications. Social media tools open new ways for global communication and arguably, community building. Its potential for effecting change—both positive and negative—is enormous. With regard to enhancing e-learning, given the wide acceptance and ease of use of social media, it is uniquely positioned to be the technology of the moment as well as of the near future, laying the groundwork for more sophisticated instructional models that integrate technology into online curricula in support of student learning. As early as 2001, Prensky was arguing that due to ever-evolving technology that surrounds them constantly, contemporary students think and process information fundamentally differently than their predecessors did. To enhance learning for these "digital natives," educators must "reconsider both our methodology and our content" (p. 3). This chapter, among others is evidence that in addition to content, educators are in fact considering methodology as they incorporate new technologies into curriculum development and design.

References

Adeoye, B. F. (2011). Culturally different learning styles in online learning environments: A case of Nigerian university students. *International Journal of Information and Communication Technology Education 7*(2), 1–12.

Adeoye, B. F., Adebo, G. M., and Olakulehin, F. K. (2011). Poverty and phobia of internet connectivity and usage among university students in southwestern Nigeria. In G. Kurubacak and T. V. Yuzer (Eds.), *Handbook of Research On Transformative Online Education and Liberation: Models for Social Equality* (pp. 461–471). Hershey, PA: IGI Global. doi:10.4018/978-1-60960-046-4.ch026.

Adeoye, B. F., and Oni, A. (2010). Socio-cultural dimensions of e-learning systems. In T.V.
Yuzer and G. Kurubacak (Eds.), *Transformative Learning and Online Education: Aesthetics, Dimensions and Concepts* (pp. 154–165). Hershey, PA: IGI Global. doi: 10.4018/976-1-61520-985-9.ch011.

Adeoye, B. F., and Wentling, R. M. (2007). The relationship between national culture and the usability of an e-learning system. *International Journal on E-Learning 6*(1), 119–146. Chesapeake, VA: Association for the Advancement of Computing in Education (AACE).

Adeoye, Y. M., Oluwole, A. F., and Blessing, L. A. (2013). Appraising the role of information communication technology (ICT) as a change agent for higher education in Nigeria. *International Journal of Educational Administration and Policy Studies 5*(8), 177–183.

Allen, I. E., and Seamen, J. (2014). *Grade change: Tracking online education in the United States.* Babson Survey Research Group and Quahog Research Group.

Assie-Lumumba, N. T. (2012). Cultural foundations of the idea and practice of the teaching profession in Africa: Indigenous roots, colonial intrusion, and post-colonial reality. *Educational Philosophy and Theory 44*(S2). doi: 10.1111/j.1469-5812.2011.00793.x.

Burke, L. (2013). Educational and online technologies and the way we learn. *International Schools Journal 32*(2), 57–65.

Buzzard, C., Crittenden, V. l., Crittenden, W. F., and McCarty, P. (2011). The use of digital technologies in the classroom: A teaching and learning perspective. *Journal of Marketing Education 33*(2), 131–139. doi: 10.1177/0273475311410845. Retrieved from http://jmed.sagepub.com.

Cole, M. T., Shelley, D. J., Swartz, L. B., and Adeoye, B. (2014a). Using digital technologies to aid e-learning: A pilot study. In B. Adeoye and L. Tomei (Eds.), *Effects of Information Capitalism and Globalization on Teaching and Learning* (pp. 22–35). Hershey: PA: IGI Global.

Cole, M., Shelley, D., Swartz, L., and Adeoye, B. (2014b). Does student and instructor use of social media facilitate online learning: A look at two universities. In R. Neves-Silva, G. Tshirintzis, V. Uskov, R. Howlett, and L. Jain (Eds.), *Frontiers in Artificial Intelligence and Applications: Vol. 262. Smart Digital Futures 2014* (pp. 553–564).

Crampton, A., Ragusa, A. T., and Cavanagh, H. (2012). Cross-discipline investigation of the relationship between academic performance and online resource access by distance education students. *Research in Learning Technology 2012, 20*:14430. doi: 10.3402/rlt.v20i0/14430.

Czerkawski, B. C. (2013). Strategies for integrating emerging technologies: A case study of an online educational technology master's program. *Contemporary Educational Technology 4*(4), 309–321.

Dorner, D. G., and Gorman, G. E. (2006). Information literacy education in Asian developing countries: Cultural factors affecting curriculum development and programme delivery. *IFLA Journal 32*(4), 281–293. doi: 10.1177/0340035206074063.

Fabris, C. (2015). Professors know about high-tech teaching methods, but few use them.
Retrieved from http://chronicle.com/blogs/wiredcampus/professors-know-about-high-tech-teaching-methods-but-few-use-them/55777?cid=wc&utm_source=wc&utm_medium=en

Frantzen, D. (2013). Is technology a one-size-fits-all solution to improving student performance?: A comparison of online, hybrid and face-to-face courses. *Journal of Public Affairs Education 20*(4), 565–578.

FTI Consulting (2015). U.S. postsecondary faculty in 2015: Diversity in people, goals and methods, but focused on students. Retrieved from http://postsecondary.gatesfoundation.org/wp-content/uploads/2015/02/US-Postsecondary-Faculty-in-2015.pdf.

Grosseck, G. (2009). To use or not to use web 2.0 in higher education? *Procedia Social and Behavioral Sciences 1*(2009), 478–482. doi: 10.1016/j.sbspro.2009.01.087.

Hamerling, J. A. (2012). Best practices in undergraduate clinical laboratory science online education and effective use of educational technology tools. *Lab Medicine 43*(6), 313- 319. Retrieved from www.labmedicine.com.

Hedberg, J. G. (2011). Towards a disruptive pedagogy: Changing classroom practice with technologies and digital content. *Educational Media International 48*(1), 1–16.

Hofstede, G. (1997). *Cultures and organizations: Software of the mind.* New York, NY: McGraw-Hill.

Kaplan, A. M., and Haenlein, M. (2010). Users of the world, Unite! The challenges and opportunities of social media. *Business Horizons 53*, 59–68.

Krentler, K. A., and Willis-Flurry, L. A. (2005). Does technology enhance actual student learning? The case of online discussion boards. *Journal of Education for Business, 80*(6), 316–321. doi:10.3200/JOEB.80.6316–321.

Lowe, B., D'Alessandro, S., Winzar, H., Laffey, D., and Collier, W. (2013). The use of Web 2.0 technologies in marketing classes: Key drivers of student acceptance. *Journal of Consumer Behavior 12*, 412–422. doi: 10.1002/cb.1444.

Lwoga, E. (2012). Making learning and Web 2.0 technologies work for higher learning institutions in Africa. *Campus-Wide Information Systems 29*(2), 90–107.

Majid, N. (2014). Integration of Web 2.0 tools in learning a programming course. *The Turkish Online Journal of Educational Technology 13*(4), 88–94.

McCabe, D. B., and Meuter, M. L. (2011). A student view of technology in the classroom: Does it enhance the seven principles of good practice in undergraduate education? *Journal of Marketing Education 33*(2), 149–159. doi: 10.1177/0273475311410847. Retrieved from http://jmed.sagepub.com.

Nielsen, J. (1993). *Usability engineering.* Boston, MA: Academic Press ISBN0–12–518405–0.

Nihuka, K. A., and Voogt, J. (2011). Instructors and students' competencies, perceptions, and access to e-learning technologies: Implications for e-learning implementation at the Open University of Tanzania. *International Journal on E-Learning 10*(1), 63–85. Chesapeake, VA: Association for the Advancement of Computing in Education (AACE). Retrieved from http://www.editlib.org/j/IJEL/v/10/n/1.

Nijhuis, C., Pieters, J., and Voogt, J. (2013). Influence of culture on curriculum development in Ghana: An undervalued factor? *Journal of Curriculum Studies 45*(2), 225–250. doi: 10.1080/00220272.2012.737861.

Picciano, A. G., Seamen, J., and Allen, I. E. (2010). Educational transformation through online learning: To be or not to be. *Journal of Asynchronous Learning Networks 14*(4), 17-35.

Prensky, M. (2001, October). Digital natives, digital immigrants. *On the Horizon 9*(5), 1–6.

Saeed, N., Yang, Y. and Sinnappan, S. (2009). Emerging web technologies in higher education: A case of incorporating blogs, podcasts and social bookmarks in a web programming course based on students' learning styles and technology preferences. *Educational Technology and Society 12*(4), 98–109.

Siemens, G., and Conole, G. (2011). Special Issue—Connectivism: Design and delivery of social networked learning. *The International Review of Open and Distance Learning. 12*(3), i-iv.

Thiele, A. K., Mai, J. A., and Post, S. (2014). The student-centered classroom of the 21st century: Integrating Web 2.0 applications and other technology to actively engage students. *Journal of Physical Therapy Education 28*(1), 80–93.

Tsai, C., Shen, P., and Chiang, Y. (2013). Research trends in meaningful learning research on e-learning and online education environments: A review of studies published in SSCI-indexed journals from 2003 to 2012. *British Journal of Educational Technology 44*(6). doi: 10.1111/bjet.12035.

Tung, L. C. (2013). Improving students' educational experience by harnessing digital technology: ELGG in the ODL environment. *Contemporary Educational Technology 4*(4), 236–248.

Tunks, K. W. (2012). An introduction and guide to enhancing online instruction with Web 2.0 tools. *Journal of Educators Online 9*(2), 67.

Tuten, T., and Marks, M. (2012). The adoption of social media as educational technology among marketing directors. *Marketing Education Review 22*(3), 201–214. doi: 10.2753/MER1052–800822030.

Varela, D., and Westman, L. (2013). Active learning and the use of technology, or how one online popular culture course changed how we teach everything else. *Interdisciplinary Humanities*, 42–53.

Veletsianos, G. (2010). A definition of emerging technologies for education. In G. Veletsianos (Ed.), *Emerging Technologies in Distance Education* (pp. 3–22). Athabasca, AB: Athabasca University press. Retrieved from http://www.aupress.ca/books/120177/ebook/01 Veletsianos 2010#sthash.nxVP3yOs. dpuf.

Wankel, C., and Blessinger, P. (2013). Increasing student engagement and retention in e-learning environments: Web 2.0 and blended learning technologies. *Cutting Edge Technologies in Higher Education.* Bingley, England: Emerald.

Chapter 15

A Survey on Augmented Reality-Based Education

Javid Iqbal
and
Manjit Singh Sidhu

Introduction

Augmented reality can be defined as a technology where real world objects can be seen superimposed by virtual machine generated entities in order to provide interaction between real world and virtual world objects. AR has been employed in numerous areas including construction, automobile, aerospace, military, medicine and education to bridge the gap between visualization and information retrieval. In recent years, development of human-computer interaction has improved significantly. AR acts a connecting bridge by eventually combining the real world and virtually generated objects together and allows the user to view it seamlessly in the same environment. This enhances the usability and information retrieval which human beings cannot perceive with their regular senses.

Researcher such as Azuma (1997) tended to present a definition of AR based on a system that fulfills three basic criteria: (1) combination of real and virtual, (2) interactive in real time and (3) 3D registration of virtual and real objects, where registration refers to the accurate alignment of real and virtual objects. A similar definition is proposed by other researchers (Hollerer and Feiner, 2004; Kaufmann, 2003; Zhou, Duh, and Billinghurst, 2008), who define AR based on its features which the real and computer-generated information are combined in a physical world, interactively in real time, and display virtual object intrinsically align to real world orientation. AR provides effective means of enabling human beings to enhance their information retrieval capabilities that cannot be perceived with regular sense. Eventually, AR acts a bridging link between the real world entities and virtual objects to be viewed on the same platform seamlessly.

Dance Learning Technology

Mastering new gestures and postures is an important aspect of many physical activities such as dancing, martial arts, exercise, aerobics and sports. Learning and expertising in these skills can be challenging, time consuming and needs repetitive practice. People who could hardly find time to learn dance or do any physical activity find it difficult to attend classes or training sessions.

Though there could be no replacement for expert coaching, self-learning is found to be more desirable for in house practices. Information technology has enabled physical training to be practiced at house in order to supplement professional training. The video Web sites such as YouTube provides enormous training videos that are informative for aspects like exercise, surgery, academics and a lot more to mention. But these videos when utilized as a training medium do not provide feedback or guidance that can serve to be as a personalized motivation for the users.

Dancing to music is the most enjoyed artistic skill of human beings. The research on e-learning had taken place from 1993 when online system delivered tutorial used many software's modules to create a virtual classroom environment for students and teachers, various learning activities may be adopted depending upon the content of learning. The evolution of dance video games, fitness training with gaming console embedded with AR technology, visualization methods and sensing capabilities have made physical movement much more fun learning. In the past decade, there have been many literature work focusing on the analysis of facial recognition and motion synthesis, but still faces many challenging issues in the field of computer vision and learning technology.

AR Applications

The emergence of AR technology had been gaining the attention of researchers and educators since it serves to be an interesting and exciting alternative way for creating a teaching medium. AR provides an interactive interface for the users to work and interact with the real world and virtual world entities simultaneously. AR is considered as a medium that combines the aspects of ubiquitous computing, tangible computing and social computing. AR applications in education field have been employed in many learning scenarios which include chemistry, biology, astronomy, mathematics, automotive engineering and dance. Utilizing dance for education purpose improves creative thinking among students, increases their information perception skills in concrete subject domain and understanding of spatial spaces.

This advantage of AR in educational sector has enabled the significant transformation of traditional classroom training towards computer vision based e- learning strategies. This chapter reviews the AR based dance learning standards and the transformation of learning methods in educational sector.

Motivation

Education is defined to be a mode for life-long enlighten experience. As long as self- learning is concerned, medium such as books and Internet serves to be as most important pillars. On the other hand, training an individual often depends on the needs of the learner level of understanding and grasping capabilities. The objective of this chapter is to review the AR-based education and learning methods that had been utilized in educating an individual. Since AR involves visual perception of real world entities with interactive virtual information; a learner tends to find this method of learning more interactive than traditional classroom teaching. Education based on AR technology involves gaming consoles which drives the learner to gain knowledge with inspiration and enthusiasm. This also provides the learner with self-confidence, self-motivation and ways for constant improvement through constructive feedback. The authors in this chapter aim to provide insight into AR-based educational perspective and the influence of computer vision on learning technologies.

From traditional learning towards Computer vision

The process of educating an individual has emerged from classroom teaching, motor learning, e-leaning towards computer vision from past few decades as depicted in figure 1. During the 1990s the advancements of technology gave way to computer based training. The following decades eventually have witnessed the proliferation of information sharing, embedded systems, artificial intelligence, neural networks and cloud computing.

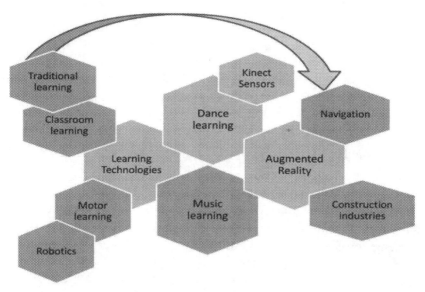

Fig1. The Learning paradigm and significant transformation
|(Refer Appendix 1)

The theory of educational technology defines three major philosophical frameworks namely behaviourism, cognitivism, and constructivism. The learning methods that included traditional learning and classroom teaching followed the behaviourism framework and then Cognitivism took its own pace of development for problem solving methods with computer based training. The AR based learning can be related to Constructivism framework of educational technology as it involves the learner's own information seeking techniques. Albors, and Carrasco (2011) who was one of renowned educational researcher suggested that learning can be integrated into different phases namely Acquisition, apprehension, generalization, motivation and retaining, in order to assure learning efficiency.

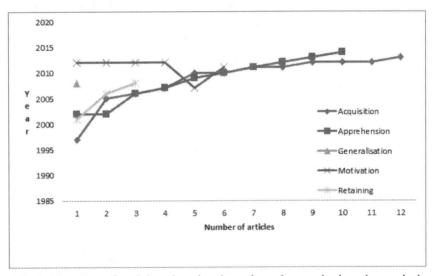

Fig.2. Number of articles showing learning phases during the period 1990–2014

Fig 2 shows the number of articles grouped under different phases of learning technology during the year 1990–2014. Fig 3 shows percentage of articles according to learning stages. Majority of the articles which are around 35% consists of Knowledge acquisition as the learning element. Acquisition learning stage includes learners encoding storage phase which eventually gives way for stimulation and guidance process as suggested by Albors and Carrasco (2011). This review also emphases that 29% of the articles are based on Apprehension learning phase as depicted in Fig 2. Apprehension stage gains learner's attention and selective perceptions

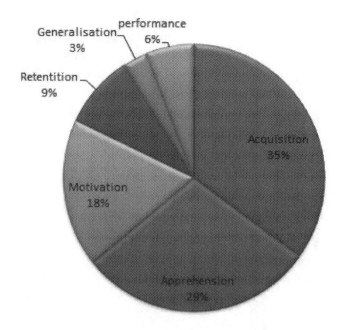

Fig. 3: Percentage of articles according to learning stages (Refer Appendix 2)

The above result indicates that last few decades have focused upon knowledge acquisition and apprehension as a primary source of delivering education. Computer vision can be defined as a process of acquiring, analyzing, processing and understanding images in order to provide useful information. The above classification of articles was done based upon the fact that the authors in the past decades have emphasized acquisition and apprehension as effective means of educating an individual. Both these learning stages are the key elements in today's computer vision approach. The results have clearly indicated that there is a major transformation from traditional learning towards computer vision in educational sector during the last two decades.

AR in navigation and construction industry

Augmented reality is employed in both indoor and outdoor spaces for navigation purposes. Some of the applications are Across Air, Google Googles, Google Sky map, In Road Augmented Driving, AR invader and

AR compass. These applications enable the user to make use of location information, seek traffic updates, routes and gain insights into new places of interests. Wang (2014) had presented an AR navigation system with automatic marker free image registration using 3D image over relay and stereo tracking for dental surgery. AR is also employed in architecture, construction, engineering and facility management (ACE/FM). Rankohi (2013) provided an expanded foundation for future research by presenting a statistical review of AR technology in the ACE industry.

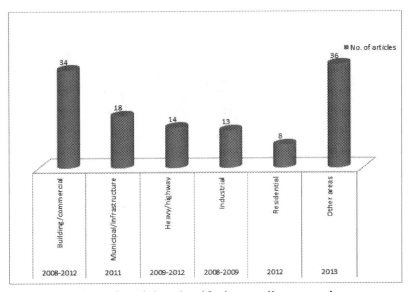

Fig. 4: Number of Articles classified according to various sectors

Fig. 4 shows that 34 articles have employed AR in building / commercial sector during the year 2008–2012. Municipal/ Infrastructure, highway, industrial and residential construction sectors have 18, 14, 13 and 8 articles from the year 2008 to 2012 respectively. Other areas include visualization, education, archeology, social development, games, physical training, military, cultural heritage communication, and elderly people training and shopping.

Augmented Reality in Educational Sector

Augmented reality as defined earlier is a technology that makes use of computer vision techniques to collaborate computer generated virtual objects with real time environment in order to increase or to enhance what can be visualized by the human user (Carmigniani, 2011). In the educational sector, AR can be used in dance education, chemistry, biology, astronomy, mathematics, computer graphics, etc. AR in chemistry education is used for the exploration of physical models of amino acids (Fjeld, 2002). In biology AR based learning system is used to get insights into the interior of human organs on a detailed basis with easier understanding (Gillet, 2004). In astronomy AR technology is applied for better understanding of seasonal changes in light and temperature, rotation and revolution of the sun and earth (Brett, 2002). AR is also used as a visualization tool in computer graphics laboratories and in computer aided design lectures (Kaufmann, 2008 and Chen, 2006). In mathematics AR is used for teaching calculus and algorithms (Dunleavy, 2009). In the field of dance education Kinect sensors are employed to train the students in physical movements as well as master them in their skills.

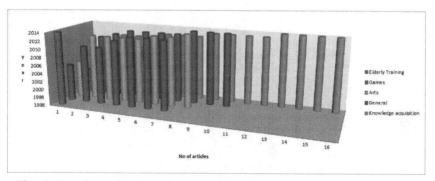

Fig. 5: Number of articles according to learning applications during 1990–2014 (See Appendix 3)

Figure 5 illustrates the number of articles for different types of learning applications in various aspects in the past two decades. Arindam (2014) presented insights acquired from experiments performed from hand held devices in outdoor locations for visualization of occluded objects. Jungong (2013) presented a compressive review of recent Kinect-based computer vision algorithms and applications. Cadavieco and Thiengtham (2012) had presented an analysis of educational experiences related to the use of mobile device in the class room and development of template matching based on AR toolkit for Thai alphabet learning on mobile AR. Such contributions where AR has made its major impact in gaining knowledge and educating an individual has been grouped under the learning application of knowledge acquisition in this review. AR technology has also been employed in entertainment and elderly training for educational purposes. There are seventeen articles (40%) which focus upon knowledge acquisition as its primary aim. There are eight articles (19%) that deal with improving the artistic skills of a learner based on AR technology. There is only one article (2%) for elderly training which emphasizes on exercise biking with virtual environments. There are 7 articles (16%) that provide assistive learning environments based on ubiquitous computing. There are 10 articles (23%) that deal with general AR based learning aspects and outcomes as depicted in (Fig 6).

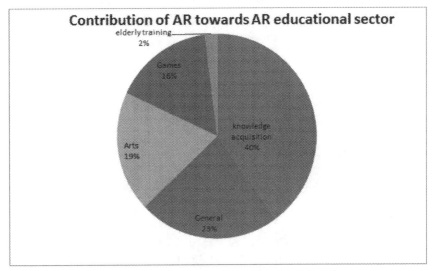

Fig.6.Contribution of AR Towards educational sector
(Refer Appendix 3)

Fun-filled AR based dance learning Using Kinect

Computer vision and motion sensing technologies have enabled the users to actively, physically and mechanically interact with the digital environment in varied ways. The hybrid combination of traditional art forms and advanced computer vision techniques have made the authors in the last decade to drive out to AR based dance leaning systems. Isabelle (2014) describes Cha learn gesture data set that is user dependent, small vocabulary and one-shot learning using Kinect camera. Silva (2013) developed a prototype to monitor fall risk while playing a game using smartphone accelerometer. Rukun (2012) introduced a novel method for synthesizing dance motion that follows the emotions and the contents of a piece of music. Yang (2012) presented an automatic dance lesson generation system which is suitable in a learning-by-mimicking scenario where the learning objects can be represented as multi-attribute time series data. Jacky (2011) proposed a new dance training system based on motion capture and virtual reality technologies. Forouzan (2004) presented multimedia information repository for cross cultural dance studies like East Indian dance with the use of two 3D Vicon motion capture systems. Marcin (2012) proposed techniques for novel Human Pose Coestimation for joint pose estimation over multiple persons in a common, but unknown pose. Kuramoto (2013) had proposed a visualization method of velocity and acceleration of teacher's motion for the learner to understand more clearly and easily. Anderson (2013) discussed a novel system YouMove that allows users to record and learns physical movement sequences. The Kinect-based recording system is designed to be simple, allowing anyone to create and share training content. The corresponding training system uses recorded data to train the user using a large-scale AR mirror. The system trains the user through a series of stages that gradually reduce the user's reliance on guidance and feedback. This also discusses the design and implementation of YouMove and its interactive mirror. The authors have presented a user study in which YouMove was shown to improve learning and short-term retention by a factor of 2 compared to a traditional video demonstration. While the presented implementation uses a half-silvered mirror as a display, the software could also run as a traditional video-based AR system. The Kinect has difficulty tracking movements that cause large amounts of occlusions. This would be more accessible to users, but does not provide the real-time feedback that the mirror does. It would be interesting to better understand any learning difference between a mirror and video based system on various devices (large screen, small screen, etc.). The addition of social features and richer inclusions of gaming technologies could also greatly help YouMove. One can imagine online yoga, dance or martial arts classes, with competition from online peer groups, but more work is needed to achieve this.

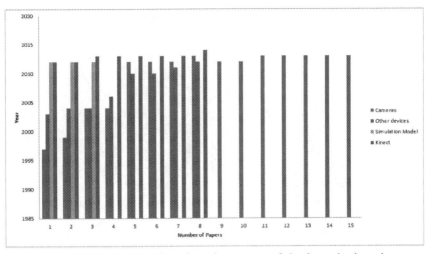

Fig.7. Number of articles showing the usage of devices during the year 1990–2014 in AR based dance learning. (Refer Appendix 4)

Figure 7 presents the number of articles that have used various devices for AR based dance learning in the past few decades. There are 8 articles that have used cameras for AR based training. There are 3, 15, 8 articles that have used simulation models, other devices and Kinect cameras respectively for AR based dance learning systems.

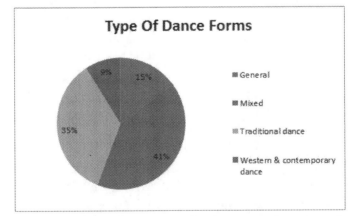

Fig.8. Percentage of articles contributing towards various dance forms. (Refer Appendix 5)

Figure 8 illustrates the percentage of articles in this review that have contributed towards learning of different dance forms. There are 41% of articles that deal with studies related to mixed physical movements. Traditional and contemporary dance forms have 35% and 9% respectively. The rest of the 15% of the articles focus on general aspects like motion recognition, Dance notation system and any gesture detection.

Benefits of AR in Educational Sector

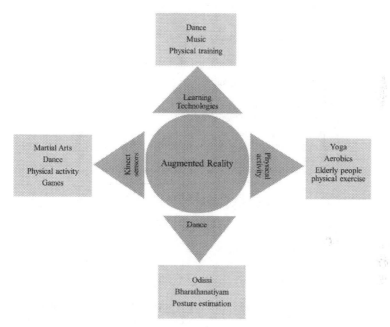

Fig.9. Cyclic taxonomy of this review

Figure 9 represents cyclic taxonomy view of this review where AR based learning technologies were reviewed for dance education that have employed Kinect sensors. There are numerous advantages of AR in educational sector some of which are interactivity, easier understandability and attractive way for learning and teaching according to past studies. Other benefits include stimulation of conceptual thinking, constructivism, receiving greater sense of 3D space, Theoretical understanding, perception enhancement and fun based interaction as depicted in Figure 10. The features of AR based learning include safer, cheaper, user friendly, animated, timely, easily understandable, responsive, seamless integration with other media, interesting, reliable, efficient and portable.

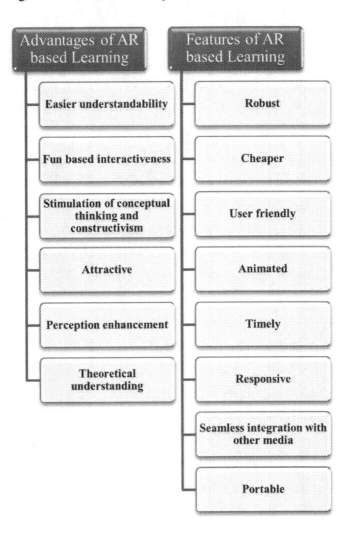

Fig.10. Advantages and Features of AR based Learning

Conclusion

Augmented reality has made distinctive contributions towards learning experiences. The developments in computer vision technology have led researchers to enhance and assess AR learning aspects. In the course of this review there are many interesting facts that can be derived cyclic taxonomy of which is illustrated in Figure 9. The evolution of learning paradigm from traditional methods towards computer vision, relative to the theory of education technology has showed a significant transformation and has eventually led researchers and technologists to adopt AR as one of the promising direction for multimedia learning technology. The insights for AR technology in navigation and construction industries provide an essential need for comprehensive systems, integration of multiple platforms, user friendly interfaces, defect detection and seamless integration at an affordable cost. The AR technology can be used for group as well as individual learning which motivates the learner in every possible way. The advantage of AR in educational sector includes opportunity to visualize digital information, observe the finer details of subjects and possibilities to examine the virtual information perceptively as many times as needed. Hence AR has made its major contribution towards knowledge acquisition in educational sector during the last few decades. There has also been emergence of AR based dance learning technology with tracking capabilities such as Microsoft Kinect. This technology has been employed for Fun-filled dance learning with responsive interactions and self-motivating feedback. It can further be concluded that the traditional dances which have already faded away [as in Figure 8] with history, can be explored and the cultural heritage can be preserved with computer vision technology. The Kinect V2 sensor is seen as a future direction for AR based motion recognition and gesture detection, for which the work is underway.

References

Albors, G. J., and Carrasco, J. C. R. (2011). *New Learning Paradigms: Open Course versus Traditional Strategies. The Current Paradox of Learning and Developing Creative Ideas*, Springer book on Social Media Tools and Platforms in Learning Environments, 53–79.

Anderson, F., Grossman, T., Matejka, J., and Fitzmaurice, G. (2013). *YouMove: Enhancing Movement Training with an Augmented Reality Mirror,* 26th Annual ACM Symposium on User Interface Software and Technology, 311–320.

Arindam, D., and Sandor. C. (2014). *Lessons learned: Evaluating visualizations for occluded objects in handheld augmented reality,* Elsevier International Journal of Human-Computer Studies, 704–716.

Azuma, R. T. (1997). *A Survey of Augmented Reality,* Teleoperators and Virtual Environments, 355–385.

Billinghurst, M., Piumsomboon, T., and Bai, H. (2014). *Hands in Space- Gesture Interaction with Augmented- Reality Interfaces,* IEEE journal on Computer Graphics and Applications, 77–80.

Brett, E. S, and Nicholas, R. H. (2002). Using Augmented Reality for Teaching Earth-Sun Relationships to Undergraduate Geography Students, IEEE International Augmented Reality Toolkit Workshop.

Cadavieco, J. F., Maria, F. and Costales, A. F. (2012). *Using Augmented Reality and m-learning to optimize student's performance in Higher Education,* World Conference on Educational Sciences.

Carmigniani, J., Furht, B., Anisetti, M., Ceravolo, P., Damiani, E., and Ivkovic. M. (2011). *Augmented reality technologies, systems and applications,* Multimedia Tools and Applications, 341–377.

Chen, Y. C. (2006). *A study of comparing the use of augmented reality and physical models in chemistry education,* Proceedings of ACM international conference on Virtual reality continuum and its applications.

Dunleavy, M., Dede, C., and Mitchell, R. (2009). *Affordances and Limitations of Immersive Participatory Augmented Reality Simulations for Teaching and Learning,* Journal of science education and technology, 7–22.

Fjeld, M. and Benedikt, M. V. (2002). Augmented Chemistry: An Interactive Educational Workbench, IEEEInternational Symposium on Mixed and Augmented Reality.

Forouzan, G., Vissicaro, P., and Park, Y. (2004). *A Multimedia Information Repository for Cross Cultural Dance Studies,* Multimedia Tools and Applications, 89–103.

Gillet, A., Sanner, M., Stoffler, D., Goodsell, D. and Olson, A. (2004). *Augmented Reality with Tangible AutoFabricated Models for Molecular Biology Applications,* ACM International Conference on Visualization.

Gordon and Rowland. (2004*). Shall We Dance? A Design Epistemology for Organizational Learning and Performance,* Springer Journal on Educational Technology Research and Development, 33–48.

Isabelle, G., Athitsos, V., Jangyodsuk. P. and Escalante, H. J. (2014). *The ChaLearn Gesture dataset (CGD 2011),* Springer Journal of Machine Vision and Applications.

Jacky. C. P., Leung, H., Tang, J. K. T. and Komura, T. (2011). *A Virtual Reality Dance Training System Using Motion Capture Technology,* IEEE transactions on Learning Technologies.

Jungong, H., Shao, L., Xu, D., and Shotton, J. (2013*). Enhanced Computer Vision with Microsoft Kinect Sensor: A Review,* IEEE Transactions on Cybernetics.

Kaufmann, H., and Schmalstieg. D. (2002). *Mathematics and geometry education with collaborative Augmented Reality,* ACM SIGGRAPH conference.

Kuramoto, I., Nishimura, Y., Yamamoto, K., Shibuya, Y. and Tsujino, Y. (2013). *Visualizing Velocity and Acceleration on Augmented Practice Mirror Self-Learning Support System of Physical Motion,* International Conference on Advanced Applied Informatics.

Marcin, E., and Ferrari, V. (2012). *Human Pose Co-Estimation and Applications,* IEEE.

Moazami, F., Bahrampour, E., Azar, M. R., Jahedi. F and Moattari, M. (2014) *Comparing two methods of education (virtual versus traditional) on learning of Iranian dental students: a post-test only design study,* BMC Medical Education.

Rankohi, S., and Waugh, L. (2013*). Review and analysis of augmented reality literature for construction Industry,* Springer Journal on Visualization in Engineering.

Rukun, F., Xu, S., and Geng, W. (2012). *Example-Based Automatic Music-Driven Conventional Dance Motion Synthesis,* IEEE transactions On Visualization and Computer Graphics.

Silva, P. A., Nunes, F., Vasconcelos, A., Kerwin, M., Moutinho, R. and Teixeira, P. (2013).

Using the Smartphone Accelerometer to Monitor Fall Risk while Playing a Game: The Design and Usability Evaluation of Dance! Don't Fall, Foundations of Augmented Cognition, 754–763.

Stephen, F. (1997). *Situated Learning Theory versus Traditional Cognitive Learning Theory: Why Management Education Should Not Ignore Management Learning,* Springer journal on Systemic Practice and Action Research, 727–747.

Thiengtham, N. and Sriboonruang, Y. (2012). *Improve Template Matching Method in Mobile Augmented Reality for Thai Alphabet Learning,* International Journal of Smart Home. transactions on Pattern Analysis and Machine Intelligence.

Wang, J., Suenaga, H., Hoshi, K., Yang, L., Kobayashi, E., Sakuma, I., and Liao. H. (2014*).

Augmented Reality Navigation With Automatic Marker-Free Image Registration Using 3D Image Overlay for Dental Surgery,* IEEE Transactions On Biomedical Engineering.

Yang. Y, Leung, H., Yue, L., and Deng, L. (2012). *Automatic Dance Lesson Generation,* IEEE Transactions on Learning Technologies.

Appendix 1

Table 1. Applications of AR

No	Year	Author	Application	Types of learning paradigm
1.	2010	Bruin, E. D. et al.,	• Motor • Cognitive • Rehabilitation	Motor learning
2.	2011	Ivanova, M. et al.,	• Web-based applications • AR applications utilized in different learning scenarios	Learning and teaching computer graphics
3.	2013	Han, J. et al.,	• Video analysis applications • Health-care applications • Indoor video surveillance • Activity of a human • Kinect in gaming applications	Understanding Kinect
4.	2013	Wang, J. et al.,	• Dental surgery	Medical
5.	2013	Rankohi, S. et al.,	• Visualization and simulation applications for construction	Architecture, engineering, construction, and facility management (AEC/FM) industry

Appendix 2

Table 2. Learning stages

Title	Stages	Year
Situated Learning Theory Versus Traditional Cognitive Learning Theory: Why Management Education Should Not Ignore Management Learning	Acquisition	1997
Reviewing Reality: Human Factors of Synthetic Training Environments	Retaining	2001
The Use of E-Mail and In-Class Writing to Facilitate Student–Instructor Interaction in Large-Enrollment Traditional and Active Learning Classes	Apprehension	2002
Learning Environments and Learning Styles: Nontraditional Student Enrollment and Success in an Internet-Based Versus a Lecture-Based Computer Science Course	Apprehension	2002
Structural Learning of Graphical Models and Its Applications to Traditional Chinese Medicine	Acquisition	2005
Applying the Agent Metaphor to Learning Content Management Systems and Learning Object Repositories	Acquisition	2006
Learning Online: A Comparative Study of a Situated Game-Based Approach and a Traditional Web-Based Approach	Retaining	2006
Changing the Mindset: From Traditional On-Campus and Distance Education to Online Teaching and Learning	Apprehension	2006
Students' Conceptions of Constructivist Learning: A Comparison between a Traditional and a Problem-based Learning Curriculum	Performance	2006

Web3D Technologies in Learning, Education, and Training: Motivations, Issues, Opportunities	Motivation	2007
Technology in Dance Education	Apprehension	2007
Breaking the Traditional E-Learning Mould: Support for the Learning Preference Approach	Acquisition	2007
Lessons from the E-Learning Experience in South Korea in Traditional Universities	Retaining	2008
Traditional and Emerging it Applications for Learning	Generalisation	2008
ICT Use, Educational Policy and Changes in Pedagogical Paradigms in Compulsory Education in Denmark: From a Lifelong Learning Paradigm to a Traditional Paradigm?	Apprehension	2009
Can We Get Better Assessment from a Tutoring System Compared to Traditional Paper Testing? Can We Have Our Cake (Better Assessment) and Eat It Too (Student Learning during the Test)?	Apprehension	2010
Comparing Learning Results of Web-Based and Traditional Learning Students	Acquisition	2010
Use of Virtual Reality Technique for the Training of Motor Control in the Elderly	Acquisition	2010
Learning by Interest: Experiences and Commitments in Lives with Dance and Crafts	Motivation	2011
A Methodology for Integrating Traditional Classroom Learning with Contemporary Online Learning	Apprehension	2011
New Learning Paradigms: Open Course Versus Traditional Strategies. The Current Paradox of Learning and Developing Creative Ideas	Acquisition	2011
Semi-Automatic Analysis of Traditional Media with Machine Learning	Acquisition	2011
An In-service Training Course (INSET) on ICT Pedagogy in Classroom Instruction for Greek Primary School Teachers	Acquisition	2012
Personality and Learning Styles Surrounded by W3 Software: The Macao Portuguese School Case	Apprehension	2012

A Service-oriented Web Application for Learner Knowledge Representation, Management, and Sharing Conforming to IMS LIP	Acquisition	2012
Support for the Teacher in a Technology-enhanced Collaborative Classroom	Motivation	2012
Using Computer-assisted Instruction to Enhance Achievement of English-Language Learners	Motivation	2012
Using Technology Pedagogical Content Knowledge Development to Enhance Learning Outcomes	Acquisition	2012
mLearning: Anytime, Anywhere Learning Transcending the Boundaries of the Educational Box	Motivation	2012
A Quest for Meta-learning Gains in a Physics Serious Game	Performance	2012
Mobile Phones in Education: Challenges and Opportunities for Learning	Motivation	2012
A New Pedagogical Design for Geo-informatics Courses Using an E-training Support System	Acquisition	2013
An Interaction Model between Human and System for Intuitive Graphical Search Interface	Apprehension	2013
Comparing Two Methods of Education (Virtual versus Traditional) on Learning of Iranian Dental Students: A Posttest-only Design Study	Apprehension	2014

Appendix 3

Table 3. AR towards education sector and learning applications

Title	Type of journal published	Year	Sector	Web-based/ standalone AR technology/ Handheld systems	Type of learning application (games, elderly training, arts, knowledge acquisition, general)
Lessons learned: Evaluating visualizations for occluded objects in handheld augmented reality	International Journal of Human-Computer studies	2014	Engineering	Handheld AR systems	Knowledge acquisition
Audio-augmented paper for therapy and educational intervention for children with autistic spectrum disorder	International Journal of Human-Computer studies	2014	Education	Web-based	Arts
Delivering Educational Multimedia Contents through an Augmented Reality Application: A Case Study on its Impact on Knowledge Acquisition and Retention	Turkish Journal of Educational Technology	2013	Education	Web-based	General

Fun Learning with AR Alphabet Book for Preschool Children	International Conference on Virtual and Augmented Reality in Education	2013	Entertainment	Standalone AR system	Games
Enhanced Computer Vision with Microsoft Kinect Sensor: A Review	IEEE Transactions on Cybernetics	2013	Engineering	Standalone AR system	Knowledge acquisition
Using Augmented Reality and m-learning to optimize students performance in Higher Education	Procedia—Social and Behavioral Sciences	2012	Education	Handheld AR systems	Knowledge acquisition
Improve Template Matching Method in Mobile Augmented Reality for Thai Alphabet Learning	International Journal of Smart Home	2012	Education	Standalone AR system	Knowledge acquisition
Augmented reality and mobile learning: the state of the art	World Conference on Mobile and Contextual Learning	2012	Education	Web-based and standalone AR systems	Arts
Orchestrating TEL situations across spaces using Augmented Reality through GLUE!-PS AR	IEEE Transactions on Learning Technology	2012	Education	Web-based AR system	Arts
Augmented Reality Interfaces for Assisting Computer Games University Students	IEEE Transactions on Learning Technology	2012	Entertainment	Web-based AR system	Games
Enhancement of Learning and Teaching in Computer Graphics Through Marker Augmented Reality Technology	International Journal on New Computer Architectures and Their Applications	2011	Education	Standalone AR system	Knowledge acquisition
Augmented Reality and Education Current Projects and the Potential for Classroom Learning	New horizons for Learning	2002	Education	Standalone AR system	Knowledge acquisition

Applications of Augmented Reality technology for archaeological purposes	IEEE workshop	2002	Architecture and archeology	Standalone and handheld AR system	General
Creating Next Generation Blended Learning Environments Using Mixed Reality, Video Games and Simulations	Springer TechTrends Journal in Learning and Instruction	2005	Entertainment	Web-based and standalone AR systems	Games
Collaborative Learning through Augmented Reality Role Playing	International Conference on Computer support for collaborative learning	2005	Education	Web-based AR system	Arts
Augmented Reality in Education	Science Center To Go workshop	2011	Education and Architecture	Web-based and standalone AR systems	Knowledge acquisition
Design Principles for Augmented Reality Learning	Springer TechTrends Journal	2014	Education and Navigation	Web-based AR system	Knowledge acquisition
Using Motion Sensing for Learning: A Serious Nutrition Game	Springer-Verlag	2013	Entertainment	Web-based AR system	Games
Augmented reality enriches hybrid media	International Academic MindTrek Conference	2012	Entertainment	Web-based AR system	General
Augmented Heritage-situating Augmented Reality Mobile Apps in Cultural Heritage Communication	International Conference on Information Systems and Design of Communication	2013	Architecture and archeology	Web-based AR system	Arts
Asynchronous Immersive Classes in a 3D Virtual World: Extended Description of vAcademia	Springer-Verlag	2013	Education and Entertainment	Web-based AR system	Knowledge acquisition

Augmented Reality Navigation With Automatic Marker-Free Image Registration Using 3D Image Overlay for Dental Surgery	IEEE transactions on Biomedical Engineering	2014	Navigation for Dentistry	Web-based AR system	General
Augmented Exercise Biking with Virtual Environments for Elderly Users: A Preliminary Study for Retirement Home Physical Therapy	Workshop on Virtual and Augmented Assistive Technology	2014	Entertainment	Standalone AR system	Elderly training
Augmented Reality Learning Experiences: Survey of Prototype Design and Evaluation	IEEE TRANSACTIONS ON LEARNING TECHNOLOGIES	2014	Education	Web-based AR system	Knowledge acquisition
Augmented Reality Technology and Art: The Analysis and Visualization of Evolving Conceptual Models	International Conference on Information Visualisation	2012	Entertainment	Web-based AR system	Arts
An augmented reality-based authoring tool for E-learning applications	Multimedia tools and Applications	2014	Education	Web-based AR system	Knowledge acquisition
Augmented reality technologies, systems and applications	Multimedia tools and Applications	2011	Engineering	Web-based and standalone AR systems	General
Implementing digital game-based learning in schools: augmented learning environment of 'Europe 2045'	Multimedia Systems Springer	2010	Entertainment	Web-based AR system	Games

Affordances and Limitations of Immersive Participatory Augmented Reality Simulations for Teaching and Learning	Journal of science education and technology	2009	Education	Web-based AR system	Knowledge acquisition
Design and implementation of an augmented reality system using gaze interaction	Multimedia tools and Applications	2014	Application	Standalone AR system	General
What Teachers Need to Know About Augmented Reality Enhanced Learning Environments	TechTrends	2013	Education	Web-based AR system	Knowledge acquisition
Designing Kinect games to train motor skills for mixed ability players	Springer	2013	Games	Standalone AR system	Games
A classification of eLearning tools based on the applied multimedia	Springer	2013	Education	Web-based and standalone systems	General
Human Pacman: A Mobile Augmented Reality Entertainment System Based on Physical, Social and Ubiquitous Computing	Springer	2010	Education	Web-based AR system	Games
Hands in Space Gesture Interaction with Augmented-Reality Interfaces	IEEE Computer Graphics and Applications	2014	Entertainment	Web-based and handheld AR systems	General
Expected user experience of mobile augmented reality services: A user study in the context of shopping centres	Springer	2011	Retail	Handheld AR systems	General

A new typology of augmented reality applications	ACM	2012	Topology	Web-based handheld and standalone AR systems	Knowledge acquisition
3D Video and Its Applications	Springer	2012	Applications	Web-based systems	Arts
SMART: scalable and modular augmented reality template for rapid development of engineering visualization applications	Springer journal on Visualization in Engineering	2013	Education	Web-based AR system	Arts
Using Virtual Reality and Augmented Reality to Teach Human Anatomy	Thesis	2011	Education	Web-based AR system	Knowledge acquisition
A State of the Art Report on Kinect Sensor Setups in Computer Vision	Springer journal	2013	Education	Standalone AR system	Knowledge acquisition
Review and analysis of augmented reality literature for construction industry	Springer journal on Visualization in Engineering	2013	Engineering	Web-based and handheld AR systems	General

Appendix 4

Table 4. AR-based dance-learning devices

Title	Year	Type of journal	Kind of physical movement or dance form	Devices used
The ChaLearn gesture dataset (CGD 2011)	2013	Machine Vision and Applications	ChaLearn gesture	RGB-D Kinect camera
MUSE: Understanding Traditional Dances	2014	IEEE Virtual Reality	Middle Eastern indigenous dance, Al Ardha	Kinect motion sensor
Identification of Odissi Dance Video Using Kinect Sensor	2013	IEEE	Indian classical dance 'Odissi'	Kinect sensor
Stochastic Regular Grammar-based Learning for Basic Dance Motion Recognition	2013	International Conference on Advanced Computer Science and Information Systems	Dance motion recognition	Kinect depth sensor camera
A multi-modal dance corpus for research into interaction between humans in virtual environments	2013	Springer Journal on Multimodal User Interfaces	Salsa	Nintendo Wii or Microsoft Kinect

Using the Smartphone Accelerometer to Monitor Fall Risk while Playing a Game: The Design and Usability Evaluation of Dance! Don't Fall	2013	Springer-Verlag	Dance! Don't Fall game	Smartphone
Desirability of a Teaching and Learning Tool for Thai Dance Body Motion	2013	Springer-Verlag	Thai dance	LabanEditor and Thai Dance Notation System
A Novel Technique for Space-Time-Interest Point Detection and Description for Dance Video Classification	2013	Springer-Verlag	Indian classical dance	STIP detector and descriptor
Study on the Synchronous E-Learning Platforms for Dissemination of Traditional Dance	2013	IEEE	Traditional dance	Screen (150 inches), a projector, speakers, a mike, webcams (video resolution was 1600 x 1200)
Example-Based Automatic Music-Driven Conventional Dance Motion Synthesis	2012	IEEE Transactions On Visualization and Computer Graphics	Chinese, Tibetan, Uighur, and Mongolian dance	Optical 3D motion capture device, desktop PC with an Intel Core i7 Duo 2.66 GHz CPU, 4 GB memory, and an NVIDIA GTX285 graphics card
Real-time Body Motion Analysis For Dance Pattern Recognition	2012	IEEE	Any dance pattern	Dynamic stereo vision sensors
Framework for teaching Bharatanatyam through Digital Medium	2012	IEEE Fourth International Conference on Technology for Education	Bharatanatyam	No device used
Music similarity-based approach to generating dance motion sequence	2013	Multimedia Tools and Applications	Any dance form	Motion capture device

Automatic Dance Lesson Generation	2012	IEEE Transactions on Learning Ttechnologies	Latin and hip hop dances	Marker-based optical 3D motion capture system with several cameras
Learn2Dance: Learning Statistical Music-to-Dance Mappings for Choreography Synthesis	2012	IEEE Ttransactions on Multimedia	Any dance form	Simulation models
Creating Tangible Cultural Learning Opportunities for Indigenous Dance with Motion Detecting Technologies	2012	IEEE International Games Innovation Conference	Indigenous dances	Xbox Kinect
Production of Body Model for Education of Dance by Measurement Active Quantity	2012	IEEE Conference on Computer Electronics	Any dance form	Simulation model
Interactions and systems for augmenting a live dance performance	2012	International Symposium on Mixes and Augmented Reality	Any dance form	The motion-tracking system, IR LED footstrap, IR camera, static FireWire digital camera (Point Grey Dragonfly 2 DR2-HICOL), XSens MVN motion capture suit, NAO humanoid robot
A Virtual Reality Dance Training System Using Motion Capture Technology	2011	IEEE Transactions On Learning Technologies	Any dance form	3D graphics, motion matching, motion database, and motion capture system.
DanVideo: an MPEG-7 authoring and retrieval system for dance videos	2010	Multimedia Tools and Applications	Labanotation and Benesh dance notation	Video Annotator, MPEG7 Instance Generator, Parser, Visualizer, MPEG-7 Query Generator, MPEG-7 Tree Generator, and Query Processor
Interactive Theater Experience with 3D Live Captured Actors and Spatial Sound	2010	Springer	Any dance form	Wireless network, content marker, syncronisation server, render client, tracking system

Augmented Reality Environment for Dance Learning	2003	IEEE	Macedonian dance	Prototype model
Compression and recognition of dance gestures using a deformable model	2004	Springer	Any dance form	Hidden Markov models (HMMs)
A Multimedia Information Repository for Cross Cultural Dance Studies	2004	Multimedia Tools and Applications	East Indian dance	Two 3D Vicon motion capture systems, one with sixteen cameras and one with four cameras.
Analysis and Synthesis of Latin Dance Using Motion Capture Data	2004	Springer-Verlag	Latin dance	Eight digital cameras (640 × 480 pixels, 60Hz) and real-time capture software, motion graph editor
Expressive Footwear for Computer-Augmented Dance Performance	1997	International Symposium on Wearable Computers	Any dance form	Two piezoelectric pads, sensors, dual-axis accelerometer, 8-bit A/D converter, analog multiplexer, battery, low-power RF transmitter
Augmented Performance in Dance and Theater	1999	International Journal of Dance and Technology	Any dance form	Video camera, Silicon Graphics computers, large projection screen, phased array microphone
Indian Classical Dance Classification by Learning Dance Pose Bases	2012	IEEE	Indian Classical Dance-Bharatnatyam, Kathak, and Odissi	Pose descriptor, online dictionary learning technique, Learning SpaceTemporal 3D dictionary
Results and Analysis of the ChaLearn Gesture Challenge 2012	2013	Springer-Verlag	Any dance form	Kinect camera
Gestures	2012	Safari online book	Any dance gesture	Microsoft Kinect SDK

Video-as-Data and Digital Video Manipulation Techniques for Transforming Learning Sciences Research, Education, and Other Cultural Practices	2006	Springer	Any dance form	Software analysis tool
Human Pose Co-Estimation and Applications	2012	IEEE Transactions an Pattern Analysis and Machine Intelligence	Any dance form	Simulation model
Visualizing Velocity and Acceleration on Augmented Practice Mirror Self-Learning Support System of Physical Motion	2013	International Conference on Advanced Applied Informatics	Physical motions	Augmented Practice Mirror (APM), wire human model
Shall We Dance? A Design Epistemology for Organizational Learning and Performance	2004	Springer	Any dance forms	No device used

Appendix 5

Table 5. Different dance forms or physical movement

Kind of physical movement or dance form	Dance forms
Dance motion recognition	General
Dance! Don't Fall game	General
Labanotation and Benesh dance notation	General
Any dance gesture	General
physical motions	General
Any dance pattern	Mixed
Any dance form	Mixed
Any dance form	Mixed
Any dance form	Mixed
Any dance form	Mixed
Any dance form	Mixed
Any dance form	Mixed
Any dance form	Mixed
Any dance form	Mixed
Any dance form	Mixed
Any dance form	Mixed
Any dance form	Mixed
Any dance form	Mixed
Any dance forms	Mixed
ChaLearn gesture	Traditional
Middle Eastern indigenous dance, Al Ardha	Traditional
Indian classical dance "Odissi"	Traditional
Thai dance	Traditional
Indian classical dance	Traditional
Traditional dance	Traditional

Chinese, Tibetan, Uighur, and Mongolian dance	Traditional
Bharatanatyam	Traditional
Indigenous dances	Traditional
Macedonian dance	Traditional
East Indian dance	Traditional
Indian Classical Dance—Bharatanatyam, Kathak, and Odissi	Traditional
Salsa	Western
Latin and hip hop dances	Western
Latin dance	Western

Chapter 16

Web 2.0 as an Innovation: Perception, Attitude, and Adoption in Tertiary Institutions in Lagos State, Nigeria

Nathan Emanuel N.
and
Blessing F. Adeoye

Introduction

The development of the Internet into the highly versatile, dynamic and democratized medium today has brought with it incredible transformations and opportunities in practically all fields of human activity. A particular transformation in the field of education is the use of Web 2.0. According to Selwyn (2008), the past five years have seen growing excitement within the educational community about Web 2.0 technologies. Web 2.0 is an umbrella term for a host of recent Internet applications such as social networking, Wikis, folksonomies, virtual societies, blogging, multiplayer online gaming and 'mash-ups' Whilst differing in form and functions, all these applications share a common characteristic of supporting Internet-based interaction between and within groups, which is why the term 'social software' is often used to describe Web 2.0 tools and services.

The current generation of students entering universities and colleges use Web 2.0 applications (Lenhart and Madden 2005; 2007). Educators suggest that Web 2.0 tools should be integrated into higher education because digital natives expect to learn with new technologies and because higher education should prepare students for the workplace of the future (Alexander, 2006; Prensky, 2001; Roberts, Foehr and Rideout, 2005; Strom and Strom 2007). The use of Web 2.0 tools has grown considerably in the education sector in the last few years; they are being emphatically and overwhelmingly adopted by the conventional age students and the "digital netizens." A netizen is a person who is, literally, a citizen of the Internet. A netizen understands this fact, and uses the Internet to communicate. The Internet and in particular Web 2.0 have changed the process of teaching and learning and information process. Web 2.0 tools have brought about a radical evolution in education which makes teaching and learning look impossible without them (Hargadon, 2009).

On-line learning is extremely interesting to observe in the educational sector, as an enhanced efficiency at this level is further on naturally disseminated in all segments and fields of activity. Moreover, taking into account all the great advantages of using Web 2.0 in providing high quality, modern educational services and catalysing learning processes, it is of utmost importance for the future of education and the development of generations to come. This is the dawn of a new era pertaining entirely to "digital natives" (Mason and Rennie, 2007), as today's children are using Web 2.0 technologies comfortably and efficiently and they will continue to do so ever more naturally.

The status of integration of Web 2.0 is the underlying theme of this chapter, which will range across social networking and several Web 2.0 tools that will enhance teaching and learning. Selwyn (2008) pointed out that Web 2.0 marks a distinct break from the Internet applications of the 1990s and early 2000s, facilitating 'interactive' rather than 'broadcast' forms of exchange, in which information is shared 'many-to-many' rather than being transmitted from one to many. Web 2.0 applications are built around the appropriation and sharing of content amongst communities of users, resulting in various forms of user-driven communication, collaboration and content creation and recreation.

Statement of the Problem

The needs for exploring the Internet for teaching and learning materials as an alternative for the grossly unavailable materials in some of the high institutions requires the resourcefulness of both the teachers and students in making use of Web 2.0 tools to ensure that teaching and learning is not retarded. The lack of materials calls for institutions of higher learning to accelerate growth in the usage of Web resources to enhance delivery of educational content in general. With the increased awareness as well as the desire for more flexible delivery alternatives, educational institutions will be able to move faster and extend the Web 2.0 tools to their campuses.

Despite interest in using Web 2.0 tools for educational purposes, students lacked the experience in using them for learning. No studies were identified that survey the awareness of Web 2.0 in teaching and learning in tertiary institutions in Lagos State or the benefits of any one Web 2.0 technology over others for a particular discipline in tertiary institutions in Lagos State. The purpose of this study was to survey the awareness and the status of integration of Web 2.0 in teaching and learning in tertiary institutions in Lagos State.

Research Questions

To fulfil the purpose, this study seeks to answer the following research questions:

1. What are the current technological tools used by teachers in teaching and learning?

2. What is the teachers' awareness and perception of the pedagogical benefits of Web 2.0?

3. What is the students' awareness on use of Web 2.0 tools in learning?

4. What are the challenges associated with the use of the Web 2.0 during teaching and learning?

Review of Related Literature

Parker and Chao (2007) investigated the application of Web 2.0 in Education and reported that Web 2.0 tends to complement, enhance and add new collaborative dimensions to the classroom. It is a potential for knowledge sharing among academics in higher education and hence improve knowledge creation as well as innovation in the academic universe of discourse. Collins and Allan (2009) asserted that Web 2.0 technologies provided teachers with new ways to engage students in a meaningful way. "Children raised on new media technologies are less patient with filling out worksheets and listening to lectures" because students already participate on a global level. Collins and Allan (2009) also stated that the lack of participation in a traditional classroom stems more from the fact that students receive better feedback online.

Traditional classrooms would only require students to do assignments and when they are completed, that would be all; but Web 2.0 shows students that education is a constantly evolving entity. Whether it is participating in a class discussion or participating in a forum discussion, the technologies available to students in a Web 2.0 classroom does increase the amount of their participation (Collins and Allan, 2009). Web 2.0 tools are needed in the classroom to prepare both students and teachers for the shift in learning and teaching. According to Collins and Halverson (2009) the self-publishing aspects as well as the speed with which their work becomes available for consumption allows teachers to give students the control they need over their learning. This control is the preparation students will need to be successful as learning expands beyond the classroom.

Integrating technology into instruction tends to move classrooms from teacher-dominated environments to student-centred. While it is still important for teachers to monitor what students are discussing, the actual topics of learning are being guided by the students themselves. Web 2.0 calls for major shifts in the way education is provided for students. Students in a Web 2.0 classroom are expected to collaborate with their peers. By making the shift to a Web 2.0 classroom, teachers are creating a more open atmosphere where students are expected to stay engaged and participate in the discussions and learning that is taking place around them (Russell and Sorge, 1999).

Donnison (2004) conducted a survey on the ways in which teaching and learning could benefit from the inclusion of Web 2.0 applications in specific contexts or disciplines of higher education. The survey also aimed at determining which Web 2.0 technologies were currently used by undergraduate students in different disciplines on-campus and which Web 2.0 technologies undergraduates find most beneficial for learning in their respective disciplines. The research provided insight into the undergraduate perspective to educators seeking to integrate Web 2.0 tools in their teaching in different disciplines, as well as those calling for increased inclusion of new technologies in higher education. The research identified several benefits of Web 2.0 technologies to learners in higher education (Alexander, 2006; Elgort, Smith, and Toland 2008; Lamb, 2004).

Further studies by Ellison and Wu (2008), Farmer, Yue, and Brooks (2008), Hall and Davison (2007), Williams and Jacobs (2004), and Xie, Ke, and Sharma (2008) have focused on one particular tool, for example, blogs, within a certain discipline. The researchers reported that blogs encourage students to read and provide peer feedback, and also enhance reflection and higher-order learning skills. Wikis were found not only to improve students' writing skills but also to engage students and facilitate collaborative learning in various disciplines (Luce-Kapler 2007; Parker and Chao, 2007). Podcasting has been used successfully at institution-wide or in specific disciplines like language learning, chemistry or psychology in higher education (Chinnery, 2006; Miller 2006; and Woodward, 2007). Although many studies have been conducted regarding the use of Web 2.0, however, no studies were identified that compared the use of any one of the components Web 2.0 technology across disciplines or that identified the benefits of any one Web 2.0 technology over others for a particular discipline.

Kvavik, Caruso and Morgan (2004) surveyed 4374 students across 13 institutions, the researchers found that respondents used technology mainly for word processing (99.5%), surfing the Internet for pleasure (99.5%) and e-mailing (99.5%). Only 21% had created their own content on the Web. Likewise, in a survey of teenagers, Lenhart and Madden (2007) found that 59% read blogs daily, but only 28% had created their own online journals or blogs. However, Sandars and Schroter (2007) surveyed 3000 medical students' familiarity with and use of Web 2.0 technologies and found out high familiarity, but low use of most Web 2.0 technologies except for social networking tools (Facebook, YouTube). Despite interest in using Web 2.0 tools for educational purposes, students lacked the experience in using them for learning or teaching. Based on these findings, students' familiarity and prior use of Web 2.0 for teaching and learning were taken into account in redesigning the curriculum.

Downes (2005) offered an early review of the potential of these Web 2.0 technologies for learning. Both outline more open, participatory and hierarchical structures in teaching methods. Reviewing the use of social media like blogs and wikis, Bruns and Humphreys (2007), also argued that the production of content by the user requires a shift in changing teaching methods towards approaches that support community building through collaboration, hierarchical structures of engagement, mentoring, fostering creativity and critical literacy capacities. Siemens (2009), considering this from the perspective of networked learning and connectivism, reflected on the role of academic teaching methods: Given that coherence and lucidity are keys to understanding our world and how do educators teach in networks. For educators, control is being replaced with influence. Instead of controlling a classroom, a teacher now influences or shapes a network.

Methodology

The research design adopted for this study was the descriptive survey research design, which involved the assessment of respondents' opinion using sampling techniques (Wolman, 1973). This method is preferred because it afforded the opportunity of studying details in a large group by selecting from representative sample. This survey design assisted the researcher in obtaining information from a sample of respondents considered to be the exact representative of the entire student population.

Population of the Study

The population of this study involved randomly selected lecturers and undergraduates in the University of Lagos and lecturers and undergraduates in Yaba College of Technology, Yaba. Lagos. These two institutions were selected for this study because they are both in the Lagos metropolis, which has different ethnic groups and it is believed students and lecturers in the schools are already familiar with the use of information technology.

Sample and Sampling Technique

The sampling technique adopted for this study was the simple random sampling technique. A sample population of 250 respondents, comprising of 200 undergraduates, and 50 lecturers from the University of Lagos and Yaba College of Technology, both in Yaba local government areas of Lagos State participated in this study.

Instrumentation

The researchers used questionnaire for data collection. The questionnaire was generated by the researchers with contributions from experts in Measurement and Evaluation and Educational Technologist, both in the University of Lagos and College of Education, Akoka respectively. Questionnaire was used for data collection because it gives the respondents the freedom to pick his or her responses. Hatf (1972) and Ilogu (2005) indicated that questionnaire is an instrument for securing answers to questions by using a form which the respondent fills himself. The questionnaire was divided into two sections. Section A and B. Section A contained questions on the demographic features of the respondents, while the second part contained statements designed to gather information needed to achieve the specific objective of the study.

Validity and Reliability of the Instrument

This instrument was validated to ascertain whether it measures what it tends to measure. Therefore, to ascertain the validity of this instrument, the instrument was validated by experts in educational technology. It was further presented to the project supervisor for proper vetting and modification before the final approval was given for the retention of the questionnaire and administration. Thus the construct, face and content validity of the questionnaire were assured. A pilot test was also carried out with 20 respondents from students at Federal College of Education, Akoka, using a 5-point Likert scale type questionnaire and after two weeks the second administration followed to ascertain its internal consistency. The simple percentage was used for the analysis.

Administration of the Instrument

The researchers visited the selected schools and administered the questionnaire with the help of the officers of the institutions. The purpose of the study was explained. A promise to treat all pieces of information supplied in strict confidence was made before administering the questionnaire to ensure consistency in response. At the end, the completed questionnaires were collected for analysis by the researcher.

The instrument used for the study were scored and recorded. The scores from the respondent on the questionnaire were tallied and summed up accordingly. The items in the questionnaire were coded for easy scoring of the responses. The responses were scored in descending order for positive items and ascending order for negative items.

The items were rated in a four point Likert scale thus:

Descending order	Points	
Strongly agree	SA	4
Agree	A	3
Undecided	U	0
Disagree	D	2
Strongly disagree	SD	1

Ascending order	Points	
Strongly agree	SA	1
Agree	A	2

Undecided	U	0
Disagree	D	3
Strongly disagree	SD	4

Data Analysis

The data analysis was based on the questionnaire as the research questions were analysed using the frequencies of the answered questions from the respondents, which were afterwards translated into simple percentages in a tabular form for an easy interpretation.

Table 1. Distribution of student respondents by gender

Sex	f	%
Male	131	52.8%
Female	118	47.2%
Total	250	100%

Tables 1 revealed that 52.8% of the students' respondents were males; of the total study sample of respondents, while the remaining 42.7% were females.

Table 2. Distribution of lecturers' respondents by gender

Sex	f	%
Males	32	64
Female	18	36
Total	50	100%

Table 2 revealed that 64% of the lecturers' respondents were males, and 36% were females.

Table 3. Distribution of the lecturers' respondents' qualification

Qualification	f	%
H.N.D/B. Ed./B.Sc.	26	52
M.Ed./M.Sc.	20	40

Ph.D.	4	8
Total	50	100

Table 3 revealed that 52% of the lecturers had HND/B.Ed./B.Sc., 40% had M.Ed./M.Sc., 8% had a PhD. This showed that lecturers with HND/B.Ed./B.Sc were more than the rest.

**Table 4. Distribution of the
lecturers respondents by years of experience**

Experience	f	%
2–5 yrs	12	24
6–10 yrs	22	44
11–15 yrs	16	32
Total	50	100

Table 4 revealed that 24% of the respondents had between 2–5 years of experience, 44% of the respondents had between 6–10 years of experience while the remaining 32% had between 11–15 years of experience. This showed that lecturers with 6–10 years were more.

Analysis of Data from Yaba College of Technology Students

What are the current technological tools used by teachers in teaching and learning?

Table 5. Yaba College of Technology students' view on the awareness of current technological tools used during classroom instructions

Statements	SA	A	U	D	SD
I use a computer system and laptop during classroom interactions with my teachers.	7.2%	16.7%	3.1%	37.5%	35.4%
I have been taught with a projector during lectures.	14.9%	22.3%	7.4%	26.6%	28.7%
I have been taught with PowerPoint slides.	13.3%	33.3%	14.6%	25.3%	37.3%

I am familiar with the use of Wikipedia materials.	16.8%	56.8%	6.3%	9.5%	10.5%
I have visited blog sites for materials.	18.8%	49.2%	8.6%	18.3%	8.6%
I have posted on a Twitter page.	17.0%	39.0%	8.0%	20.0%	16.0%
I am familiar with Web 2.0 tools like blogs.	13.5%	27.1%	17.7%	21.9%	19.7%
I use Web 2.0 tools like Facebook to communicate with my teachers and other classmates.	28.7%	30.1%	8.2%	32.9%	26.0%
I am familiar with the use of iPod/podcast.	22.3%	34.0%	8.5%	25.5%	9.5%
I get more materials from the Internet/Web when studying.	31.9%	56.3%	7.4%	4.2%	0.0%
Wikis enhance my learning.	24.2%	53.5%	13.1%	7.01%	3.0%
I search YouTube for further studies	22.6%	34.0%	11.3%	21.6%	10.3%

Thirty-seven percent (37%) of Yaba Tech respondents disagree that they use computer system during classroom interactions with their teachers. The trend is similar for projectors in which 28.7% strongly disagree that they have been taught with projectors during lectures. Similarly 37.3% strongly disagree that they have been taught using PowerPoint slides, but 56.8% of the respondents agreed to be familiar with the use of Wikipedia materials, while 46.2% have visited blog sites for materials. Thirty nine percent of the respondents have posted on twitter page, 27.1% is familiar with Web 2.0 tools like blogs, and 32.9% disagree to have used Web 2.0 in any form with their teachers and classmates.

Thirty four percent of the participants agreed to be familiar with the use of iPod/podcast and 56.3% agreed to be getting materials from the Internet, and 53.5% of the respondents agreed that Wiki enhances their learning while 34.0% search YouTube for further studies. In analysing the responses, a higher percentage of student agreed that they are aware of currently technological tools.

Reasons for not using Web/Internet materials

Table 6. Yaba College of Technology students' reasons for not using Web/Internet materials

Statements	SA	A	U	D	SD
Wiki materials are confusing	5.5%	12.0%	32.9%	26.3%	23.1%
No Internet facilities in my school	6.5%	10.8%	6.5%	31.5%	44.6%
No educational technology resource like computer systems in my school	5.2%	7.3%	5.2%	36.5%	45.8%
Using social networking tools in the school makes class interaction difficult	9.2%	6.1%	14.3%	24.5%	45.9%
Web 2.0 tools like iPod are not common in the school	14.0%	32.3%	15.1%	23.7%	15.1%
No awareness of Web/Internet usage in my school	7.4%	11.6%	6.3%	36.8%	37.9%

Thirty-two percent of the respondents were undecided on whether Wiki materials are confusing while 26.3% disagreed that Wiki materials are confusing. Forty-four percent of the respondents strongly disagreed that there is no Internet facilities in their school. The trend is similar for educational technology resource like computer systems in which 45.8% strongly disagreed that they do not have it in their schools. Similarly, 45.7% disagreed that using social networking tools in their school makes class interaction difficult. Thirty-two percent of the respondents agreed that Web 2.0 tools like iPods are not common in the school which 37.9% strongly disagreed that there is no awareness of Web/Internet usage in their schools. Analysis revealed that one of the problems of not using Web/Internet materials is because these facilities are not readily available in the institutions.

What is the teachers' awareness and perception of the pedagogical benefits of Web 2.0?

Table 7. Yaba College of Technology perceptions of the benefits of using Web materials/Internet facilities

Statement	SA	A	U	D	SD
Using e-mails to communicate makes learning easier	34.8%	47.8%	6.5%	6.5%	4.3%
The Internet facility enhances learning and communication	54.0%	42.5%	0%	2.3%	1.2%
The social networking tool do not create effective communication amongst students	10.5%	10.5%	3.5%	29.1%	46.5%
The wikis enables fast and slow learners to learn at their own pace.	27.3%	38.6%	23.9%	10.2%	8.0%
The Internet creates opportunities for students to source for study materials.	53.1%	38.4%	4.1%	3.1%	1.0%
I interact with my teachers online.	2.1%	10.4%	12.5%	33.3%	41.7%
Studying online is boring.	6.3%	14.7%	15.8%	27.4%	35.8%

Table 7 revealed that 54% agreed that using e-mails to communicate makes learning easier. 46% strongly disagree that social networking tools does not create effective communication amongst students, while 38.6% of the respondents agreed that Wiki enable fast and slow learners to learn at their own pace. Similarly, 53.1% of the respondents strongly agreed that the Internet creates opportunity for students' to source for study materials. Forty-one percent strongly disagreed that they interact with their teacher online, while 35.8% strongly disagreed that studying online is boring. Despite a higher number of respondents agreeing that the Internet and other Web 2.0 tools are useful in teaching and learning, its usage is low.

Analysis of Data Collected from University of Lagos Students

The first set of questions was intended to survey the awareness of current technological tools used during classroom instructions in University of Lagos.

What is the students' awareness on use of Web 2.0 tools in learning?

Table 8. University of Lagos students' view on the awareness of current technological tools used during classroom instructions

Statements	SA	A	U	D	SD
I use computer system during classroom interactions with my teachers.	0.9%	14.3%	8.9%	52.7%	23.2%
I have been taught with a projector during lectures.	14.2%	30.1%	4.4%	32.7%	18.6%
I have been taught with PowerPoint slides.	8.5%	34.2%	2.6%	35.9%	18.8%
I am familiar with the use Wikipedia materials.	15.7%	54.7%	7.8%	16.5%	5.2%
I have visited blog sites for materials.	20.7%	45.6%	8.6%	17.2%	7.8%
I have posted on a Twitter page.	20.0%	24.3%	7.8%	33.9%	13.9%
I am familiar with Web 2.0 tools like blogs.	12.9%	23.3%	10.3%	35.3%	18.1%
I use Web 2.0 tools like Facebook to communicate with my teachers and other classmates	21.5%	41.1%	6.5%	20.6%	19.6%
I am familiar with the use of iPod/ podcast	19.0%	39.6%	9.5%	22.4%	9.5%
I get more materials from the Internet/ Web when studying	41.9%	47.0%	1.7%	6.0%	3.4%
Wikis enhance my learning	21.4%	49.1%	12.5%	12.5%	4.5%
I search YouTube for further studies	13.8%	25.9%	13.8%	33.6%	12.9%

Fifty three percent of the respondent disagree that they use computer system during classroom interactions with their teachers. The trend is similar for projectors in which 32.7% disagreed that they have been taught with projectors during lectures. Similarly, 35.9% disagreed that they have been taught using PowerPoint slides, while 54.7% of the respondents agreed to be familiar with the use of Wikipedia materials, 45.6% have visited blog sites for materials. 33.9% disagreed that they have posted on Twitter page, 35.3% disagreed to be familiar with Web 2.0 tools like blogs, 41.1% agreed to have used Web 2.0 to communicate with their teachers and classmates. 39.6% Agreed to be familiar with the use of iPod/Podcast and 47.0% agreed to be getting materials from the Internet. 49.1% of the respondents agreed that Wiki enhances their learning while 33.6% disagreed that they search YouTube for further studies. Analysis of the responses indicated that higher percentage is aware of the current technological tools which enhance teaching and learning.

What are the reasons for not using Web/Internet materials?

Table 9. University of Lagos students' response on reasons for not using Web/Internet materials

Statements	SA	A	U	D	SD
Wiki materials are confusing	4.5%	11.8%	29.0%	36.6%	18.1%
No Internet facilities in my school	4.3%	4.3%	0.9%	29.3%	61.2%
No educational technology resource like computer systems in my school	2.6%	3.4%	5.2%	28.4%	60.3%
Using social networking tools in the school makes class interaction difficult	6.2%	4.4%	5.3%	32.7%	51.3%
Web 2.0 tools like iPods are not common in the school	15.6%	22.0%	16.5%	26.5%	19.3%
No awareness of Web/Internet usage in my school	5.4%	2.7%	9.0%	34.2%	48.6%

Thirty-seven percent of the respondent's disagreed that Wiki materials are confusing. That is, the Wiki materials have been very helpful in their studies. 61.2% strongly disagreed that there is no Internet facilities in their school. It means that there are Internet facilities in their school which encourage Web 2.0 and other technological usage. Similarly 60.3% strongly disagreed that there is no educational resources like computer systems in their school. 51.3% strongly disagreed that using social networking tools in the school makes class interaction difficult while 26.5% also disagree that Web. 2.0 tools like iPods are not common in their schools. 48.6% of the respondents also strongly disagreed that there is no awareness of Web/Internet usage in their school.

What are the benefits of using Web materials/Internet facilities?

Table 10. University of Lagos students' perception of benefits of using Web materials/Internet facilities

Statement	SA	A	U	D	SD
Using e-mails to communicate makes learning easier.	38.8%	46.6%	4.3%	7.8%	2.6%
The Internet facility enhances learning and communication.	56.9%	35.3%	2.6%	5.2%	0%
The social networking tools do not create effective communication amongst students.	8.8%	11.5%	8.0%	34.5%	37.2%
The wiki enables fast and slow learners to learn at their own pace	28.3%	33.6%	22.1%	9.4%	6.2%
The Internet creates opportunity for students to source for study materials	52.6%	38.6%	4.4%	2.6%	1.8%
I interact with my teachers on-line	13.4%	15.2%	11.6%	31.3%	28.6%
Studying online is boring	12.9%	7.8%	12.1%	29.3%	37.9%

Forty-seven percent of the respondents agreed that using e-mails to communicate makes learning easier. Fifty-seven percent of the respondent strongly agreed that Internet facility enhances learning and communication. Thirty-three percent strongly disagreed that social networking tools does not create effective communication amongst students; while 33.6% of the respondents agreed that Wikis enable fast and slow learners to learn at their own pace. Similarly, 52.6% strongly agreed that the Internet creates opportunity for students to source for study materials 31.3% disagree that they interact with their teachers on-line while 37.9% strongly disagreed that studying online is boring. The result here indicated that a higher percentage of students are aware of the benefits of Internet and Web materials in teaching and learning.

Analysis of Lecturer's Responses

What is teachers' awareness of current technological tools used during classroom instructions by the lecturers?

Table 11. Teachers' awareness of current technological tools used during instruction by the lecturers

S/n	Statements	SA	A	U	D	SD
1	I use computer systems during classroom instructions	26.3%	23.5%	0%	35.3%	11.8%
2	I use projector during class room instructions	15.8%	10.3%	15.8%	42.1%	15.8%
3	I use PowerPoint presentation for lecturing	23.5%	11.8%	5.9%	52.9%	5.9%
4	I create blogs pages for references	11.8%	5.9%	17.6%	52.9%	11.8%
5	Internet Facilities have been useful for my lecturing	50.0%	44.4%	0%	0%	5.6%
6	When lecturing, I refer my student to my Twitter page	0%	22.2%	27.8%	38.9%	11.1%

Thirty-five percent of the lecturers disagreed that they use computer systems during classroom instructions. The trend is similar for projector in which 42% disagreed that they use projector a\during classroom instructions. Similarly, 53% disagreed that they use power points presentation for lecturing. 53% also disagreed to have created blogs pages for references. While 45% agreed that Internet facility has been useful for lecturing. Forty percent disagreed that when lecturing, they refer their student to their Twitter page. The analysis revealed that teachers are aware of current technological tools used in teaching and learning, however, they might not have used them during interactions.

Awareness/Use and perception of the pedagogical benefits of Web 2.0 Tools

Table 12. Awareness/use and perception of the pedagogical benefits of Web 2.0 tools

S/N	Statements	SA	A	U	D	SD
7	I use iPods/podcasting for lecturing my students	5.6%	0%	33.3%	44.4%	16.7%
8	I use Web 2.0 tools like Facebook to communicate with my students	5.6%	27.8%	11.1%	33.3%	22.2%

9	I am familiar with the use of YouTube for lecturing	5.9%	41.2%	11.8%	26.3%	11.8%
10	I get more materials from the library than the Internet	25.0%	25.0%	6.3%	12.5%	31.3%
11	I create Wikis for further references when lecturing	12.5%	6.3%	37.5%	37.5%	6.3%
12	Uploading Video/Audio podcast when teaching is common	12.5%	25.0%	12.5%	31.3%	18.6%

Forty-four percent of the lecturers disagreed that they use iPods/podcasting for lecturing. This is because iPod usage is not popular in the school. Thirty-three percent also disagreed that they use Web 2.0 tools like Facebook to communicate with their students. Forty-two percent agreed that they are familiar with the use of YouTube for lecturing; 32% strongly disagreed that they get more materials from the library than the Internet. Thirty-eight percent disagreed that they create Wiki for further references when lecturing; 33% disagreed that they upload video/audio podcast when teaching. The result above revealed that teachers are aware of the benefits of Web 2.0 tools, but a lower percentage uses these tools in teaching and learning.

Reasons for not using Web/Internet materials

Table 13 shows the results for the reasons of not using Web/Internet materials

S/N	Statement	SA	A	U	D	SD
13	Wiki materials are confusing	5.9%	17.6%	23.5%	41.2%	11.8%
14	No Internet Facilities in my school	0%	0%	0%	33.3%	66.7%
15	No educational technology resource like computer systems in my school	0%	12.5%	0%	44.4%	50.0%
16	Using Social Networking tools in the school makes class interaction difficult	12.5%	11.1%	11.1%	16.7%	55.6%
17	Web 2.0 tools like iPods are not common in the school	11.1%	33.3%	38.9%	11.1%	12.5%
18	No enough training on the use of PowerPoint	22.2%	12.5%	12.5%	44.2%	22.2%
19	No constant power supply in my school	33.9%	22.2%	0%	33.3%	11.1%

Forty-one percent of the lecturers disagreed that wikis are confusing while 66.7% strongly disagreed that there are no wiki facilities in their school. Fifty percent strongly disagreed that there is no educational technology resource like computer systems in their school; 55.6% strongly disagreed that using social networking tools in the school makes class interaction difficult. Thirty-three percent of the lecturers agreed that Web 2.0 tools likes iPods are not common in the school, while 44.2% disagreed that there is no enough training on the use of PowerPoint. Thirty-three percent strongly agreed that there is no constant power supply in their school. The results revealed that a high percentage of teachers do not use these tools due to the lack of constant power supply.

Benefits of using Web materials/Internet Facility

Table 14. The benefits of using Web materials/Internet facilities

S/n	Statement	SA	A	U	D	SD
20	Communicating with e-mails with students make learning easier	38.9%	38.9%	0%	22.2%	0%
21	The Internet facility enhances teaching and interactions	66.7%	27.8%	0%	0%	12.5%
22	The social networking tools do not create effective communication amongst students and teachers	5.9%	11.8%	5.9%	47.1%	26.3%
23	The wikis enable fast and slow learners to learn at their own pace	17.6%	23.5%	23.5%	17.6%	11.8%
24	The Internet creates opportunity for teachers to source for lecturing materials	66.7%	27.8%	0%	0%	5.5%
25	Being online enables the teacher to reach more learners	44.4%	27.8%	11.10%	11.1%	5.5%

Thirty-nine percent of the respondent strongly agreed that communicating with e-mails with students makes learning easier. The trend is similar where 67% strongly agreed that the Internet facility enhance teaching and interaction. Forty-seven percent disagreed that the social networking tools does not create effective communication amongst students and teachers; 23% agreed that the Wiki enables fast and slow learners to learn at their own pace; 67% strongly agreed that the Internet creates opportunity for teachers to source for lecturing materials; 45% strongly agreed that being online enables the teachers reach more learners.

Discussion

Data was collected and analysed. The first question on the awareness of current technological tools used by students and teachers during classroom instructions, revealed that both student and teachers are aware of educational technological tools, like laptop, iPad, projector, video recorder, the Internet, video-cast, and Facebook. However, these tools have not been used by the teachers during teaching. Thirty six percent of students' response from Yaba College of Technology indicated that they did not use computer systems during classroom interactions with their teachers, while 42% of students' response from the University of Lagos also indicated that they did not use computer during lectures. Twenty-eight percent of the participants from Yaba College of Technology and 32% from the University of Lagos respectively indicated that they did not use projectors during lectures. Thirty-five percent of the lecturers disagreed that they use computer systems during classroom instructions while 42% disagreed that they use projector during class lessons. The research revealed that despite the awareness of these educational technology tools, they were not integrated in classroom teaching at both institutions sampled. Therefore, it is of uttermost importance that more awareness and usage of these technological tools be encouraged by emphasizing its importance on a consistence basis.

The second and third research questions emphasized on teachers and students awareness and use of Web 2.0 tools during teaching and learning. The research revealed that both the students and teachers are aware of Web 2.0 tools like blogs, Wikipedia materials, iPads, Google, Yahoo, YouTube, e-mails, Web sites, audio podcasting, RSS, video podcasting, etc., of Web. 2.0, but do not use them for teaching and learning purposes except for leisure and personal entertainment. Students in Yaba College of Technology (56.8%) and University of Lagos (54.7%) respectively are familiar with the use of Wikipedia materials while 46.2% and 39.0% of students in Yaba College of Technology have visited blog sites and have also posted on Twitter pages. They have used iPad/podcast (34.0%) and have retrieved materials from the Internet for learning. Similarly, the lecturers have visited several blogs; though they have not created their own blogs.

The forth questions indicated the effect/benefits of the use of Web 2.0 tools on teaching and learning. The opinions of the students and teachers showed in table 7,10 and table 14 respectively, showed that Web 2.0 and other technological is useful in teaching and learning. 48% of students from Yaba College of Technology and 46.6% of students from the University of Lagos respectively agreed that e-mails made communication and learning easier. Some lecturers as well required students to e-mail their assignments to them. 54% of students from Yaba College of Technology and 56% of students from the University of Lagos strongly agreed that Internet facility enhances learning and communication amongst teachers and students. 46.5% and 37.2% of the students' response agreed that social networking tools like Facebook, YouTube, yahoo-chat, and Twitter create effective communication amongst students. Students are able to communication amongst themselves through this medium. They get more materials on the Internet when studying or carrying out a given assignment (47.0%) and wikis enables fast and slow learners to learn at their own pace (38.6%) and (33.6%). Students are able to go on the Internet to get additional material on their own for further studies as the net is a good source for materials for both learners and teachers.

What are the challenges associated with the use of the Web 2.0 during teaching and learning?

The fifth research question deals with the challenges associated with the use of the Web 2.0 and the reasons for not using the Web/Internet and other technological tools to enhance learning. The response from the students (32.5%) revealed that learners (11.6% and 5.4%. See table 6 and table 9) do not use Internet facilities because they do not have access to it in their schools. The usage can be improved when the Internet facility is available at the school.

On educational technology resources in the schools, 7.3% and 5.4% (see tables 6 and 9) of the students indicated the usage is limited by lack of educational technology resources like the computer systems. On the other hand, 12.5% of the lecturers indicated that there are not enough educational technology resources and this limited their usage. Resources like iPod/podcast/video cast are not common in the schools (33.3%) and this pose a challenge in using them. Eighteen percent of the participants indicated that they do not use Wikis because it appears confusing. Twenty two percent of the lecturers opined that one of the problems of using this technology is the inadequate professional training by the teachers on the usage and lack of awareness in the schools. This is consistent with the finding of Muodumogu (2003) that the lack of commitment and training on the part of the teachers affected the use of instructional materials and many educational technology tools.

Others blamed the nonusage of technological tools on the lack of power supply (33.3%) in the schools environment, inadequate provision of the facilities by the necessary agencies, laziness of the lecturers to prepare PowerPoint slides for their lessons etc. There is no budget for the schools to provide these amenities/materials to enhance learning and when not available, there will be no usage. This can be supported with the finding of Muodumogu (2003) that lack of funds to procure the materials and difficulty involved in preparation, storage, and movement of instructional materials hinders their learners/teachers from using them.

Conclusion

Based on the findings of this study, it was concluded that the importance of Web 2.0 and other educational technological tools in the schools cannot be over emphasized. The usage presently is at a low level as some teacher and students are not aware of the benefits of Web 2.0 in teaching and learning. The analysis revealed that Web 2.0 tools are of great importance in teaching and learning in our tertiary institutions as these tools were revealed to be potent in achieving teaching and learning efficiency. Colleges, polytechnics and Universities should be assisted in integrating the use of Web 2.0 tools to enhance teaching and learning however, this should start by an awareness of its benefits.

References

Alexander, B. (2006) "Web 2.0: A new wave of innovation for teaching and Learning?" *Educause,* 41(2), pp 32–44.

Chinnery, G. M. (2006) "Emerging technologies—Going to the mall: Mobile assisted language Elearning," *Language Learning and Technology,* 10(1), pp 9–16.

Downes, S. (2005). 'E–learning 2.0' Association for Computing Machinery. Available online http://www.elearnmag.org/subpage.cf m?section=articlesandarticle=29–1 Accessed 05/05/2012.

Elgort, I., Smith, A. G., and Toland, J. (2008) "Is wiki an effective platform for group coursework?" *Australasian Journal of Educational Technology,* 24(2), pp 195–210.

Ellison, N., and Wu, Y. (2008) "Blogging in the classroom: A preliminary exploration of student attitudes and impact on comprehension," *Journal of Educational Multimedia and Hypermedia,* 17(1), pp 24.

Farmer, B., Yue, A. and Brooks, C. (2008) "Using blogging for higher order learning in large cohort university teaching: A case study" *Australasian Journal of Educational Technology,* 24(2). pp 123–136.

Hall, H. and Davison, B. (2007) "Social software as support in hybrid learning environments: The value of the blog as a tool for reflective learning and peer support." *Library and Information Science Research,* 29(2). pp 163–187.

Hargadon, S. (2009). *Educational networking: The important role Web 2.0 will play in education* (White paper). Retrieved December 12, 2010 from http://audio.edtechlive.com/lc/ EducationalSocialNetworkingWhitepaper.pdf. Accessed on March 22, 2007.

Kvavik, R. B., Caruso, J. B. and Morgan, G. (2004). ECAR Study of students and information technology 2004: Convenience, connection, and control. Boulder, CO: EDUCAUSE Center for Applied Research. Available: http://www.educause.edu/ir/library/ pdf/ers0405/rs/ers0405w.pdf. Accessed on March 22, 2007.

Lamb, B. (2004) "Wide open spaces: Wikis, ready or not," EDUCAUSE Review, 39(5). pp 36–48.

Lenhart, A. and Madden, M. (2005) "Teen content creators and consumers" Pew Internet and American life Project, Available online at http://www.pewinternet.org/pdfs/ PIP Teens Content Creation.pdf. Accessed on January 16 2012.

Luce-Kapler, R. (2007) "Radical change and wikis: Teaching new literacies," *Journal of Adolescent and Adult Literacy*, 51(3). pp 214–223.

Mason, R., and Rennie, F. (2007). Using web 2.0 for learning in the community. *The Internet and Higher Education, 10*(3), 196–203.

Miller, D. B. (2006) "Podcasting at the University of Connecticut: Enhancing the educational experience" *Campus Technology*. Available online at http://campustechnology.com/news article. asp?id=19424&typeid=156. Accessed January 14, 2012.

Parker, K. R. and Chao, J. T. (2007) "Wiki as a teaching tool" *Interdisciplinary Journal of Knowledge and Learning Objects, 3.*

Prensky, M. (2001) "Digital natives, digital immigrants," *On the Horizon,* 9(5). pp 1–6.

Roberts, D. F., Foehr, U. G., and Rideout, V. (2005). *Generation M: Media in the lives of 8–18 year-olds.* Menlo Park, CA: Kaiser Family Foundation.

Russell, M. and Tao, W. (2004). Effects of handwriting and computer-print on composition scores: a follow-up to Powers, Fowles, Farnum, and Ramsey.

Practical Assessment, Research and Evaluation, 9(1). Available online at http://PAREonline.net/getvn.asp?v=9andn=1. Accessed January 14, 2012.

Sanchez, C. A., Wiley, J., and Goldman, S. R. (2006). Teaching students to evaluate source reliability during Internet research tasks. In S. A. Barab, K. E. Hay, and D. T. Hickey (Eds.). *Proceedings of the seventh international conference on the learning sciences* (pp. 662–666). Mahwah, NJ: Erlbaum.

Sandars J. and Schroter S. (2007) "Web 2.0 technologies for undergraduate and postgraduate medical education: an online survey," *Postgraduate Medical Journal*, 83(986). pp 759–62.

Siemens, G. (2009) 'Connectivism and Connective Knowledge' Available online at http://ltc.umanitoba.ca/connectivism/?p=189. Accessed 05/15/2012.

Strom, R. D. and Strom, P. S. (2007). *New directions for teaching, learning, and assessment,* Netherlands: SpringerVerlag.

Toland, J., and Klepper, R. (2008). Business-to-consumer electronic commerce in developing countries. In Mehdi Khosrow-Pour (Ed.), *Encyclopaedia of Information Science and Technology* (2nd ed., pp. 489–494). Hershey, PA, Information Science Reference.

Williams, J. B. and Jacobs, J. (2004) "Exploring the use of blogs as learning spaces in the higher education sector," *Australasian Journal of Educational Technology,* 20 (2). pp. 232–247.

Woodward, J. (2007) *Podcasts to support workshops in Chemistry,* Available online at https://breeze.le.ac.uk/lfconimpalajonny/ Accessed on March 22, 2012.

Xie, Y., Ke, F. and Sharma, P. (2008) "The effect of peer feedback for blogging on college students' reflective learning processes," *The Internet and Higher Education,* 11(1). pp. 18–25.

SECTION 6

Integration of Educational Technology in Health Education

Chapter 17

Integrating Electronic Learning Technology and Tools into Medical Education and Training

T. Oluwatobiloba Olatunji
&
Adekunle Olusola Otunla

Introduction

With the advent of information communication technology (ICT) and its wide applications and tools at all levels of education, medical instruction has taken new directions. This chapter explores the concept, tools, and application of existing innovative educational technologies to medical teaching and training. The chapter also deals with and highlights existing local and global efforts in integration, the implementation process using various methodological and pedagogical approaches, and their attendant challenges while proffering some solutions. Electronic learning (e-learning) has immense benefits that are specific to medical education and training because the core of medical teaching, training, and instruction lies in practical hands-on skills.

E-learning approaches afford the students the privilege of providing individualised virtual learning environments where the student interacts with simulated, standardized, or typical patients with varied accessibility to learning materials such as notes, videos, audio, etc. in the form of clinical instructions. E-learning provides opportunities for self-administered, comfortably paced learning approaches and processes, which could be repeated as many times as possible with no discomfort to the simulated patient. The listed fundamental steps to the integration and implementation processes was recommended such as content development, content structure standardization, content organization/management, and content distribution based on any chosen learning medium.

E-learning is a generic term that is closely associated with much earlier technological innovations such as Web-based learning (WBL), online learning (OL), distributed learning (DL), computer-assisted instruction (CBI), or Internet-based learning (IBL). It refers to the use of Internet tools and technologies to enhance knowledge and performance. Historically, there have been two common e-learning modes, open and distance education (ODE) and computer-assisted instruction (CAI). Open and distance education uses information and communications technology tools and devices to deliver instruction to learners who are at remote locations from a central site (Bernard, Abrami, Lou & Borokhovski, 2004).

Computer-assisted instruction, also called computer-based learning (CBL) and computer-based training (CBT), uses computers to assist in the delivery of learning through the use of stand-alone multimedia computers, programmed instruction (PI) packages, and self-paced learning and teaching packages (Ferguson and Wilson, 2001; Smith 2004). A related term or concept is multimedia learning (MML), which refers to the use of two or more media such as text, graphics, animation, audio, or video to produce engaging content that learners access via computer. Another innovative strategy is the term "blended learning," which is a fairly new approach in education but a concept familiar to modern educators. It is an approach that combines e-learning technology with traditional instructor-led training, where, for example, a lecture or demonstration is supplemented by an online tutorial (Otunla, 2013a).

Studies on Integration of Computers in Medical Teaching and Training

Medical education refers to "a process of teaching, learning and training of students with an ongoing integration of knowledge, experience, skills, qualities, responsibilities and values which qualify an individual to practice medicine" (IIME, 2001). This definition implies that medical education is a branch of education, but the field of medicine is dynamic as new discoveries constantly emerge on preventive methods, diagnostic techniques, and patient management. Medical education is learning related to the practice of being a medical practitioner, either the initial training to become a physician or additional training thereafter.

Within the last one and a half decades, changes in health care delivery and advances in medicine and especially public health due to advances in technology, telemedicine, bio-technology, medical informatics, health informatics, and nursing informatics, among other emerging fields of studies, have placed increased demands and challenges on medical teachers who are mostly practicing consultants. Medical teachers always find it difficult to stop an ongoing medical lecture to bring the student back to a point that might have been missed. With e-learning, it is possible to pause, rewind, and fast forward a lecture to listen over and over again till the learner is satisfied with content assimilation.

Thus, Ruiz, Mintzer and Leipzung (2006) concluded that as a result of professional demands on today's medical educators, they are facing different challenges than their predecessors in teaching tomorrow's physicians. Moreover, traditional instructor-centred teaching is yielding to a learner-centred model that puts learners in control of their own learning. A recent shift towards competency-based curricula emphasizes the learning outcome, not the process, of education (Gibbons, and Fairweather, 2000). Other issues are related to problems associated with finding time to teach new concepts, themes, and emerging health hazards that are products of climatic changes, ozone layer depletion, bio-terrorism, and so on; all these are limitations to covering conventional course content as contained in the medical school curricula. Further, patient rights and privacy issues represent the major restraint for the traditional apprenticeship approach. As patients become more aware of their rights and hospital bills are on the increase, patients find it more difficult to consent to hands-on practice with several medical students enthusiastic to perform their first vaginal or abdominal examination.

E-learning has gained popularity in the past one and a half decade; however, its use is highly variable among medical schools (Chumley-Jones, Dobbie, and Alford, 2002; Izet, 2008). E-learning technologies offer learners control over content, learning sequence, pace of learning, time, and often media, allowing them to tailor their experiences to meet their personal learning objectives. In diverse medical education contexts, e-learning appears to be at least as effective as traditional instructor-led methods such as lectures. Students do not see e-learning as replacing traditional instructor-led training but as a complement to it, forming part of a blended-learning strategy (Johnson, Hurtubise, and Castrop, 2004). E-learning can be used by medical educators to improve the efficiency and effectiveness of educational interventions in the face of the social, scientific, and pedagogical challenges created within the learning environment (Otunla, 2012, Otunla, 2013b). Self-directed, self-paced learning is perhaps the most important reason because learners have the liberty, first to choose their most convenient style or mode of learning and, second, to progress through the content in their preferred order, answer evaluation questions, and assess their performance in a cycle till they have achieved the required level of assimilation.

Effectiveness and Efficiency of E-Learning in Medical Education

The effectiveness of e-learning has been demonstrated primarily by studies in higher education, government, corporate, and military environments. However, these studies have limitations, especially because of the variability in their scientific design. Often they have failed to define the content quality, technological characteristics, and type of specific e-learning intervention being analysed. In addition, most have included several different instructional and delivery methodologies, which complicate the analysis while other studies compared e-learning with traditional instructor-led approaches (Rosenberg, 2001; Ruiz, Mintzer, and Leipzig, 2006). Yet three aspects of e-learning have been consistently explored: product utility, cost-effectiveness, and learner satisfaction. Utility refers to the usefulness of the method of e-learning. Several studies outside of health care have revealed that most often, e-learning is at least as good as, if not better than, traditional instructor-led methods such as lectures in contributing to demonstrated learning.

For instance, Gibbons and Fairweather (2000) cited several studies from the pre-Internet era, including two meta-analyses that compared the utility of computer-based instruction to traditional teaching methods. The studies used a variety of designs in both training and academic environments, with inconsistent results for many outcomes. Yet learners' knowledge, measured by pretest and posttest scores, was shown to improve. Moreover, learners using computer-based instruction learned more efficiently and demonstrated better retention. Other reviews of the e-learning (specifically Web-based learning) literature in diverse medical education contexts reveal similar findings. Chumley-Jones, Dobbie, and Alford (2002) reviewed 76 studies from the medical, nursing, and dental literature on the utility of Web-based learning. About one-third of the studies evaluated knowledge gains, most using multiple-choice written tests, although standardised patients were used in one of the studies. In terms of learners' achievements in knowledge, Web-based learning was equivalent to traditional methods. Among two other studies evaluating learning efficiency, only one demonstrated evidence for more efficient learning via Web-based instruction.

Further, Chumley-Jones, Dobbie, and Alford (2002) reported that a substantial body of evidence in the nonmedical literature has shown, on the basis of a sophisticated cost analysis, that e-learning can result in significant cost savings, sometimes as much as 50%, compared with traditional instructor-led learning. Savings are related to reduced instructor training time, travel costs, and labour costs, reduced institutional infrastructure, and the possibility of expanding programs with new educational technologies. Only one study in the medical literature evaluated the cost-effectiveness of e-learning as compared with text-based learning. The authors found the printing and distribution of educational materials to be less costly than creating and disseminating e-learning content. Studies in both the medical and nonmedical literature have consistently demonstrated that students are very satisfied with e-learning. Learners' satisfaction rates increase with e-learning compared to traditional learning, along with perceived ease of use and access, navigation, interactivity, and a user-friendly interface design (Otunla, 2012). Interestingly, students do not see e-learning as replacing traditional instructor-led training but as a complement to it, forming part of a blended-learning strategy.

Process Overview of E-Learning Materials and Content Development

Creating e-learning material involves several components; once content is developed, it must be managed, delivered, and standardized. Content comprises all instructional material, which can range in complexity from discrete items such as texts, audio, video and digital images, and objects as well larger instructional modules. A digital learning object is defined as any grouping of digital materials structured in a meaningful way and tied to an educational objective. Learning objects represent discrete, self-contained units of instructional materials assembled and reassembled around specific learning objectives, which are used to build larger educational materials such as lessons, modules, or complete courses to meet the requirements of a specified curriculum (Otunla, 2013a). E-learning typical examples include tutorials, case-based learning, hypermedia, simulations, and game-based learning modules. Content creators use instructional design and pedagogical principles to produce learning objects and instructional materials. Content management includes all the administrative functions (e.g., storing, indexing, cataloging) needed to make e-learning content available to learners. Examples include portals, repositories, digital libraries, learning management systems (LMSs), search engines, and e-portfolios. Johnson, Hurtubise, and Castrop (2004) described the learning management system (LMS) as an Internet-based software and platform that facilitates the delivery and tracking of e-learning across an institution. LMSs can serve several functions beyond delivering e-learning content; it can simplify and automate administrative and supervisory tasks, track learners' achievement of competencies, and operate as a repository for instructional resources twenty-four hours a day. Learning management systems that are mostly familiar to medical educators are Moodle®,

WebCT®, or Blackboard®, but there are more than 200 commercially available systems, a number that is growing rapidly. E-learning content delivery may be either synchronous or asynchronous; synchronous delivery refers to real-time, instructor-led e-learning where all learners receive information simultaneously and communicate directly with other learners (Otunla, 2013c). Examples include teleconferencing (audio, video, or both), Internet chat forums, and instant messaging. With asynchronous delivery, the transmission and receipt of information do not occur simultaneously. The learners are responsible for pacing their own self-instruction and learning. The instructor and learners communicate using feedback technologies such as e-mail, blogs, wikis, short messages / chats, Facebook, Twitter etc. but not in real time (Otunla, 2013c). Therefore, a variety of approaches and tools could be adopted for asynchronous delivery; these include e-mail, online bulletin boards, newsgroups, and Weblogs.

E-learning Formats, Tools, Resources, and Innovative Approaches

There are many innovative e-learning formats, tools, resources, and innovations that are applicable to medical teaching, training, and instructions (Rosenberg, 2001; Ward, Gordon, Field and Lehmann, 2001). Some of these may be as simple as what medical teachers could develop themselves or as complex as those designed using special authoring software and requires professionals for their design and development. Some of them are discussed as follows:

- **Portable Document Format (pdf)**: This is the most common and most basic form of electronic learning in medical education where lecture notes are typed out, clinical pictures, surgical procedures, and laboratory procedures are prepared and simply formatted into a PDF document and made available to learners for self-study.
- **PowerPoint Presentations Slides:** This is another most common and most basic form of electronic learning in medical education where lecture notes, clinical pictures, video of surgical procedures and laboratory procedures are developed and formatted into slideshows and made available to medical students for personal revision and self-study.

Slides and enhanced PowerPoint could be formatted into PDF and embedded or hyperlinked to relevant online medical documents, content, charts, diagrams, journals, clinical procedures in videos, and so on.

- **Video/Audio:** Video tutorials simplify steps involved in learning clinical skills like physical examination, patient communication, and surgical procedures. This format involves storing the procedure or conversation electronically using video or audio recorders, editing the data, using basic or professional software to highlight important areas, as well as removing noise and unwanted areas, adding texts, images, objects, arrows, illustrations, diagrams, charts, etc. to follow or explain the procedure, and exporting to commonly readable formats like AVI, MP4, WMV, MPG, and so on.

- **Mobile Applications**: With the growing popularity of mobile platforms like iOS by Apple, Android by Google, and the proliferation of several mobile applications, medical educators are also exploring this medium of knowledge transfer by developing games, flash cards of practice questions (based on specialty), mobile versions of popular textbooks like the Oxford handbook, anatomy atlases, and so on. It is still a recent medium, but extensive work has already been done to create interactive material, study aids, mnemonics, medical dictionaries and so on make them available on mobile platforms.

- **Mannequin Simulation:** It refers to the use of electronically preconfigured mannequins (life-like dolls) to mimic specific disease conditions or trauma cases like airway obstruction, myocardial infarction, chest trauma, and so on. Learners are then allowed to practice various management skills after demonstration sessions by the tutors. Procedures like physical examination can be practiced over and over again on the low-fidelity mannequins or dolls. High-fidelity mannequins can be otherwise be programmed to simulate several conditions including choking, obstructed labor, and seizures and even respond to administration of intravenous fluids and other management modalities. This is brought about by advanced programming techniques integrating software with hardware interfaces. Mannequins can also be used to practice dental extractions, biopsies, cesarean sections, and several others.

- **Games**: Games and simulations could be used on both desktop and mobile operating systems; medical games have widespread application, particularly to test retention and assimilation of learning content. It is also used to test decision-making quality. Commonly, a word puzzle, time-based test, or competition can be administered via a game interface. Learners score points and get rewards as they progress through various levels of difficulty. Several exist, some of which are Workout Trainer, the Urinary Anatomy game, Grey's Anatomy.

- **Virtual Reality:** This is very similar to how pilots are trained and race drivers prepare; virtual reality, which is closely related to games or an advanced application of gaming, refers to the creation of specific virtual environments that are created such as a road traffic accident, mass casualty, postpartum hemorrhage, bleeding, ectopic pregnancy. It is created through computer graphics and accessed via head mounted display (HMD), data gloves, or specialised input devices. This mainly tests decision-making skills and allows users to practice their hands-on skills. Scenarios can be repeated several times and difficulty levels adjusted to the learner's abilities. Such scenarios may be difficult to come by or even rare, but students have the opportunity to be exposed to it in special labs where these devices exist. Management of other medical conditions like HIV/AIDS, abortions, and psychiatric counseling are other existing applications of this technology to help doctors test the results of their decisions through complex algorithms developed based on specific cases managed and several possible outcomes following the progress of several decisions made during management.

- **Robotics:** Also referred to as robotic surgery, computer-assisted surgery, or robotically assisted surgery, this is the method of performing surgery using very small tools attached to a robotic arm remotely controlled by a surgeon. Apart from its widespread use in minimally invasive surgical intervention and advanced assisted open surgery, its application in surgical training is important. Using 3D generated images of live patients (through MRI or CT scans of areas of interest), it is possible to perform virtual surgery as practice before the actual procedure. A virtual three-dimensional theatre is created, and surgeons can use buttons, keyboards, or other specialised control equipment to cut, ligate, retract, repair, and so on prior to the main procedure or as regular scholarly practice.

E-learning Integration Requirements and Implementation Phases

Political will remains a major limitation in taking steps towards integration and adoption of electronic learning in medical education technology especially in sub-Saharan countries in Africa. Constitutional support in form of laws, policy statements, and decrees by the governing and academic boards of medical schools or colleges of medicine must provide legal backing for the implementation and financing of electronic medical education. Without the legal backing, any effort will result in isolated, uncoordinated, unstandardised content at best. Next to it is engagement of highly qualified information technology (IT) staff and provision of IT infrastructure to support any e-learning modality chosen or adopted. It also implies that IT consultants must be appointed, and IT staff must help in guiding the decision making to commence an IT project that it feasible and sustainable. IT equipment like computers, servers, backup storage, recording, and editing equipment must be acquired or leased to facilitate easy content creation and management. High-speed Internet facilities must be provided to enable easy and hitch-free content distribution.

The acceptance and cooperation of medical teachers and students alike must be sought through conduct of preliminary or baseline survey on staff (both teaching and nonteaching) and students' competence, skills, and knowledge in deployment and use of technology in medical teaching and training. Data gathered from the baseline will guide and facilitate easy execution and adoption of necessary modalities for innovative e-learning technology in medical education. Collaboration with local and foreign organisations and institutions who have implemented such technologies is also crucial to the success of such integration. Such collaboration shall prevent reinventing the wheel and can help avoid several problems that may stall the execution of the e-learning process. The collaboration can provide resources, finances, equipment, and ad hoc staff that can ease the process. Finance is the major hurdle for most medical schools looking to adopt electronic education. University funds, grants, donations, fundraising events, special student levies/fees, and partnerships with large ICT firms are some of the ways medical schools are surmounting this challenge.

Assessment methods have to be predefined, and the indexes must drive or determine technologies to be selected. Each medical college or institution must determine what works best within their unique local environment, what will produce the desired learning outcomes, and how to measure, evaluate, or assess them. The integration of e-learning into existing medical curricula should be the result of a well-devised plan that begins with a needs assessment and concludes with the decision to use e-learning. Although some institutions have tried to use e-learning as a stand-alone solution to updating or expanding their curricula, the best approach is to begin with an integrated strategy that considers the benefits and burdens of blended learning before revising the curriculum. E-learning offers students tools and materials for self-instruction and collaborative learning at both undergraduate medical education and postgraduate medical education levels. For instance, the Accreditation Council for Graduate Medical Education (ACGME) in the United States of America has established six core competencies towards which e-learning can be applied.

E-learning materials suited for each of these competencies can be integrated into the education of residents and fellows, replacing lectures and other synchronous methods of instruction. Asynchronous e-learning can be effectively used during demanding clinical care rotations, especially when duty hours are limited yet curriculum requirements remain high. In continuing medical education, physicians with daily clinical obligations can attend medical e-conferences using e-learning. The complexity and breadth of medical education content, together with the scarcity of experts and resources in e-learning, make the creation of centres of excellence in e-learning a reasonable proposition.

A recommended stepwise integration plan for tropical medical schools is provided as follows (Ferguson and Wilson, 2001):

- **Policy statement:** Policy formulation or adoption is the first step in integration; the university college of faculty must officially announce its decision with a detailed policy document to all stakeholders: staff, students, parents, and members of the public.
- **Needs assessment:** An assessment must follow to document and establish the demand for the intervention. This can be administered via a questionnaire, manually or electronic, to students and teaching staff to assess their knowledge, attitudes, acceptance, and concerns for integration. This can be compared with results for other institutions or one-year post execution results or, most importantly, used to drive decision making.
- **Enlightenment Workshops:** Seminars and workshop must be conducted for teachers and students to bridge knowledge gaps and highlight the numerous benefits of using e-learning or blended learning in terms of cost, time, learning, distance, and so on. Committees must be set up to carry out the assignment of implementation; it may include and ICT committee that handles the technical aspects, an academic committee that selects the specific topics or areas to be covered and learning outcomes, and a student feedback committee, which handles tests and samples the material first-hand before distribution and collates and coordinates student input to the process. Process design must then ensue with contribution of all committees and external collaborators, taking into consideration long-term sustainability, cost, adoption, and distribution channels.
- **Resource audit:** The audit will highlight existing infrastructure based on process design decisions; this is to ascertain the necessary human and material resources available to implement the solution. It will also further highlight staffing and resource needs.
- **Equipment purchase:** Equipment that is required should be identified based on the audit so that costs are not duplicated. All necessary nonexistent resources must be purchased on

a long-term basis starting with the most urgently required in view of available funding. Purchasing all equipment at once may be daunting and stall the project; it may also lead to pilferage and wasting of some of the equipment. Purchase process must be well supervised and the input of experts sought to avoid wastages and mismanagement of scarce resources.

- **Pilot study:** A pilot study should be launched to test the process from content creation to management and distribution.
- **Feedback:** Results from the pilot study should involve instructors, collaborators, and the end users (students) at this stage to highlight improvements, errors, gaps, and all necessary inputs to amend and perfect the process.
- **Process redesign and standardisation**: The academic committee will finalise certain standards based on the pilot redesign, which will form a template for all others with some room for specialty-specific adjustments.
- **Content creation schedule:** This must be agreed upon by the academic committee starting with the specialty with least resistance to serve as a standard for the rest. Once a start point has been determined, a schedule must be developed to cover as many departments/units/specialties as desired in the first phase of the project. Medical schools might begin with pediatrics, surgery, anatomy, etc., and others follow based on the local terrain.
- **Content creation:** This phase marks the beginning of several cycles of producing e-learning contents such as slides, video/audio, scenarios, games, and other formats as earlier enumerated from department to department. The content collection stage involves collation of all materials developed to a central source where content is organised into learning modules).
- **Standardisation** for consistency will be checked and effected.
- **Content Packaging and Distribution:** This will take place via a learning management systems adopted by the institution; medical education lab, or volumes of digital video discs based on the local terrain.

- **Feedback:** Feedback is again necessary as e-learning content will further be upgraded and updated from time to time to keep the contents relevant and user-oriented.

Challenges and Opportunities

Resistance to change remains the foremost setback for integration of e-learning in medical education in most tropical medical colleges. Lecturers usually complain of the additional workload to their already tight schedules in the short term without consideration the immense benefits in the long term. Lack of political will in among institutions administrators is another roadblock, especially where those in positions of authority fail to consider or accept the benefits of technology use. Lack of finance is a major hurdle as resources are scarce and must be distributed among other needs. An e-learning project is cost intensive, but this can be mitigated by breaking it into phases over years, adopting some of the practical solutions already stated earlier. Fear of truancy remains a major deterrent to adoption, and this is averted by strict attendance monitoring, spread-out continuous assessment, in-class level of contribution per student, problem-based learning, and blended learning. Internet connectivity remains a major excuse, but since almost every student is accessing Facebook via the Internet, the same mode could be adapted for learning purposes. Another innovation is the laboratory or CD-based distribution systems that require local cabled networking in form of an intranet.

Conclusion

In conclusion, design, development, and use of e-learning technologies are viable alternatives to insufficient human materials for clinical teachings. Therefore, course content modules, anatomical teachings, patients' case history, clinical diagnostic procedures, theatre routines, laboratory experiments, and practices could be developed into e-learning materials in the form of blended learning and used as supplements to face-to-face interactions and for medical teaching, training, and instructions.

Recommendations

For medical colleges to integrate e-learning formats, tools, and resources into their undergraduate and postgraduate curriculum, the following recommendations are put forward.

- Mentoring of young and junior medical academic staff and consultants by senior medical teachers and professors for computer technology adoption and diffusion in undergraduate medical teaching and training must be encouraged.
- ICT infrastructures and capacity building in computer integration should be intensified by medical colleges and institutions in Nigeria.
- Nongovernmental organizations and donor agencies should come to the assistance of the medical colleges to expand existing computer and ICT facilities possibly through public-private-partnerships (PPP), build, operate, and transfer (BOT) schemes among other partnership programmes.

References

Bernard, R., Abrami, P. L., Lou, Y., and Borokhovski, E. (2004). How does distance education compare with classroom instruction? A meta-analysis of the empirical literature. Rev Educ Res. 74:379–439.

Chumley-Jones, H. S., Dobbie A. and Alford C. L. (2002) Web-based learning: sound educational method or hype? A review of the evaluation literature. Acad Med. 77 (10 suppl): S86–S93.

Ferguson, J. D. and Wilson, J. N. (2001). Process redesign and on-line learning. International Journal of Educational Technology. Retrieved from: http:// IJET@lists.uiuc.edu

Gibbons, A., and Fairweather, P. (2000) Computer-based instruction. In: Tobias S, Fletcher J (eds). Training and Retraining: A Handbook for Business, Industry, Government, and the Military. New York: Macmillan Reference USA, 410–42.

Institute for International Medical Education (2001). *Global minimum essential requirement in medical education, document approved for cure committee members,* 2, 7, 19 and 32.

Izet, M. A. (2008). E-learning as new method of medical education. Informatica Medica, Johnson, C. E., Hurtubise, L. C. and Castrop J. (2004). Learning management systems: technology to measure the medical knowledge competency of the ACGME. Med Educ. 38:599–608.

Otunla, A. O. (2012). Utilization Patterns in and impact of computer-mediated Learning on Medical Undergraduate Students' Pre-clinical Skills. Unpublished Doctoral Dissertation, University of Lagos, Akoka-Lagos, Nigeria.

Otunla, A. O. (2013a): Embedded Assessment Strategies for E-Learning Technologies. In Adams O. U. Onuka (Ed.) Learning. Society for the Promotion of Academic and Research Excellence (SPARE). Ibadan; Pp. 36–47.

Otunla, A. O. (2013b): University Lecturers' Information Literacy Skills In Relation To Computer-Mediated Professional Development. Information Systems, Development Informatics and Business Management. Vol. 4 No. 3 September, 2013 pages 71–76. Available @http://cisdijournal.net/uploads/ V4N3P11-CISDI - Otunla.

Otunla, A. O.(2013c): Operational processes for Computer-Mediated Learning in Management of Large Surgical, Theatres, Laboratory and Teaching Sessions in Medical and Health Sciences. In U. M. O. Ivowi (Ed.), Seeking Total Quality Education in Nigeria: A Book of Reading in honour of Prof. T. D. Baiyelo. Lagos; Foremost Educational Services Ltd. Pp. 313–327.

Rosenberg, M. (2001). E-Learning: Strategies for Delivering Knowledge in the Digital Age. New York: McGraw-Hill.

Ruiz, J. G., Mintzer, M. J. and Leipzig, R. M. (2006) The Impact of E-Learning in Medical Education. Academic Medicine. 81 (3) pp 207–212.

Smith, R. (2004) Guidelines for Authors of Learning Objects (http://www.nmc.org/guidelines/NMC/20LO/20Guidelines.pdf). The New Media Consortium, Austin, TX.

Ward, J. P., Gordon, J., Field M. J. and Lehmann, H. P. (2001). Communication and information technology in medical education. Lancet.;357:792–96.

Chapter 18

Applications of Innovative Educational Technology Tools in Teaching and Learning of Health Education

Georgy O. Obiechina

Introduction

Health education does not only need new ideas and innovations that shatter the performance expectations to meet the new millennium goal of education but also make a meaningful impact on students and teachers. These new innovations must grow large enough to serve millions of health education students and teachers. The current technologies have a positive transformative role to play in shaping the future of health education in Nigeria's educational institution especially at colleges of education and universities. Health education is among the main catalysts for the growth and progress of any society. The Joint Committee on Health Education and Promotion Terminology (JCHEPT) (2011) defined health education as any combination of planned learning experiences using evidence-based practices or sound theories that provide the opportunity to acquire knowledge, attitudes and skills needed to adopt and maintain healthy behaviour. It is also a process that informs, motivates, and helps people to adopt and maintain healthy practices and lifestyles, advocates environmental changes as needed to facilitate goal, and conducts professional training and research to the same end (Marcus, 2012).

Health education does not only impart knowledge, skills, values and belief, but also is responsible for building human capital which breeds, drives and sets technological innovation and economic growth of any nation. In this era, information, communication and education stand out as very important and critical input for growth and healthy survival of any nation. Many learning environments have looked to technology in their efforts to redesign teaching and learning programmes. While technology integration has long been a key area of concern in education, the intersection of technology with its rapidly transforming educational landscape is framing the nature of technology in education in profound new ways. The new technologies are provoking a reconceptualisation of teaching and learning, while also serving as catalysts for transformation and innovation.

Traditional and nontraditional methods of teaching and learning also constitute stepping stone in health education.

Traditional Method of Teaching in Health Education

Traditional education also known as back-to-basics, conventional education or customary education. It refers to long-established method found in schools that society traditionally used. The teaching method promotes the adoption of progressive educational practices, a more holistic approach which focuses on individual students' needs and self-expression. Traditional teacher-centred use of lecture and discussion method, aimed at learning and memorization of information. The method includes the use of blackboard, chalk, slate, exercise book, whiteboard and marker, transparent slides and overhead projector, computer programmes, multimedia projector, PowerPoint, video among others. Health education students make use of textbooks, attend lectures, write notes as lecturer dictates, formulate question and do oral presentation. However, recent developments in health education revealed the adoption of the use of technology which allowed a number of important changes both for the students and the educators.

Nontraditional methods of teaching and learning in health education

Nontraditional methods of teaching are commonly known as innovative or modern teaching methods that involves the use of technology, animation, special effects and are generally learners' self-direct and interactive in nature; for example the use of computer, e-learning, videos and media to enhance learning. Nontraditional methods influence communication and retention of important concepts.

In the twenty-first century, health education has embraced a new meaning and identity. The slow but obvious evolution in health education sees a drastic change of how learners' critically think and learn through the use of various teaching methods (Kidd, Kinsley, and Morgan, 2012). With evolution of education, there is also a shift in the use of traditional teaching methods from lecturer style of teaching to nontraditional teaching methods such as demonstration by lecturer, use of overhead projector (OVH), viewing of prerecorded demonstration on video tapes. Other aspects of nontraditional method include the use of virtual environment, e-learning, information and communication technology (ICT) in education, learning technology, multimedia learning, technology-enhanced learning (THL), computer-based instruction (CBI), online education, network learning. These help to engage students in various ways.

Metamorphosis of Health Education into Health Promotion

Health education is essentially a social process through which individuals acquire knowledge, skills, and dispositions relating to improvement and maintenance of their health and that of others. It is an instrument of values transmission, a conscious and unconscious effort at behaviour modification to the end that individuals or groups understand, accept and adjust to healthful living standard. Health education provide channels for delivering programmes, provide access to specific populations in an existing communication system for diffusion of programmes, and facilitate development of policies and organizational change to support positive health behaviour. Major settings that are relevant to contemporary health education include schools, communities, worksites, health-care settings, homes, the consumer marketplace, and the communications environment.

Health Education is an area within professions which encompasses environmental health, physical, social, emotional, intellectual and spiritual health. Their responsibilities as verified through the 2010 health educator job analysis project includes the following and serve as the basis for conducting examination for students.

1. Assessing individual and community needs for health education
2. Plan health education strategies, interventions and programmes
3. Implement health education strategies, intervention and programmes
4. Conduct evaluation and research related to health education
5. Administer health education strategies, interventions, and programmes
6. Serve as a health education resource person
7. Communicate and advocate for health and health education

WHO in collaboration with other organization held their first international conference which resulted in the Ottawa Charter for Health Promotion 1986. WHO (2005) defined health promotion as the process of enabling people to increase control over their health and its determinants and thereby improve their health. To reach a state of complete physical, mental and social wellbeing, an individual or group must be able to identify and to realize aspiration to satisfy needs and to change or cope with the environment. The main aims of health promotion are prevention of illness, restoration of the sick, and rehabilitation. Health promotion techniques have had its origins in health education with early promotion focusing on changes to an individual's health by modifying individual behaviours. It is also embedded in other disciplines.

The scopes of health education techniques broaden to include wellness and holistic approach. The conference of wellness held in 2013 by National Wellness Institute (NWI) in Wisconsin noted that wellness is a very dynamic subject that has to be observed throughout one's lifetime for an individual to attain good health. Wellness is more of a lifestyle with integration of seven different components namely physical, emotional (mental), intellectual, social, environment, occupational and spiritual health that expands ones potential to live and work effectively and to make a significant contribution to social life (Corbin et al, 2011). It is the positive component of optimal health.

Holistic is a wellness approach that addresses the body, mind and spirit or the physical, emotional, mental and spiritual aspects of an individual (Elizabeth and Martin, 2007). A holistic approach encourages the individuals to engage in self-care and educate themselves about their health. It urges them to be active participant in their treatment and health care, rather than giving all the power to a health-care provider.

Innovative Application

Innovation is a new, sustainable approach that has led to an overall improvement in the students' experience, which is supported by evidence. Consequently, the student, individual and institutional perceptions and interpretations of innovation are subjective and vary. Practice that may be viewed and described as innovative by one student group or institution may be common place in another. However, since innovative practice is time bound and operating within a rapidly changing environment, it may be viewed as both innovative and common practice simultaneously, across institutions.

Kamen (2011) stated that technology ushers in fundamental structural changes that can be integral to achieving significant improvements in productivity. Moore (2011) noted that in 2005, only 9% of students performed adequately in mathematics. With introduction of innovation in schools after three years, math scores have gone up to 62% and violence and behaviour issues have decreased dramatically. Students have reported increased motivation to do school work because the technology, and the work, is more stimulating—resulting in a demonstrated greater investment in their educational journey

There are several definitions of innovation in the literature, for example, the Advisory Committee on Measuring Innovation in the 21st Century Economy (2007) defines innovation as the design, invention, development and/or implementation of new or altered products, services, processes, systems, organizational structures, or business models for the purpose of creating new value for customers and financial returns for the firm. Innovation is the implementation of a new or significantly improved product (good or service) or process, a new marketing method, or a new organizational method in business practices, workplace organization, or external relations (UNESCO Institute for Statistics, 2005).

Health education needs to break away from traditional ways of thriving and nurturing an environment of innovation. Innovation in health education generally can be considered as a process that brings together various health education ideas in a way that they have an impact to promote the health of the society. It could be linked to positive changes in efficiency, productivity, quality, competitiveness, among others. True health education innovations are those products, processes, advocacy, strategies and approaches that improve significantly, health behaviour and lifestyle of an individual. Health education technology is the effective use of technological tools in learning. Such tools include media, machines and networking hardware, as well as considering theoretical perspectives for their effective application (Selwyn, 2011). Health education technology has become an important part of society today where every individual is mindful of their lifestyle and good health.

This chapter anchors on some of the educational technology tools in teaching and learning in health education. These tools are e-learning, instructional technology, information, and communication technology (ICT) in education, learning technology, multimedia learning, technology-enhanced learning (THL), computer-based instruction (CBI), online education, network learning, distance learning, virtual learning environment, virtual library, interactive whiteboard (IWB), and information, education and communication (IEC).

E-learning in Health Education

E-learning in health education is the teaching and learning that is facilitated through information, communication and technology (ICT) both in and outside the classroom setting. E-learning is education via electronic media normally on Internet. It is a broad term used in all form of educational technology, computer based instruction, online education, multimedia learning, technology enhanced learning, and virtual learning. However, e-learning can be defined as the use of Internet technologies to deliver a wide range of solutions that enhance knowledge, skills and performance to people. These have become an important component of today's teaching and learning process in health education institutions. Although e-learning tools have been used in many settings for long, evidence of their use in health education is spare. It is also designed to fit the needs of every type of learner lifestyle and the students.

Broadbent (2010) identifies four types of e-learning in education namely: informal, self-paced, and leader-led and through performance support tools. In informal e-learning, a learner could access a Web site or join an online discussion group to find relevant information. Self-paced e-learning on the other hand refers to the process whereby learners' access computer based or Web-based training materials at their own pace. Leader-led e-leaning as the name suggests refers to an instructor, tutor or facilitator leading the process. This type of learning can further be divided into two categories: learners accessing real-time (synchronous) learning materials and learners accessing delayed learning materials (asynchronous). The fourth and last type of e-learning described is through the use of performance support tools which refers to materials that learners can use to help perform a task (normally in software) such as using a wizard.

Asynchronous e-learning is a student's-centred teaching method that uses online learning resources to facilitate information sharing outside the constraints of time and space among a network of people (Broadbent 2010). The online resources used to support asynchronous learning include e-mail, electronic mailing lists, threaded conferencing system, online discussion boards among others. The students require being actively involved with and taking more responsibility for their own learning. However, students are required to;

(a) Become proficient with the technology required for the course.

(b) Use new methods of communication with both students and instructors.

(c) Strengthen student's independence through collaboration with their peers.

Synchronous learning refers to a learning environment in which everyone takes part at the same time. In this method, students are watching a live streaming of a class, take part in a chat with instructors participating in a class via a Web conference tool such as Blackboard Collaborate, Adobe Connect, WebEx, and Skype, among others.

E-learning is an effective tool that develops regional capacity and fulfills the needs of health education students and their teachers in developing countries including Nigeria (Nartker, Stevens, Shumays, Kalowela, Kiss, and Potter, 2010). It is convenient, effective and least expensive means of delivering digital resources to students. Frehywot, Yianna, Zahra, Nadia, Heather, Hannah, Selam, Kristine, Abdel, and James (2013) noted that students who participated in pure e-learning environment such as emersion through simulation, multimedia programmes or distance learning courses, also reported positive feedback and noted that these tools facilitated their understanding of curriculum material. E-learning can also occur in or outside of the classroom. It is obvious that more health educators are experimenting with this innovative learning approach to complement traditional teaching in our Nigeria universities.

Interactive Whiteboard (IWB) in Health Education

This is a large instructional tool that allows computer images to be displayed onto a board using a digital projector. The computer desktop is projected onto the board's surface where users control the computer using a pen, finger, stylus, mouse, among others. The board is typically mounted to a wall or floor stand, and can be used in a variety of settings including classroom and seminars at all levels of health education. An interactive whiteboard is a powerful visual tool for teachers to create lessons that incorporate video, moving diagrams and materials and to keep students engaged. (Painter, Whiting and Wolters, 2010). Items can be dragged, clicked and copied, also lecturer can handwrite notes which can be transformed into text and saved.

The multimedia capability of interactive whiteboards also allows health educators to convey information in many ways. The teacher can show the students a structure or object of part of the body from Google

Earth, PowerPoint graphs on city population. These can be annotated, adding labels to demonstrate the key aspects. Students can interact in a variety of ways with whiteboard including writing, manipulating objects with the help of handheld devices and saved (Beauchamp and Parkinson, 2012). According to Glover and Miller (2010), manufacturers of IWB technology have been setting up various online supports to communities for teachers and educational institutions in deploying the use of the IWB in learning environment. Such Web site regularly contributes research findings and administers free whiteboard lessons to promote widespread use of IWB in classrooms. Moss et al. (2007) reported that IWB technology led to consistent gain across all key stages and subjects with increasingly significant impact in the classroom.

The features available when using an interactive whiteboard include;

i) Add annotations.
ii) Highlight text.
iii) Add notes and drawings and then save them to be printed out and shared, or added to a virtual learning environment.
iv) Show pictures and educational videos to the whole lecture theaters. You can label parts or highlight elements of an image.
v) Demonstrate the contents available on a Web site in a teacher-directed activity.

Teleteaching in Health Education

The advancement in technology and telecommunication has generally created special opportunities in tele-education. Teleteaching in health education involves the use of information, communicators and technology (ICT) for delivering lectures to students. It aims at equal access to health education irrespective of geographical location of the student in need. Networks for tele-education enable the integration of distributed health educators competence and contribution to improve the quality of education (Rheuban and Sullivan, 2005).

The use of specifically designed networks for tele-education contributes to the continued mastery of the students. For the optimal performance of tele-education application, networks and communication tools used should be appropriate for the health education student applications. Grashchew, Roelofs, Rakowsky, and Schlog (2008) stated that participation of industrial partners in integrating the communication networks and getting satellite transmission capacity available, several telemedical services such as offline access to archived data, live consultation of experts, teleteaching among others have become available.

The purpose of teleteaching include;

a) Connect to other universities by a special Internet service. Teleteaching should act as a channel for prompt, long distance interaction with experts for all students.
b) Increase collaboration between developed and developing country's universities.
c) Offer teaching support for all health educators.
d) Increase knowledge and skill of educators and students especially in the present era of globalization.
e) Identification of new areas for research collaboration between Nigeria and other African and developed Universities.

Virtual Library in Health Education

A virtual library also known as a digital library or an electronic library may be defined as the online facility provided by a conventional library to read books and access other facilities. It may also be known as a Web site which offers links to various sites with a large store of information in a catalogued or archived form. It is a collective manner to the entire number of online books and other literary material related to any subject available on the Internet.

Virtual Health Library (VHL) is an online digital library made specifically for information on health education. It is meant to promote the development of a network of sources of scientific and technical information with universal access on the Internet. VHL is a model for the management of information and knowledge. It also includes the cooperation and convergence between institutions, system, networks and initiatives of producers, intermediaries and users, in the operation of networks of local, national, regional and international information sources favouring open and universal access (Eisenberg, 2012). Universities that offer a distance learning education facility should have virtual library for their students.

Distance Learning in Health Education

Distance Learning (DL) is an apparent alternative to traditional methods in health education. Applications of distance learning in health education have closely reflected the evolution of Internet networking communication and information technology. Distance learning was introduced long ago, evaluated and accepted in many disciplines, mostly faculties of education. The introduction of DL to health education students occurred much later. As a result of remarkable achievement in technology and the increasing need for continuing updated knowledge, DL today has become an important alternative to traditional method of education. Anneroth (2003) stated that DL is the only feasible ways to help the USA's 500,000 public health workers to meet new challenges. It is a very good tool to serve the growing demand for postgraduate health education students.

The evolution of technology has dominated the DL methodologies including their courses. These allow the innovation application tools such as teleconference, teleteaching, virtual library, IWB, video application, virtual learning environment, and virtual classroom, among others. The virtual classroom is the powerful combination of a variety of media and resources with the Internet as a backbone to students (Cravener, 2004). Distance learning in health education provides a broader method of communication within the realm of education. With the many innovation tools that technological advancements have to offer, communication appears to increase in distance learning amongst students and their health educators as well as students and their classmates (Borje, 2005). However, DL uses asynchronous and synchronous method interchangeably to meet the need of the students. Several tools are used to facilitate the synchronous sessions, such tools include adobe connect, a proprietary Web-conferencing tool, ustream.tv (a free video-streaming service) and Skype video conferencing conferencing. Participants also engage in a number of asynchronous activities such as reading, reviewing and critiquing course reading through participant blogs. Through both the asynchronous and synchronous activity, personal learning network grow as individual freely connects with those interested in and collaboration. Social interactions become authentic, dynamic and fluid.

Information, Education, and Communication (IEC)

Information, education, and communication (IEC) combines strategies, approaches, and methods that enable teaching individuals, families, groups, organizations and communities to play active roles in achieving, protecting and sustaining their own health. Embodied in IEC is the process of learning that empowers students and other individuals to make decisions, modify behaviours and change social condition. The influence of underlying social, cultural, economic, and environmental conditions on health education are also taken into consideration in the IEC processes. Identifying and promoting specific behaviours that are desirable are usually the objectives of IEC efforts.

Health information can be communicated through many channels to increase awareness and assess the knowledge of different populations about various issues, products and behaviours. Channels might include interpersonal communication such as individual discussions, demonstration, counselling sessions or group discussions and community meetings and events. It could be through mass media communication such as radio, television, information technologies concept maps, flowcharts, and other forms of one-way communication, such as brochures, leaflets and posters, visual and audio visual presentations and some forms of electronic communication.

Education on the other hand involves teaching someone, using formal system of school, college, university or passing knowledge of a particular subject. Communication is the successful way of sharing of thoughts, feelings, ideas, and information with others through speech, writing, printing, electronic media, signals among others for health promotion and wellbeing of people (Obiechina 2012).

Although, communication service provision is essential for transmitting information and building trust with the students, communication with other individuals and groups within the community is also vital in health. It is through such communication networks that lecturers can pass information to students on user's needs, priorities and concerns. It also helps lecturers to better understand the specific setting and context in which they are working, which will be useful in the later development of IEC approaches, messages and materials to the students.

The constraints to use of technological innovation in health education

There are a lot of constrains militating against the implementation of new innovation in health education. It is obvious that there are inadequate infrastructure, including software facilities, lack of finance, inadequate dedication and space in health education system. Inadequate infrastructure is compounded by inconsistence power supply and irregular Internet service.

There are also shortages of facilities such as bandwidth which disrupts the rate of data flow in digital network. Personnel in information, communication technology system is insufficient. Incentives and will to adopt new innovations are lacking among some health educators. Inadequate finance in academic institution and no visible plans to expand the already overstretched facilities in health education.

Conclusion

The current innovative technologies have a positive transformative role to play in shaping the futures of health education tutor and their students in all the tertiary institution in Nigeria. It provides self-directed learning to both student and lecturer as facilitator or guides. There is cooperative learning in small student groups, acquisition of critical thinking, speaking, and writing skills among students. It linked health educators and students to positive changes in efficiency, productivity, quality, competitiveness, knowledge and skill to compete with the global world. That is, it provide opportunities for students to connect with a variety of meaningful ways by using e-learning, virtual library, interactive lectures, engaging assignments, hands-on lab/field experiences, and other active learning strategies. Innovation application provides students with experiences, preconceptions, or misconceptions by using pretests, background knowledge to probe, write or verbal activities designed to reveal students thinking about a subject. It offer support to students by the of concept maps, flowcharts, outlines, comparison tables to make the structure of their knowledge clear. It facilitates long-term group projects, class discussions, and group activities to enhance the social side of learning.

References

Anneroth, G., M. (2003). Worldwide survey on distance learning in dental Education. *International Dentist Journal,* 44 (2) 506–510.

Beauchamp G., and Parkinson, J. (2012). Beyond the wow factor: developing interactivity with the interactive whiteboard. *Journal of Health Science,* 21 (2) 91–101.

Borje, H. (2005). The evolution, principles and practices of distance education. *Journal of Computer-Mediated Communication,* 11 (3) 222–230.

Broadbent, B. (2002). ABCs of E-learning. Reaping the Benefits and Avoiding the Pitfalls.

Cravener, P., G. (2004) Education on the web: A rejoinder. *Computer Innovative Technology for computer professionals* (3) 107–108.

Corbin, C., B., Welk, G. J., Corbin, W., R., and Welk K. A. (2011). *Concepts of Fitness and Wellness.* A comprehensive lifestyle approach (9th ed). Mc Gracehill companies, America, New York.

Elizabeth, A. and Martin, M. A. (2007). Oxford concise medical dictionary. *Oxford University Press* (8th ed.) Great Britain.

Eisenberg, M., L. (2012). Information literacy: essential skills for the information age. *Journal of Library and Information Technology,* 28 (2) 37–45.

Frankelius, P. (2009). Questioning two myths innovation literature. *Journal of High Technology Management Research,* 20 (1) 40–51.

Frehywot, S. Yianna, V., Zahra, T., Nadia, M., Heather, R., Hannah, W., Selam, B., Kristine, K., Abdel, K., K., and James, S. (2013) E-learning in medical education in resource constrained low- and middle-income countries. *Journal of Human Resources for Health,* 11 (4) 152–157.

Glover, D., and Miller, D., H. (2010). The Interactive Whiteboard Expansion: a literature survey. Technology, pedagogy and Education. *Journal of Information Technology for Teacher Education,* 10 (3) 257–261.

Graschew, G., Roelofs, T. S., Rakowsky, S., and Schlog, P. M. (2008). Disaster emergency logistic telemedicine advanced disaster satellites system. Telemedical services for disaster emergencies. *International Journal of Risk Assessment and Management* vol. 9, pp. 351–366.

Joint Committee on Terminology (2011). Report of the 2010 joint committee on health education and promotion terminology. *American Journal of Heath Education,* 32 (2) 89–103.

Kame, D. (2011) Education Innovation: what it is and why we need more of it. The office of Innovation and Improvement at the U.S. Department of Education.

Kidd, L. I., Knisley, S. J. and Morgan, K. I. (2012). Effective of a second life simulation as a teaching strategy for undergraduate mental health nursing students. *Journal of Psychosocial Nursing.* (50): 7: 28–37

Marcus, A. (2012). Health education and health promotion. Planning, implementing and evaluating health promotion programme. (5th ed.) San Francisco Donatelle.

Moore, J. L., Dickson-Deane, C., and Galyen, K. (2011). E .Learning the solution for individual learning. *The internet and Higher Education,* 14

Moss, G., Jewitt, C., Levaaic, R., Armstrong, V., Cardini, A., and Castle, F. (2007) The Interactive Whiteboards, Pedagogy and Pupil Performance Evaluation. Schools Whiteboard Expansion (SWE) Project, London Challenges 12 (1) 92–98.

Nartker, A. J., Stevens. L., Shumays, A., Kalowela, M. Kiss B., D., and Potter, k. (2010). Increasing health worker capacity through distance learning: a comprehensive review of programmes in Tanzania. *Journal of Public Health Medicine,* 8 (3) 30–35

Obiechina G. O. (2012). Promoting Health through Effective and Improved Consumer Health Education. *Journal of Education in Developing Areas* (JEDA) 20 (1) 411–420

Painter, D., Whiting, E., and Walters, B. (2010). The use of an Interactive Whiteboard in promoting interactive teaching and learning. *Journal of Computer Assisted Learning,* 21 (2) 102–106.

Rheuban, K., S., and Sullivan, E. (2005). Opportunities and challenges of e- health and telemedicine via satellite. *European Journal of Medical Research,* 110 (2): 58–66.

Salge, T., O., and Vera, A. (2012). Benefiting from public sector innovation: The Moderating role of customers, Learning Orientation. *Journal of Education for Business,* 72 (4) 550–560.

The Advisory Committee on Measuring Innovation in the 21st Century (2007). Federal register (72) 18627–18628.

UNESCO Institute for Statistic (2005). *The measurement of scientific and technological activities. (3rd* ed) pg 34.

World Health Organisation (2005). Promoting Mental Health: Concepts, emerging evidence, and Practice. *A report of the WHO Geneva.*

Chapter 19

Innovative Applications of Educational Technology Tools in Nonformal and Informal Health Education

Ekenedo, Golda O.

Introduction

In the last ten years, education has benefited from a real e-revolution. Technology ushered in fundamental structural changes that can be integral in achieving significant improvements in teaching and learning. Technology-enhanced learning has evolved both from enhancements to earlier generations of face-to-face teaching and enhancements to earlier generations of distance education. For example, many graduate programs have deliberately designed programs for working adults, which are predominantly offered online but also include short-term face-to-face residencies. The nature of health education is such that the use of technology is core in realizing its goals. Health education targets the general public with its learning content. Invariably, all segments of the society irrespective of age, culture, literacy level, occupation, social class, and gender need to be reached with health education. Technology provides other avenues for presenting health information through that invite rather than discourage people with literacy or language issues to learn about health matters.

Health education is important in promoting individual health, community health, as well as the proper use of health-care facilities and services. The role of health education in total health care cannot be overemphasized; in fact, it is central to health care. Well-delivered health education promotes health literacy which according to American Medical Association (1999) is believed to be a stronger predictor of health outcomes than social and economic status, educational level, gender and age. "Health literacy" is a new concept in the field of health which is defined by WHO (2008) as "the cognitive and social skills which determine the motivation and ability of individuals to gain access to, understand, and use information in ways that promote and maintain health."

The definition of health literacy by WHO emphasizes three factors: obtain health information; understand health information; and use health information. In other words, for health education to successfully lead to health literacy, appropriate health information must be given in such a way that it must be understood as well as ensure its use by the recipient. Persons with limited health literacy can hardly make take informed decisions concerning their health and well-being and can also be at serious social disadvantage in terms of their ability to read and understand written medical instructions, including medication dosages and understanding results of medical tests and diagnosis, locate health providers and services offered, share personal information such as health history, provide self-care in chronic illnesses, or understand how to take medicines. Unfortunately, a study by Atulomah and Atulomah (2012) has shown that health literacy in Nigeria is low. They found, for example, that 43.3% of their respondents indicated needing help in explaining clearly what their illnesses are and required frequent reminder on how to take medications prescribed. These parameters of health literacy are critical for those with chronic illnesses. There is, therefore, the need to improve health literacy, and there is no other way to do it other than through health education. Since health education promotes health literacy and health literacy promotes health status, there is a need for efficiency and innovation in the process of health education in order to achieve effective results, hence the need to apply technological innovations in teaching and learning about health. As noted by Atulomah and Atulomah (2012), in technology resource constrained settings, education can become, compromised resulting to an ineffective venture.

Theoretical Background

A number of theories have been put forward to explain how individuals learn in a virtual environment; and provide guide for developing effective online and multimedia resources. Three of such theories are discussed here to provide bases for innovative application of educational technology tools such as e-learning, online and multimedia resources in nonformal and informal health education. First is the theory of connectivism propounded by Siemens (2004). Siemens proposes the need for a new theory to address the change in how people think and handle knowledge as a result of the increased uses of technology in all aspects of daily living. He argues that the "know-how and know-what is being supplemented with the know-where (the understanding of where to find knowledge needed)" (Hodges, 2014). Tenets of the theory include that learning may reside in nonhuman appliances; that nurturing and maintaining connections is needed to facilitate continual learning; accuracy (accurate, up-to-date knowledge) is the intent of all connectivist learning activities. Connectivism provides the needed shift in learning skills and activities to provide a successful and up-to-date learning environment through the use of online tools and resources (Siemens, 2005).

The second, transactional distance (TD) theory, was developed by Michael Moore in 1997. Moore (1997) uses Dewey's concept of transaction as describing the role between individuals and the learning environment to establish a theory based on the unique learning environment created by distance education. When learners and teachers are separated by time and physical space (as in distance education), the potential for misunderstanding between these individuals is increased. This increase in the potential for miscommunication that can occur in distance education is what Moore (1997) defines as transactional distance. Online education by its very nature has the potential for a less predetermined structure thus allowing for the flow of dialogue between learners and instructors to increase and in turn lessen the amount of transaction distance present.

The third theory, cognitive theory of multimedia learning (CTML), was designed by Professor Richard E. Mayer. According to Mayer

(2009), people learn more deeply from words and graphics than from words alone. This he called the multimedia principle, and it forms the basis for using multimedia instruction—that is, instruction containing words (such as spoken text or printed text) and graphics (such as illustrations, charts, photos, animation, or video) that is intended to foster learning (Mayer, 2009). The cognitive theory of multimedia learning is based on the assumptions that people process audio and visual input differently, that people only process limited elements at one time, and that learning occurs when learners are presented with the right kind of cognitive processing. Five cognitive processes are presented in CTML that examine how people learn from words and pictures. Selecting words when verbal material enters through the ears, selecting images when visual materials enter through the eyes, organizing the words and the images, and integrating and building the connections between the visual, audio, and prior knowledge. Online learning environments can be designed with these principles in mind so as to promote more effective learning. Cognitive theory of multimedia learning is a research-based theory that can be applied to improve the level of online health education available.

Concept of Health Education

Health education is a process of providing skill-based health information to people so that they can improve their health status and that of the community. It focuses on providing sound knowledge about health which is capable of influencing attitudes and achieving positive change in health behaviours. In other words, the goal of health education is to empower people to take informed decisions aimed at finding solutions to their health needs and problems. The World Health Organisation's definition of *health education* describes it as any combination of learning experiences designed to help individuals and communities improve their health, by increasing their knowledge, or influencing their attitudes (WHO, 2013). Health education uses persuasion to encourage individuals to adopt lifestyles that are beneficial to their health and to drop unhealthy lifestyles. It utilizes a process or method whereby individuals, families, organizations, and communities are fully involved in all phases of health education activities including needs assessments, planning, implementation and evaluation. Health education is lifelong education. It starts from the cradle and ends with life itself.

Depending on the size of the target population, health education can be classified as mass health education, group health education or individualized health education. When the public is the recipient of the health information it is *mass health education* such as in the use of billboards and television adverts and jingles to pass on health information. It is *group health education* when the target population is a small segment of the community such as happens in the workplace. Lastly, *individualized health education* is a one on one passing of information as is the case in health counseling and between patient and doctor. Health education takes place in various settings that can be formal, nonformal or informal.

Formal health education takes place in the school setting and has a formal curriculum. Put more elaborately, formal health education is that type of health education that is provided under a formal education system. A formal education system is the hierarchically structured, chronologically graded education system running from primary school through the university and including, in addition to general academic studies, a variety of specialized programmes and institutions for full-time technical and professional training (Smith, 2001). Health education is part of the Nigerian school curriculum from primary to secondary level. Also many higher institutions in Nigeria especially universities and colleges of education offer health education as a course of study; sometimes combined with physical education.

Nonformal health education derives from general nonformal education which has been defined as "any organized and sustained educational activities that do not correspond exactly to the formal education systems of schools, colleges, universities, and other formal educational institutions" (Infodev, 2010). Therefore, nonformal health education describes any organized health education activity which takes place outside the established formal school system whether operating separately or as an important feature of some broader activity that is intended to serve identifiable clienteles and health objectives. Nonformal health education takes place in the form of distance learning, workplace health education, antenatal classes, patient education, organized youth programmes, and other community-based health education that take place in various community settings such as the church or religious institutions, sports and social clubs, town halls etc. Nonformal health education has the advantage of a wider reach and has a greater potential of influencing attitude and changing health behaviour because it is learner-centred in approach and skill-based.

Informal health education is a lifelong process whereby every individual acquires healthy attitudes, values, skills and knowledge about health from daily experience and the educative influences and resources in his or her environment such as family, neighbourhood, workplace, market place, playground, the library and mass media.

In summary, formal health education is linked with schools, and training institutions; nonformal with community groups and other organizations; and informal covers what is remaining, e.g., interactions with friends, family and work colleagues. This distinction is not rigid since there may be some overlap especially between informal and nonformal (Fordham, 1993), which incidentally are the focus of this write-up. Nevertheless for each of these forms of health education, application of educational technology produces outstanding results.

Concept of Educational Technology

Quite a number of definitions of educational technology have been put forward. However, one of the earliest definitions by Mitra (1968) in Mengal and Mengal (2009) described educational technology as a science of techniques and methods by which educational goals could be realized. This concise definition by Mitra can be understood to mean that educational technology is concerned with systematic application of science and technology in the field of education for the purpose of effectively realizing set goals. Mengal and Mengal (2009) further noted that educational technology is not limited to the use of audiovisual aids and does not symbolize merely educational hardware such as sophisticated gadgets and mechanical devices used in education, rather, it also tends to utilize the results of all good, experiments and researches in the field of human learning and the art of communication and employs a combination of all possible human and nonhuman resources to achieve the desired educational objectives.

Concept of Innovation/Innovative

An innovation is something that is new, is positively different, or is better than what was there before. Innovations however, do not exist objectively or in an unchanging sense. Concepts of newness or reformation are viewed differently by different people, and to categorise something new as 'innovative' places additional meaning on its value or relevance. As such, to be 'innovative' is an affirmative description of a process. Implicit in the concept of innovation is that creativity and imagination underpin innovative capabilities. Whereas an invention is a new idea that becomes an artifact, an innovation is a new idea or set of ideas successfully applied to processes or practices (Davenport, 1993). In the context of this paper, to be innovative refers to putting in place practices that are substantively different from, or have more desired outcomes than what has gone before. Innovative application, therefore, is also viewed in this paper to include the application of existing educational technology whether new or traditional in substantively different ways as they have never been done before. It can mean new ways of applying educational technology in health education.

Use of Educational Technology Tools in Nonformal and Informal Health Education

Over the past thirty years, nonformal education initiatives globally have effectively used ICTs for mass literacy campaigns, training of health workers, and capacity building under rural community development projects (Infodev, 2010). There is no doubt that innovative educational technology, such as e-Learning and m-Learning (mobile learning), and other technologies provide unprecedented opportunities for successful health education by facilitating the reach of large audience and increasing educational productivity by accelerating the rate of learning. Nonformal health education also has a critical role to play in reaching the marginalized groups, and ICTs are a tool in the effective performance of this role. The latest ICTs are being used to develop virtual learning communities for nonformal education purposes. Virtual learning communities are learning groups with a shared interest, who are able to overcome barriers of time, geography, age, ability, culture, and social status (Blurton, 1999).

Educational technology tools used in nonformal and informal health education can be either traditional or modern educational technologies. However, most times they are used together to achieve a particular health education goal. They can be classified into print technology, audio technology, audiovisual technology and computer-based technology. Print technology include: Posters; Billboards; Flyers/ handbills/pamphlets; and Bulletin boards. Audio technology include: Radio; Cassette; and CD. Visual technology include: Pictures; Slide presentations; projected pictures; Electronic billboards and marquees. Audiovisual technology include: Overhead slide presentations; Television; and Video. Computer-based technologies include: PowerPoint presentations; Internet; Mobile phone; and Interactive whiteboard

Innovative Application of Educational Technology Tools in Nonformal Health Education

Nonformal health education takes place in several settings which include distance learning, workplace, hospital, and community.

Health education in distance learning

Distance learning is any type of learning in which the components of structured learning activities (i.e., learner, instructor, and learning resources) are separated by time and geography (Rovai, Ponton and Baker, 2008). Distance learning is usually used to promote open learning. *Open learning* implies a greater openness for learning that is frequently missing from traditional education programmes. Open learning reduces constraints and removes barriers for learners by promoting qualities such as greater learner autonomy, independence, and flexibility. With distance learning, those who are barred by time and space but who desire to acquire health education can conveniently do so at their own time and pace. Distance education provided the base for e-learning's development.

E-learning is the delivery of a learning, training or education program by electronic means. E-learning involves the use of a computer or electronic device (e.g. a mobile phone) in some way to provide training, educational or learning material (Derek Stockley 2003). E-learning can involve a greater variety of equipment than online training or education, for as the name implies, "online" involves using the Internet or an Intranet. CD-ROM and DVD can be used in e-learning to provide learning materials. E-learning has become the primary form of distance education. E-learning can be "on demand." It overcomes timing, attendance, and travel difficulties.

Distance and online learning can benefit health professionals to update their knowledge and keep abreast of latest developments and research findings in the field of health and medicine. They can also decide to explore other areas of specialty in health. Paramedics and other auxiliary health workers can be trained and certified through distance and online learning.

Educational technologies that make distance learning effective include: videos and tapes, CDs, DVDs, television, radio, the Internet, computer, and video conferencing, e-mail, telephone, or voice mails etc. Health education contents and activities are recorded in CDs and DVDs so that learners can access them and play them at their convenience at home or in the workplace. The learner and the health instructor can interact through the Internet and e-mail. Lecture materials, assignments and other instructions are posted on the Internet where learners are expected to download them read and respond to accompanying questions. Projects and assignments can be written, supervised and corrected via e-mail. Instructions and clarifications can be sought and received through telephone. Video conferencing and webinars are emerging Internet tools used to facilitate group discussion on the phone and Internet. Open education resources provide access to distance learner of health information and research resources in the Internet without restrictions or stringent requirements. Open educational resources are teaching, learning, and research resources that reside in the public domain and are freely available to anyone over the Web. They range from digital libraries to textbooks and games.

E-learning promotes virtual learning environment and is in fact also called virtual learning. Virtual learning environments integrate heterogeneous technologies and multiple pedagogical approaches.

There are several benefits to e-learning whether on its own, or to enhance existing in house training. They include the following.

i. It is cost effective and saves time: By reducing the time taken away from the office, removing travel costs and doing away with printed materials, online learning helps to save money and increase workplace productivity. It also means staff will be happier and focused.

ii. Learning at all times and anywhere: Many face-to-face courses only operate within normal office hours. By allowing staff to complete the course when and where they like, disruptions to busy working schedule are minimised. This also means that staff will be happier because they don't need to travel to specific training centres, and if they have important work to catch up on mandatory training can be done outside of office hours in exchange for lieu time.

iii. It makes tracking of course progress easy: Perhaps the most important aspect of using computers for training is that it with a well implemented Learning Management System makes it easy to track and prove progress for staff and learners. This can be essential where proof of mandatory training is required. In many cases certificates can be printed personally by the online learner after successfully completing the programme.

iv. It's discreet: Not everybody feels comfortable learning in a large group, especially if they find something hard to understand that co-workers have no problem with. E-learning allows each individual to tackle the subject at their own pace, with interactive tasks being set in place to ensure a thorough understanding throughout each module.

Applying educational technology tools in workplace health education

The workplace is one of the most important settings for health education. This is because most adults spend a greater percentage of their waking hours in their workplaces. There is, therefore, no better place to reach such population with health programmes than the workplace. Workplace health promotion benefits the employer, employee, the family and the larger society. Traditional health communication methods used in the workplace include: announcements during meetings, bulletin board notices, printed pamphlets or handbills, marquees and electronic billboards, face-to-face individual information sessions, group information sessions, audio presentations/resources, and audio-video presentations. However, there are communication methods which are considered more innovative and more in keeping with changes in the way adult learners are prone to understand, assimilate, and retain information. These include: information-based puzzles, acrostics, or limericks; self-quizzes; health risk appraisals; mail-request vehicles; trigger cards; electronic bulletin boards; fax networks; and online telephonic support (Chapman, 2002). These methods are considered nontraditional, but they may be combined with traditional methods in the design of employee health education programmes.

Nevertheless, workplace organizations and processes have been changing. Major trends that are helping shape the future of the workplace include movement from production to information and service jobs; the increased cultural diversity of work groups; the increasing computerization of the work environment; the introduction of greater flexibility in the work locations, job definitions, and role delineation; the increased pressure for greater productivity; the emphasis on business process improvement and its role in enhancing organizational effectiveness; the "re-engineering" of work; the changing expectations and capabilities of new labour entrants into the work force; and the increased globalization of markets and productive capability (Boyett and Conn, 1991 in Chapman, 2002).

These many changes require adjustments and counter strategies from health education and promotion professionals who function in the workplace settings. Hence newer technologies include: electronic mail (e-mail); proactive telephone contact; computer-based multi-media presentations; Internet Web sites; social networks, interactive information kiosk, and virtual reality applications.

Intranet

Large organizations use intranet to pass health information to workers. This is made possible through networking of computers in the workplace so that information from one computer source is made accessible to every computer in the workplace. For instance, safety tips, information on healthy lifestyles and health notices can be passed around to workers so they receive them while they sit right there on their seats without health programme providers having to go to their offices or have them gather in one place. It saves time and resources and the problem of worker absenteeism from health talks or seminar is overcome.

Health seminar, workshop, and talk

Effective health seminars, training workshops and health talks are achieved through the use of multimedia such as computer, projector, interactive whiteboard, pictures, video illustrations, tapes, CDs, and DVDs. Multimedia messages are much more effective than plain text because individuals remember 10% of what they read, 20% of what they hear, 30% of what they see, 70% of what they see and hear, and 90% of what they see, hear, and do. Laptop and digital projector can be used to present health talks in PowerPoint. A simple yet well-crafted and well-delivered PowerPoint presentation engages the learners and makes learning easier and faster. Pictures that capture important health messages can be put on slides and used for explanations during the presentation. PowerPoint has the advantage of making presentation brief while the essence of the message is still captured and learning made effective. It is important to note that time is of essence when providing health education in the workplace because workers need to get back to work as quickly as possible. PowerPoint is good for a matured learning audience like workers.

The use of interactive whiteboard in delivering health education in the workplace is another innovative way of making learning interesting and efficient. An interactive whiteboard is an instructional tool that allows computer images to be displayed onto a board using a digital projector. The instructor can then manipulate the elements on the board by using his finger as a mouse, directly on the screen. Items can be dragged, clicked and copied and the instructor can handwrite notes, which can be transformed into text and saved. They possess a lot of features which help to add interactivity and collaboration to teaching and learning and allow the integration of media content into the lecture. They appeal to different senses, thereby allowing the instructor to accommodate different learning styles. When used innovatively they can create a wide range of learning opportunities. For example, a recorded video of an industrial accident can be shown in the interactive whiteboard, and then the health educator can pause the video at a certain point and demonstrate by drawing on the board what the worker should have done to avert the accident. The interactive whiteboard can be connected to the Internet so that materials from the Internet can be imported and used during the teaching and learning interaction. The learners themselves can also interact with the whiteboard.

Health information can be recorded on tape and CD and given to workers to listen to at their own convenient time. CDs containing music and voices that aid relaxation can be used in a stress management class. Also CDs and DVDs containing rhythmic sounds and rhythmic activities can be given out to worker participants in a physical activity class. These are very practical, cost effective, time saving and convenient ways of achieving health education goals in the workplace.

Applying educational technology tools in hospital health education

The hospital is a place that provides services especially for the sick. However, it is not only the patients that need health education, the health workers and other hospital workers also benefit from health education in the hospital setting. A lot of group and individualized health education go on in the hospital; in the antenatal clinics, during patient consultation with doctor, during patient interaction with nurses, pharmacists, at the out-patient waiting room, during pre and post operation care etc.

Health talks given to expectant mothers during antenatal can be made participatory, interactive and more result oriented through the innovative use of educational technology tools. Contents requiring demonstrations can be delivered using video, slide and projector, and interactive whiteboard. Since the capacity of the human mind to retain acquired information is not 100%, participants in antenatal classes can be given audio, visual, and audiovisual resources such as CDs, tapes and DVDs to take home and play at their own convenient time. Proactive telephone contact can be used to coach and provide support for expectant mothers especially inexperienced expectant mothers. They can access Web sites where they can obtain further health information concerning pregnancy, childbirth, motherhood and child care.

Patient education occurs during consultations. Doctor may have the need to explain to a patient the cause and process of a disease and what role he or she has to play in managing the condition successfully. The doctor communicates with patients even when they are far from his or her geographical location via phone call, SMS, e-mail. Patients can be provided links to social network of people with related health problems on the Internet so that they can start interacting with them in sharing ideas about their condition. Example of such network is "Alcoholic Anonymous." People who have problems with alcoholism go online to share ideas and experiences about how to quit successfully. Through recorded CDs and DVDs, diabetic patients, for example, can learn how to self-administer insulin, choose diets and exercises that will benefit their health condition.

Applying educational technology tools in community health education

Community health education takes the form of mass education usually through enlightenment campaigns; newspaper publications; radio and television news, propaganda, adverts and discussions; posters and billboards. It also takes the form of group education involving different homogeneous groups such as commercial vehicle drivers, adolescents, school children, traders, and religious groups.

Mass enlightenment campaigns are usually done through television and radio. It could be announcements for example about immunization schedules; news item; discussions on a particular health topic with relevant health professionals as guests; news talks and editorials; jingles; and sponsored programmes and documentaries. Outbreaks of diseases are announced on radio and TV so that the public becomes aware and beware. The public through the same medium are enlightened about the disease, causes, signs and symptoms and preventive measures, as well as numbers and centres to contact for notification of suspected cases. The case of Ebola diseases outbreak in 2014 is a typical example. The disease was contained in Nigeria due largely to effective use of ICT not only by health agencies but also by individuals especially via phone SMS and Internet-based social networks.

In group health education, pictures, videos, PowerPoint presentations, CD, DVD, slide and film presentations are popular educational technology tools. The use of interactive whiteboard would bring about fantastic results if considered by community health education providers. Another form of group health education that goes on in the communities is the training of health volunteers and alternative health-care providers in communities in order to improve on their services to the people.

For instance, when WHO saw that traditional birth attendants (TBAs) were the preferred health service providers for many women in Nigeria, they decided to train them instead of trying fruitlessly to stop women from patronizing them. Similarly, to reinforce the Midwives recently deployed to local health facilities nationwide, Pan African Development Education and Advocacy Programme (PADEAP), a nongovernmental organisation, continues to deliver training in partnership with Primary Health-care centres and local Government hospitals in Katisna State to increase the number of core services providers including Community Health Extension Workers, with a focus on deploying more skilled health staff in rural areas.

Another community health education project named 'FORSHE' targets female out-of-school youths with reproductive health education and referrals, as well as leadership and life skills to improve their overall sexual and reproductive health. Young girls who have dropped out or who have never been to school, regardless of their marital status, are part of this program, where they engage in activities such as peer education, film viewing, interpersonal communications and a variety of other youth-led activities. At the end of the project, girls have a better knowledge of reproductive health, which will hopefully lead them to change unfavorable health behaviours and improve self-assurance, as well as ensure better communication with their partners and better access to sexual reproductive health services.

The fore mentioned community health education programmes are programmes that can benefit much from the use of ICT. Most of the TBAs are not so literate and would need to be exposed to a variety of learning experiences especially ones that are interactive in nature as can be obtained in the use of videos, pictures and slides presentations. They can be provided with DVDs that they can play any time they need to recapture an experience or remember a particular procedure.

Innovative Application of Educational Technology Tools in Informal Health Education

Education technologies applicable to health education in informal settings can be clustered into 4 domains along a continuum from highly-interactive to fixed information sources as follows:

1. Personal communication: telephone, cell phone, and pager;
2. Social communication: e-mail, instant messaging, chat, and bulletin boards;
3. Interactive environments: Web sites, search engines, and computers; and
4. Unidirectional sources: television, radio, and print.

There are thousands of online health education resources that are available and can be accessed through the computer and mobile phones. In many developing countries, inexpensive mobile phones have become very much available. Not only do they provide access to health information on the Internet, they allow users to download free or inexpensive apps. Health apps allow the smartphone to act as medical devices, helping patients monitor their heart rate or manage their diabetes. As a matter of fact there are thousands of health related apps available for download today on smartphones.

Health phone is a technology that has created a library of health videos, which are preloaded free of cost on popular low cost mobile phones. They provide information on a wide range of topics, ranging from breastfeeding to hand washing, and are being made available in local languages in a place like India.

Education as a Vaccine (EVA) is a youth-led NGO that successfully leverages mobile technology and digital mediums to educate young people about sexual reproductive health and HIV prevention. The organisation is co-founded by a visionary young Nigerian, Fadekemi Akinfaderin-Agarau. In collaboration with partners such as OneWorld UK and Butterfly Works, EVA implemented the Learning about Living (LaL) project. It is based on the Nigerian Family Life and HIV/Aids Education (FLHE) curriculum. LaL uses ICT to provide information for young people, both in and out of school, about sex, HIV and health. The platform consists of two components: mobile learning and multimedia.

With "MyQuestion," a segment of the programme, Nigerian youth send their questions via SMS to a short code and receive answers directly on their mobile phone. The service is managed by young, trained volunteer counselors and ensures that the youth can receive accurate information without fear of being judged or stigmatised. As we know, in traditional African society it is deemed inappropriate and socially unacceptable to discuss sex-related issues openly with young people. This innovative service creates an anonymous channel for their questions on relationships and sexual health. Even the shyest teenager would feel comfortable using the service. Since its inception in 2007, the platform has received and responded to over 500 000 SMS queries about sexual reproductive health.

The second component involves multimedia. DVDs containing fun but educational cartoons are used to deliver the FLHE curriculum for upper primary and junior secondary school students. The multimedia clips are also accessible via a Web-based portal. The engaging story line and characters that young people can relate to allows for excellent knowledge transfer.

There is a growing number of computer software that helps people deal with health matters. These include programmes that help people on restricted diet to count calories or check the fat content of foods they eat. Example is Calorie Counter and Diet Tracker which helps programme participants keeps track of how many calories they eat in a day or a week. There are some that help users with exercise programme their doctors have prescribed for them; track medications they take; and manage their stress. Pedometer is a technology device that keeps track of how many steps you take throughout a workout or the day. This could be used to help users meet physical activity goals. Sugar Stacksis a Web site that provides images of everyday food items with the grams of sugar contained in each one represented as a stack of sugar cubes. This is a great way to talk about sugar consumption with the users since it makes the word "grams" tangible for them.

Persuasive technology is an innovation that can be successfully applied in helping people learn and adopt healthy behaviours. Persuasive technology is defined as a computer system, device, or application that is intentionally designed to change a person's attitude or behaviour (Fogg, 2003). This technology uses tools (e.g., pedometer or balance board), media (e.g., video, audio, or both), and social interaction (e.g., playing with another person) to persuade individuals to adopt the behaviour without their actually knowing it. One of such persuasive technology media known as Dance Revolution (DDR) which was not developed specifically to promote healthy behaviour has changed exercise attitudes and behaviour of children and youth using principles of persuasive technology. Dance Revolution uses video, music, and a dance platform to capture interest and engage children in the activity without their being fully aware that they are exercising. The emerging field of persuasive technology has enormous potential for promoting physical activity and healthy behaviours (Fogg and Eckles, 2007; Zhu, 2008).

Awareness is growing on the use of e-books. Before the advent of the Internet, individuals would have to refer to books or go to specialized health/medical libraries to learn about their health. This was a kind of discouragement for people to learn more about their health. Now they can access that information any time of the day from their home computers and from their smartphones. The Internet is therefore a valuable source of information on how clever developers are using technology to improve health literacy.

A lot of incidental health education occur when people view television in their homes, listen to radio and read newspapers, magazines and newsletters. Many of the programmes that are broadcast on radio and TV have health lessons that people pick up unconsciously. Some radio stations have made it a policy to promote current social and health issues with slogans which they repeat as they begin and end their news bulletins. Examples are "Be considerate to other road users, drive safe to stay alive"; "People living with HIV are people just like you do not discriminate against them"; etc. Drama series, documentaries, home videos and films are all programmes where the public consciously and unconsciously pick one or two health information.

Constraints to the Application of Educational Technology in Nonformal and Informal Health Education

There are still some challenges that prevent the full application of innovative educational technology tools in Nigeria and other developing countries. They include as follows.

1. Low level of computer literacy.

   To be able to benefit maximally from current innovative technologies, there is need for a certain level of literacy in the use of computer. It is true that many Nigerians have access to and use a wide range of information and communication technologies, a larger number are yet to fully benefit from the ICT world especially older adults and those in the rural areas due to lack of skills in the use of computer.

2. Lack of infrastructure.

The use of educational technology especially in the rural communities in developing countries presents a big challenge due to lack of basic infrastructure such as electricity, and technical support. This was highlighted in a study on the use of ICT in education in seven of the E-9 countries (Bangladesh, Brazil, Egypt, India, Mexico, Pakistan, and the People's Republic of China) undertaken by UNESCO (UNESCO 2006). Most innovative technologies are dependent on electricity for optimal use. The epileptic nature of electricity supply and unreliable telecommunications network even in urban areas is a draw back to the effective use of technology in teaching and learning health education in Nigeria.

3. Low access to technology resources.

Most technology resources are expensive and many developing countries may not easily afford them. Where they can afford to provide, they are usually in short supply so that many are not able to have access to them either as an institution or as individuals.

4. Lack of ICT skilled health educators and health-care providers.

The capability of health educators and other health health-care professionals to use educational technology tools in communicating health information is an important factor in the effective use of educational technology tools to achieve the goals of health education. Even when the technologies are available some health educators are not skilled in their use and as such the effort becomes fruitless and the goals become difficult to achieve.

5.   Low literacy level.

The benefit of some technology is limited by the literacy level of the user or information recipient. Most recent innovations in educational technology involve active participation of the learner. For instance, the use of e-resources like Internet Web sites, e-mail, and apps requires the learner to be the active participant. For illiterate persons, this is almost impossible; their ability to learn to use these resources is highly limited. Again most health information that is available in the mass media is provided in English language. Only few programmes are broadcast in native languages.

Conclusions

Educational technology tools when innovatively applied in health education are very strategic in achieving the goals of health education which include reaching every individual with health information so as to change negative health attitudes, beliefs, values and practices; and promote health status and wellbeing of individuals and communities. E-learning provides opportunities for many who are constrained by time and distance to acquire training and be certified in health education through distance learning. A lot of multi-media tools have been found very useful in making health communication highly effective. Thousands of Internet Web sites and apps have been successfully applied in making health information easily accessible to all and sundry. In fact as new technologies emerge, health educators are likely to discover more innovative ways of making use of them to reach wider audience with health information as well as make health education simpler and more effective. This will happen if efforts are made to address identified constraints to the effective application of technology tools such as low computer literacy, lack of infrastructure, low access to technology tools and poor skills in the use of ICT among others.

References

American Medical Association Ad Hoc Committee on Health Literacy for the Council on Scientific Affairs. (1999). Health literacy: Report of the council on scientific affairs. *Journal of the American Medical Association,281,* 552–557.

Atulomah, B. C., and Atulomah, N. O. (2012). Health literacy, perceived-information needs and preventive-health practices among individuals in a rural community of Ikenne local government area, Nigeria. *Ozean Journal of Social Sciences* 5(3), 95–104.

Blurton, C. (1999). New directions in education. In: *UNESCO's World communication and information 1999–2000.* Paris: UNESCO

Chapman, L. (2002). Awareness strategies. In M. P. O'Donnell, *Health promotion in the workplace (3rd Ed).* USA: Delmar.

Davenport, T. H. (1993). *Process innovation; reengineering work through information technology.* Boston: Harvard Business School Press.

Fogg, B. J. (2003). Persuasive Technology: Using Computers to Change what We Think and Do. San Francisco: Morgan Kaufmann publishers.

Fogg, B. J., and Eckles, D. (2007). *Mobile Persuasion: 20 Perspectives on the Future of Behavior Change.* Stanford: Stanford Captology Media.

Fordham, P. E. (1993). 'Informal, non-formal and formal education programmes' in YMCA George Williams College *ICE301 Lifelong Learning Unit 2,* London: YMCA George Williams College.

Hodges, V. (2014). Online learning environments and their applications to emerging theories of educational technology. *https://sites.google.com/a/boisestate.edu/edtechtheories/*

online-learning-environments-and-their-applications-to-emerging-theories-of-educational-technology

Infodev (2010). *ICT in Non Formal Education. Information and Communication Technology for Education in India and South Asia.* India: Pricewaterhouse coopers.

Mayer, R. E. (2009). *Multimedia learning (2ⁿᵈ ed).* New York: Cambridge University Press.

Mengal, S. K., and Mengal. U. (2009). *Essentials of educational technology.* New Delhi: PHI Learning Private Ltd.

Moore, M. G. (1997). Theory of transactional distance. Keegan, D. (ed.) *Theoretical Principles of Distance Education.* London: Routledge.

Rovai, A. P., Ponton, M. K., and Baker, J. D. (2008). *Distance learning in higher education: A pragmatic approach to planning, design, instruction, evaluation, and accreditation.* New York: Teachers College Press

Siemens, G. (2004). Connectivism: A learning theory for the digital age. *http://www.elearnspace.org/Articles/connectivism.htm.*

Smith, 2001). What is non formal education. *http://infed.org/mobi/what-is-non-formal-education/*

Stockley, D. (2003). E-learning definition and explanation, training, learning and performance consultant. *http://derekstokley.com.au/elearning-definition.html.*

UNESCO (2006). *Using ICT to Develop Literacy.* Bangkok: UNESCO.

WHO (1998). Division of Health Promotion, Education and Communications Health Education and Health Promotion Unit. Health Promotion Glossary. World Health Organization, Geneva *www.who.int/hpr/NPH/docs/hp_glossary_en.pdf.*

WHO (2013). Definition of health education. *http://www.who.int/topics/ health education/en/*.

Zhu, W. (2008). Promoting Physical Activity Using Technology. *Research Digest series 9, no 3.*Washington DC: President's Council on Physical Fitness and Sport.

Chapter 20

Evidence-Based Strategies to Reduce Barriers to Web-Based Education of a Low-Income Prenatal Clinic Population in the Midwestern Hospital in Illinois, USA

Florence F. Folami

Introduction

Prenatal education is an essential component of prenatal care and integral to prenatal outcome. The goal of prenatal education is to introduce education early to maintain a healthy pregnancy. Many pregnant women lack the knowledge to promote prenatal health. Tailoring prenatal education for pregnant women is an important part of prenatal care (Phillips, Galli, Watson, Felix, and Lambert, 2013). Tailoring prenatal education and instructions may be based on the clinician's experience and perceived need for information rather than on the pregnant women interest because of time limitations during clinic visit. Considering patient interest and desire for specific knowledge might improve patient attention to all educational material being presented (Pugh and Revell, 2011).

The rise of Web-based education has had a transformational impact across everyday life including health care. Exponential growth in use of the Internet, especially the social media is well documented in literatures (Tang, Gu, and Whinston, 2012; PWC Health Research Institute, 2012). According to Pew Internet and American Life Project. (2010), the changing profile of patients and demands of the health-care environment have prompted health institutions to deliver or support health education using information technology. Web-based education is seen to create access to learning (Brenner and Smith, 2013) as it is independent of geographic and temporal boundaries, and increases learner control allowing for greater flexibility and autonomy. Web-based learning provides ready access of resources for the patients. It allows for increasing knowledge and effective communication between patients and their care providers (Pugh and Revell, 2011).

Web-based tools are becoming important complements for prenatal education in order to improve prenatal outcomes (Pugh and Revell, 2011). The literature tends to use the following words interchangeably for Web-based education: online education, computer assisted learning, Web-assisted learning, Web-mediated learning, virtual learning environments, online courses, and Web-based courses. Web-based education refers to the most extreme form of online education that uses streaming videos and the more advanced functionalities available in educational software where there is no actual face to face contact between the teacher and the student.

There are approximately four million births each year in the United States. According to vital statistics for the year 2010, approximately 96% of pregnant women sought some level of prenatal care in the United States (Agency for Healthcare Research and Quality, 2012). Beyond prenatal care providers, women seek information about pregnancy and delivery from a variety of source such as books, videos, Web sites, and classes (Pugh and Revell, 2011). Among available sources, it is not clear what is preferred by women seeking care at the low income prenatal care because preferences for receiving prenatal education may make a difference in increasing prenatal education.

Web-based prenatal education have the capacity to interact with the individual with much greater frequency and in the context of the behaviour as compared to classroom education (Sim, Kitteringham, Spitz, Pierro, Kiely, Drake, and Curry, 2007). The availability of Web-based prenatal education provides the potential to deliver prenatal education tailored not only to the person's baseline characteristics but also to her frequently changing behaviours and environmental contexts (Cumbo, Agre, Dougherty, Callery, Tetzlaff, Pirone, and Tallia, 2002).

Problem Statement

Integration of Web-based educational tools into patient education has been shown to increase accessibility of resources and optimize teaching (Phillips, Galli, Watson, Felix, and Lambert, 2013). It is not very clear if the use of Web-based educational modules enhance prenatal education. Phillips, Galli, Watson, Felix, and Lambert (2013) identified positive aspects of Web-based education. The researchers observed that while the use of Web-based educational program has the ability for standardization, flexibility for the patients to select the time and place for learning, and ease of accessibility, there has been little empirical research to support the contention that Web-based prenatal education actually optimize prenatal teaching and enhance learning.

Purpose

The purpose of this study was to assess the barriers to Web-based education in a low-income prenatal clinic population. This study seeks to assess the barriers to Web-based education in a low-income prenatal clinic population at a Midwestern town in the state of Illinois, United States of America. The study will also raise a variety of interesting observations regarding the impact of Web-based education on prenatal health.

Research Questions

In order to achieve the purpose of this study, the following research questions will guide the study:

1. What are prenatal women perceptions of Web-based educational programs?
2. Do prenatal women prefer Web-based educational program over the classroom lecture?
3. What are the barriers to Web-based prenatal educational program among low income population?

Review of the Literature and Theoretical Framework

Various relevant literatures were reviewed to explore the purpose of this study. Previous studies have reported that prenatal education classes provide both knowledge and emotional support to the expectant couple (Phillips, Galli, Watson, Felix, and Lambert, 2013). These classes have been shown to affect the actual labor, delivery, and postpartum care. The following literature review provides the context for the discussion of the realities and implications of assessing the barriers to Web-based education in a low-income prenatal clinic population.

The Effects of Childbirth Preparation

Childbirth commands support for women due to its painful nature and its maternal and neonatal complications (Agency for Healthcare Research and Quality, 2012). Study of Agency for Healthcare Research and Quality (2012) shows the impact of childbirth preparation classes on the complications of labor. The study emphasized that increased knowledge and skills during pregnancy prepares pregnant mothers for labor and promotes their health. Researchers who have looked at the effects of prepared childbirth on complications during the pregnancy report a lower incidence of elevated maternal blood pressure (Galli, Watson, Felix, and Lambert, 2013).

Prenatal Care Education and Birth Outcomes

A number of researchers have reported the impact of prenatal education on birth outcomes especially the pain experienced by women in labor and the amount of pain medication laboring women require (Brenner and Smith, 2013; Tang, Gu, and Whinston, 2012; Phillips, Galli, Watson, Felix, and Lambert, 2013). Women who participated in some form of prenatal education have been found to report less pain and to require less medication to control pain than women who did not have similar preparation (Phillips, Galli, Watson, Felix, and Lambert, 2013). Women who have attended prenatal education classes also require fewer anesthesia during labor (Tang, Gu, and Whinston, 2012). Women who have attended prenatal education classes tend to have a quicker and smoother recovery (Graneheim and Lundman, 2004). The benefits of prenatal educations for the mother are well documented in the literature. It is important that the pregnant woman not only begins prenatal care early, but also receives continuous care throughout her pregnancy (Agency for Healthcare Research and Quality, 2012).

Evidences Related to Education and Learning

There is a difference between education and learning. Education is the establishment and arrangement of events to facilitate learning and emphasizes the provision of knowledge and skills (Rawool and Colligon, 2008). Learning emphasizes the recipient of knowledge and skills and results in behavioural changes. The integration of an education is necessary for the successful implementation of evidence-based prenatal education among patients. Everybody has a preferred learning style (Hawk and Shah, 2007). Knowing and understanding patient's learning style may result in effective learning (Folami, 2010). In all educational processes, an awareness and understanding of learning styles is important and that learning styles could vary between students who are from similar backgrounds pursuing similar goals and professions (Hawk and Shah, 2007). Understanding learning styles can improve an instructor's ability to plan, produce, and implement educational experiences; which could ensure a greater compatibility with learners' desires and enhance their learning, retention, retrieval, and application (Webb, Joseph, Yardley, and Michie, 2010). Web-based education utilized for this study was based on the concept of the following learning styles.

Visual Learners

These learners need to see the teacher's body language and facial expression to fully understand the content of the lesson. Traditional classroom lectures have been used to successfully promote the development and use of visual learning (Folami, 2010). Web-based education means the most extreme form of online education that uses streaming videos and the more advanced functionalities available in educational software and where there is no actual face to face contact between the teacher and the student.

Auditory Learners

Auditory learners learn best through verbal lectures, discussions, talking things through and listening. These learners benefit from reading aloud and using a tape recorder (Folami, 2010). This can be in the form of small group-based education program, and can also be associated with demonstrating and modeling. Interactive workshop will cater for the auditory learners.

Web-Based Education vs. Classroom-Based Education

Web-based health educational interventions have proliferated in recent years and appear to be an effective method for delivering health education interventions in a cost-effective manner (Tate, Finkelstein, Khavjou, Gustafson, 2009). Web-based course or electronic learning (e-learning) means using the Internet to access and participate in online modules or courses (Webb, Joseph, Yardley, and Michie, 2010). Web-based courses are offered online using course management software. Such software programs use content boards to host modules, e-mail, discussion boards, chatroom, and testing tools to organize information and support online communication. The difficulties with hardware, software, and Internet connectivity were in a group of nursing students who participated in a Web-based course (Martin, Klotz, and Alfred, 2007). Problems were prominent especially at the start of the course. However, the authors found that students were pleased with the positive psychological and social outcomes of the education. The researchers evaluated graduate nursing students' perceptions of traditional classroom teaching and Web-based model approach. Classroom methods were rated significantly higher regarding content, interaction, participation, faculty preparation, and communication.

Conceptual Framework

The Health Belief Model (HBM) is used in this study as the basis for the theoretical framework. The HBM focused on the notion of individuals participating in preventative health behaviours and stresses on personal responsibility. HBM promotes that people should trust their own abilities and have a general sense of control in their lives (Abbaszadeh, Haghdoost, Taebi, and Kohan, 2007). Utilization of prenatal care is considered a preventative health action so HBM will be a good fit for this study. The manner in which the benefits minus the barriers of a preventative health action are perceived by an individual is affected by the demographic, sociopsychological, and structural variables. The perceived threat of a disease and the perceived benefits minus the barriers of a preventative action directly impact the likelihood of taking the recommended preventative health action. Beliefs have a significant effect on the health behaviours of individuals; however, beliefs can be difficult to change. The HBM has been applied to a variety of health education topics (Graneheim and Lundman, 2004; Rawool and Colligon, 2008; Al-Ali and Haddad, 2004).

Methodology

A descriptive study was conducted to assess the barriers to Web-based education in a low-income prenatal clinic population in a Midwestern hospital in Illinois. The study focused on empowering prenatal patients' population to take more active roles in their care. The study was approved by the institutional review board of the Midwestern hospital in Illinois hospital.

Setting

The study was conducted at a Midwestern hospital in Illinois. The target population for this study consisted of the pregnant women attending the low-income prenatal clinic of the hospital. The hospital is a 372-bed hospital and considered an acute care general hospital. It offers a broad spectrum of comprehensive health services, including emergency, ambulatory, acute and extended care, and a wide range of community education and wellness programs.

Limitation

A limitation of this study was the self-selection of the sample. Certainly, tailoring information for patients is an important part of counseling. Because of time limitations, tailoring prenatal information and instructions may be based on the care providers experience and perceived need for information rather than on the patient's interest.

Sample

Total of 120 subjects participated in the study. The sample size composed of patients that attended the prenatal clinic during the time of the study. The sample size represents 83% of the total patients registered at the prenatal clinic of the Midwestern hospital in Illinois.

Brief surveys that consist of name, phone number, e-mail address, educational level, and marital status were distributed during the participants' prenatal visit at the hospital. The survey was delivered as a paper form. A limited number of demographic questions were included in the survey (maternal age, gestational age, and language preference). Access to technology, media preferences for receiving prenatal information, and potential educational topics of interest to the participant were solicited. Open ended questions followed each closed ended category to elicit further answers from the participants regarding prenatal educational topics. Completed surveys were collected at the time of their prenatal visit. Patient information was entered in the Web-based educational program to monitor their access to the education. Patient files were marked with a blue "e" to indicate that the prenatal educational module had been assigned to them. Each of the eight modules was assigned to 120 participants making total of 960 modules.

Research Instrument

This is a researcher-developed instrument. It includes items related to issues that impeded or promoted the Web-based education as well as recommendations to improve the program and intention to continue using the program. A panel of nursing staff and faculty members in nursing and public health provided a content assessment and feedback to determine clarity and validity of questions for the semi structured interview. They felt that the questionnaire was comprehensive and easy to complete. A Web-based prenatal educational program that consists of   8 different modules was created. Each module included instructions about the last stage of pregnancy, childbirth, breastfeeding, nutrition, postpartum care, newborn care, cesarean birth, and smoking cessation information. Pregnant women attending the low-income prenatal clinic were informed of the Web-based prenatal education program. An informed consent was obtained from the entire participant prior to the educational program. The participants were notified through e-mail and phone after the program has been assigned to them. The participants were also reminded about the program at each prenatal visit by the care providers. The participants have access to the Web-based educational program for 2 months after the after delivery date. Weekly monitoring of the program is done by the researcher for initiation, frequency of use, and completion of each module.

Semi Structured Interviews

This is also a researcher-developed instrument. It includes items related to issues that impeded or promoted the Web-based education as well as recommendations to improve the program and intention to continue using the program. A panel of nursing staff and faculty members in nursing and public health provided a content assessment and feedback to determine clarity and validity of questions for the semi structured interview. They felt that the questionnaire was comprehensive and easy to complete. This data were collected every two weeks.

Results

The findings organized around the four research questions of the study. The significance level for this study was set at $p < 0.05$. The findings of each are presented. The software product Statistical Package for the Social Sciences (SPSS) for Windows version 11.0.1 was used for all quantitative data analysis. The analysis included overall descriptive statistics for each variable. Bivariate comparisons between primigravida and multigravida women were calculated using student's t-test for each variable. A p-value of 0.05 was considered statistically significant. All the 120 participants completed the initial survey for a completion rate of 100%. The median age of the participants was 21 years old (range 15 to 35 years) and a median gestational age of 28 weeks (range 16 to 37 weeks). The average time spent on the educational program was per participant was 5 hours. The minimum reported was 3 hours and the maximum was 8 hours. Some participants did not have any experience using the Web-based learning and others listed themselves as experts in the computer field. Additionally, the subjects were asked if they felt comfortable using the computer as a learning tool. The average on the 5-point scale was 4.75. The range was from 3 to 5. Even though some subjects reported not customarily using a computer, almost everyone was comfortable with this computer usage.

Sixty five percent (n = 120) of the participants were having their first baby (primigravida). All the participants had access to a phone, 56% to a wireless data, and 92% to both e-mail and phone. All respondents preferred to get information from an individual meeting with a nurse or doctor and use Web-based education as supplement.

Improved use of prenatal care may be possible, in part, by improving satisfaction and engaging the patients with targeted information (Agency for Healthcare Research and Quality, 2012), both of which might be accomplished with improved attention to the topics that interest the patient. It has been reported that in some cases patients disagree with their provider about their perceived risk.

Research Question 1

The first research question asked: What are prenatal women perceptions of Web-based educational programs. To answer the first research question, descriptive statistics were used to determine participants' perceptions of Web-based educational programs. Based on mean scores, the computer program were potentially rated as terrible to wonderful, frustrating to satisfying, dull to stimulating, and difficult to easy.

Research Question 2

The second research question in this study was, "Do prenatal women prefer Web-based educational program over the classroom lecture?"

A content analysis of the added comments showed subjects much preferred Web-based educational program. Sixty eight percent of the subjects started and completed the program. Participants in this study were very positive about the Web-based educational program. Eighty percent of the participants thought that the length of the time to view the program was good. They did not feel overwhelmed. This result demonstrates the beneficial use of Web-based educational program. The Web-based format provided a great deal of flexibility for the users and allowed the users to find information of particular importance to her learning needs.

Research Question 3

The third research question asked: "What are the barriers to Web-based prenatal educational program among low income population?" Semistructured interviewed revealed clear barrier to Web-based prenatal educational program. These barriers were categorized into four factors. The factors were: no Internet/Wi-Fi access (mentioned 62 times), no computer or smartphone (mentioned 32times), forgot about link (mentioned 72 times), and didn't receive (mentioned 60 times). The participants recognized that while Web-based prenatal program was essential; it was not sufficient without an integrated program or process.

Discussion

There is evidence in the literature that integration of Web-based educational tools into patient education increases accessibility of resources and optimize teaching (Phillips, Galli, Watson, Felix, and Lambert, 2013). It is now clear that the use of Web-based educational modules enhances prenatal education. This study assessed the barriers to Web-based education in a low-income prenatal clinic population at a Midwestern town in the state of Illinois, United States of America. The study also raised a variety of interesting observations regarding the impact of Web-based education on prenatal health. In an attempt to contribute to the limited body of literature, this study assessed the impact of Web-based education in a low-income prenatal clinic population and explained ways to integrate evidence-based Web education into prenatal practice. The study gave the target hospital a framework for addressing future educational methods to address prenatal education. The participants in this study identified positive aspects of Web-based education. Considering patient interest and desire for specific knowledge might improve patient attention to all educational material being presented (Martin, Klotz, and Alfred, 2007). Rawool and Colligon (2008) identified that beliefs have a significant effect on the health behaviours of individuals; however, it can be difficult to change. A number of theories have been, and continue to be, developed to explain how people learn (Doheny, Sedlak, Estok, Zeller, 2007; Caputi and Blach, 2004).

Recommendations

Improved use of prenatal care may be possible, in part, by improving satisfaction and engaging the patients with targeted information (Anderson, and Klemm, 2008). Based on the findings and conclusions of this study, this section provides a list of recommendations as detailed below. The care provider should

- Ask patients about the Web-based education at subsequent appointments
- Inform patients that free Wi-Fi is available in the waiting room area and that the educational program can be accessed and viewed while the patient is waiting to be seen.
- Poster should be displayed in each examination room and waiting area reminding patients of the Web-based educational program and how it can answer their questions.
- Play modules in the waiting room while patients are waiting to be seen
- Offer tablet/computer device while patients are waiting
- Tablet/computer devices in the exam rooms while patients are waiting
- Play a module for a set group of people

References

Abbaszadeh, A., Haghdoost, A., Taebi, M., Kohan, S (2007). The relationship between women's health beliefs and their participation in screening mammography. Asian Pacific Journal of Cancer Prevention, 8(4), 471–475.

Al-Ali, N and Haddad, J (2004). The effect of the health belief model in explaining exercise participation among Jordanian myocardial infarction patients. Journal of Transcultural Nursing, 15(2), 114–121.

Agency for Healthcare Research and Quality (2012). National healthcare disparities report. Rockville, MD: Agency for Healthcare Research and Quality.

Anderson, and Klemm (2008). The Internet: friend or foe when providing patient education? Clinical Journal of Oncology Nursing, 12(1), 55–63. doi:10.1188/08.CJON.55–63

Brenner, J., and Smith, A. (2013). 72% of online adults are social networking site users.
Washington, DC: Pew Internet and American Life Project.

Cumbo, A., Agre, P., Dougherty, J., Callery, M., Tetzlaff, L., Pirone, J., and Tallia, R. (2002). Online cancer patient education: Evaluating usability and content. Cancer Practice, 10(3), 155–161.

Doheny, M., Sedlak, C., Estok, P., Zeller, R (2007). Osteoporosis Knowledge, Health Beliefs, and DXA T-Scores in Men and Women 50 Years of Age and Older.
Orthopaedic Nursing, 26(4), 243–250.

Folami, F. (2010). Comparison of two educational methods on nurses' adoption of safe patient handling techniques. Published doctoral dissertation, Walden University, 3411942.

Graneheim, U. and Lundman B. (2004). Qualitative content analysis in nursing research: concepts, procedures and measures to achieve trustworthiness. Nurse Education Today, 24,105–112.

Hawk, T and Shah, A. (2007). Using Learning Style Instruments to Enhance Student Learning. Decision Sciences Journal of Innovative Education, 1540–4609.

Phillips, Galli, Watson, Felix, and Lambert (2013). The effectiveness of mobile-health technology-based health behaviour change or disease management interventions for health care consumers: a systematic review.

Pew Internet and American Life Project. (2010). http://www.pewInternet.org/~/media//Files/Reports/2010/PIP_Mobile_Access_2010.pdf

Pugh, M. A., and Revell, M. A. (2011). Using online materials for prenatal education: The good, the bad and the ugly. International Journal of Childbirth Education, 26(4), 9–13.

Rawool, W., and Colligon-Wayne, A (2008). Auditory lifestyles and beliefs related to hearing loss among college students in the USA. Noise Health. 10(38), 1–10.Sim, Tang, Q., Gu, B., Whinston (2012). Content contribution in social media: the case of YouTube. Journal of Nursing Education, 4476–4485.

Tate, D., Finkelstein, E., Khavjou, O., Gustafson, A (2009) Cost effectiveness of Internet interventions: Review and recommendations. *Annals of Behavioral Medicine*. 38, 40–45.

Webb, T., Joseph J., Yardley, L., and Michie, S (2010). Using the Internet to promote health behavior change: A systematic review and meta-analysis of the impact of theoretical basis use of behavior change techniques, and mode of delivery on efficacy. Journal of Medical Internet Research, 12.

SECTION 7

Blended Learning

Chapter 21

Impact of Blended Learning Approach on Critical Thinking Ability and Academic Performance of Educationally Backward Students

Kshama Pandey
and
Ms Neetu Singh

Introduction

The experience of educators confirms that there are many children who are so backward in basic subjects that they need special help. These students do not stand out as very different from their classmates expect that they are always slow on the uptake and are often teased by the other students because of their backwardness. These pupils have limited scope for achievement. Their ability to deal with abstract and symbolic materials (i.e., language, numbers, and concepts) is very limited, and their reasoning in practical situations is inferior to that of average students. These pupils differ slightly from normal students in learning ability. They are also unable to deal with relatively complex games and school assignments. They need much external stimulation and encouragement to do simple type of work. These students are known as educationally backward students. Burt (1937) has rightly pointed out that the term, "backward" is reserved for those children who are unable to cope with the work normally expected of their age group.

There are many strategies or approaches followed by teachers to help the educationally backward students so that their academic achievement can be improved and they can enjoy the learning process as others. In the present study, blended learning is considered as a treatment for educationally backward students. It is well known that developments in information technologies have reshaped people's views towards themselves and their environments, as a result of which a parallel transformation and development at the same pace have become inevitable in the field of education. This change and development in the field of education is determined by various factors. One of the most important among these factors is teachers, which is undoubtedly followed by information technologies. A teacher has a crucial function in managing information technologies and establishing a link between students and information technologies (Oral, 2004).

It is indisputably accepted that the use of information technologies in education will greatly facilitate access to and transmission of information. Today, the first thing that comes to mind about the impact of technology on education is computers and Internet use. Computers and the Internet, which have become an integral part of daily life, could not have been expected to be left out of teaching-learning environments (Deniz, 1994). Computers and Internet use have come to the foreground in the recent practices in education.

Blended-Learning Approach

A blended-learning approach combines face to face classroom methods with computer-mediated activities to form an integrated instructional approach. According to Node (2001), "Blended learning (Integrative Learning or Hybrid Learning) is the mixing of different learning environments. It combines traditional face-to-face classroom methods with modern computer mediated activities. This strategy has been noted to create a more integrated approach for learners and lecturers." The idea is to use technological based material in support of the role of face-to-face classroom learning. Young (2002) defined blended learning as a situation where online education is combined with traditional classroom-based instruction. In this instructional method, the advantages of traditional and online learning methods are supplementary for all education environments. Online educational components naturally become a part of traditional instruction method for students to experience interaction, flexibility, and harmony in classroom environment while they take all their courses online

For designing blended-learning environments, Horton (2000) proposes certain methods such as online components that combine face-to-face and online elements for a particular course and familiarize students with face-to-face sessions, online courses defined by students in class and supported by the teacher again in class, and online presentation materials to be used by teachers for in-class presentations. Azizan (2010) concluded that utilization of technology in physical classrooms offer extra resources for the students and this is expected to enhance learners' confidence and competence as well as improve the quality of learning.

When designing a blended-learning environment, the first point to be decided is to design a part of the blended subject matter as face-to-face and some as online. The common blending technique is usually half-and half, i.e., 50% consists of face-to-face activities in classroom environment and the other 50% of activities performed in an online environment (Osguthorpe and Graham, 2003). Table 1 presents the possible components that could be found in a blended environment and used in an online- and traditionally managed classroom.

Table 1. Showing the blended-learning approach

Blended Learning

Online instruction			Traditional instruction		
Instruction environment	Activities	Applications	Instruction environment	Activities	Applications
1. Computer Based Learning	1. Introduction	1. Video	1. Classroom	1. Introduction	Varies From
2. Synchronous	2. Exercise	2. Audio	2. Synchronous	2. Presentations And	One
3. Asynchronous	3. Individual	3. Presentation Tools	3. Bidirectional	Group Work	Classroom To
4. Unidirectional	Study	(PowerPoint, Flash)	Communication	3. Exercise	Another
Communication	4. Discussion	4. Communication		4. Evaluation	
5. Bidirectional	5. Homework	Tools			
Communication	6. Group Work				
	7. Analogy				
	8. Evaluation				

Source: Korkmaz, Ö. and Karakuş, U. (2009), Http://Www.Tojet.Net/ Volumes/ V8i4.Pdf#Page=52

Schools of the information age are environments that train creative and critically thinking individuals who produce information and have access to the needed information; actively employ and spread information; and actively use information technologies (Balay, 2004). Within this framework, it is an expected outcome that the methods and technologies employed in the schools of the information age positively contribute to students' critical thinking dispositions and levels (Branch, 2000). Critical thinking is defined as an individual's ability to think openly, independently, and rationally, emphasis is made on the fact that the concept does not denote debate and constant negative criticism (Külahçı 1995). According to Elder and Paul (1997), "Critical thinking is best understood as the ability of thinkers to take charge of their own thinking. This requires that they develop sound criteria and standards for analyzing and assessing their own thinking and routinely use those criteria and standards to improve its quality." It is argued that various education stages include some course content that could only be learned through thinking and without intellectual processes, students will end up in attempting to memorize most information (Paul and Elder, 2001). Nevertheless, students are expected to analyze any presented information, or know how to use it (Brad, 1994). In this context, increased academic achievement is a natural outcome of making critical thinking education a part of educational processes (Elias and Kress, 1994). Yet, individuals cannot improve their critical thinking levels on their own. Currently, it is largely a responsibility of schools to help individuals acquire the skills of critical thinking and information analysis (Kökdemir, 2003). Usta, and Özdemir (2007) studied students' opinions about blended-learning environment and their findings proved that students have generally positive opinions about blended-learning environment. The results of the study also proved that high interaction between students and instructor exist in this type of environment. This result supported the findings of Akkoyunlu and Soylu (2006) which indicated high demands for face-to-face interaction in on line learning.

Science subject has a distinct position for its possible contributions to critical thinking skills. As a discipline analyzing and synthesizing the information collected in the context of human-natural environment interaction, Science requires students to structure the learned information by questioning it using these criteria at all stages. Thus, they improve their critical thinking skills through a questioning and synthesizing approach. In the context of the above explanations, the use of blended-learning model in science courses with dimensions such as human, environment, and human-environment interaction is expected to improve students' academic performance in Science subject and to positively contribute to their critical thinking ability. Drawing upon this proposition, the present study will attempt as it's problematic to determine how blended-learning program affects student academic performance and their critical thinking ability.

Review of Related Studies

In the modern competitive world in which academic performance is considered important, scholastic backwardness causes tremendous stress for the students. Such backwardness in a student, in most of the contexts, may make the student feels secluded and also turn into problematic child. Thus it is important to teachers that they use blended-leaning environment to improve students' academic performance and it will definitely increase their ability of critical thinking. Various research studies have been conducted in the field of blended learning; the brief discussion on related studies is as follows:

Yushau, B. (2006) conducted a research study on the effects of blended-learning on mathematics and computer attitudes in precalculus algebra and found that subjects have positive attitude towards mathematics and computer. Pereira, et al. (2007) studied the effectiveness of using blended-learning strategies for teaching and learning human anatomy and the results revealed that the percentage pass rate for the subject in the first call was higher in the blended-learning group but there were no differences regarding overall satisfaction with the teaching received. Young (2008) also found significant effect of blended learning on students' outcomes but this study revealed no significance effect of blended learning on students' attitude towards science. Blended-learning models contributed more to student attitudes towards geography and critical dispositions and levels and have positive correlation between student attitudes towards geography course and their critical thinking dispositions and levels (Korkmaz and Karakuş, 2009).

According to Al-Qahtani and Higgins (2009) studied the effects of traditional, blended and e-learning on students' achievement in higher education and explored positive effect of three methods in terms of students' achievement favoring the blended-learning method. The results revealed that no significant difference was found between the e-learning and traditional learning groups in terms of students 'achievement. Al-Saai et al. (2011) studied the effect of a blended e-learning environment on students' achievement and attitudes towards using e-learning in teaching and learning at the university level and found insignificant difference between the instructional treatments in gain scores of the achievement test. However, the results in the attitudes scale showed in favor of blended e-learning approach. Blended-learning model contributed more to the students' biology achievement than traditional teaching methods did and students' attitudes towards the Internet (Hasan, 2012). Alseweed (2013) explored significant differences among the instructional approaches in the achievement test scores in favor of blended learning, in addition, the results showed significant differences in students' attitudes in favor of blended learning. Alotaibi (2013) taught the experimental group by using blended learning, while the control group was taught by using traditional method and no statistically significant differences were found between the experimental group and the control group in critical thinking skills. Blended-learning environments affect positively on students' achievement (Dennis, et al., 2014).

Therefore the trend analysis of the researches in this area showed that various studies have been conducted on blended learning, and the results of these studies showed that the blended-learning approach was positively effective in general classrooms. Thus, it may be possible that the backward students also gain benefit from this learning approach. Therefore, the main objective of the present study is to determine the impact of blended-learning approach on critical thinking ability and academic performance of educationally backward students. Thus, it sought answers to the following questions:

1. Is blended learning able to change educationally backward students' Critical Thinking Ability?

2. Is blended learning able to change educationally backward student's Academic Performance?

Thus this study will assuredly be positively benefitted in terms of backward student's academic performance and critical thinking ability.

Statement of the Problem

The issue that this chapter dealt with is the impact of the blended-learning approach on critical thinking ability and academic performance of educationally backward students.

Objectives of the Study

1. To compare the academic performance of backward students in experimental group (taught by blended-learning approach) and control group (taught by traditional lesson plans).

2. To compare the critical thinking ability of backward students in experimental group (taught by blended-learning approach) and control group (taught by traditional lesson plans).

3. To study the effect of blended-learning approach on critical thinking ability of educationally backward students.

4. To study the effect of blended-learning approach on academic performance of educationally backward students.

Hypothesis of the Study

Ho 1. There exists no significance difference in the academic performance of the backward students in experimental and control group.

Ho 2. There exists no significance difference in the critical thinking ability of the backward students in experimental and control group.

Ho 3: There exists no significant effect of blended-learning approach on academic performance of educationally backward students.

Ho 4: There exists no significant effect of blended-learning approach on critical thinking ability of educationally backward students.

Delimitation of the Study

The proposed study was delimited as under:

1. The study was delimited to the students who are educationally backward in Science subject.

2. The students who were appearing in IX class were considered.

3. The present study was confined to B. D. Jain Girls Inter College Agra, affiliated to the UP Board.

Variables of the Study

In this study, the independent variable is blended-learning approach, the dependent variables are critical thinking and academic performance and the controlled variables are educational backwardness, class and subject.

Methodology

This study was intended to determine the impact of blended learning on critical thinking and academic performance of backward students. Keeping in view the nature of the study, the Experimental Research Method with a pretest, posttest and control group was employed.

Table 2. Exhibiting the method of the study

S.No.	Groups	Sample Size	Pretest	Treatment	Posttest
1.	Experimental Group	40	• Academic Performance	Blended Learning Based Lesson Plans	• Academic Performance
2.	Control Group	40	• Critical Thinking Ability	Traditional Lesson Plans	• Critical Thinking Ability

Selection of Sample

The researcher used Purposive method of sampling, for the selection of schools in the present study from Agra city, affiliated to the UP Board. Then backwardness of the students was determined through the school results. The students of class IX were selected, and forty students were comprised into experimental group and forty students into control group.

Tools of the Study

By observing the nature of the study following tools were developed:

1. Self-developed Academic performance Test in Science subject: Total 30 items were selected in the final draft of the tool on the

basis of expert's opinion and item analysis of the items. The test-retest reliability was measured i.e., 0.73.

2. Self-developed tool used for Critical Thinking Inventory. The dimensions for critical thinking inventory were as Analyticity, Open-Mindedness, Inquisitiveness, evaluating information, Truth-Seeking, Systematicity and Self-Confidence. Total 33 items were selected in the final draft on the basis of expert's opinion. The Cronbach's alpha reliability was 0.69 for the inventory.

Procedure of the Study

The present study was carried out in the following phases, which has following stages:

Phase 1: Construction Phase

The primary phase in the method of the present study was construction phase, which involved the subsequent stages, that was as follows:

Stage 1: Development of Academic Performance Test.

Stage 2: Development of Critical Thinking Inventory.

Stage 3: Development of Blended Learning based lesson plans and Traditional lesson plans.

Phase 2: Identification and Implication Phase

The second phase was identification and implication phase. This phase has involved following subsequent stages.

Stage 1: Identification of Educationally Backward Students from their school results.

Stage 2: Identified Educationally Backward Students were randomly distributed in control group and experimental group for carrying out the study.

Stage 3: Applying the Critical Thinking Ability Inventory on both experimental and control groups.

Stage 4: Implication of Blended Learning based Lesson Plans on experimental group.

Stage 5: Implication of traditional lesson plans on control group.

Stage 6: Administration the Academic Performance Test and Critical Thinking Inventory on both experimental and control groups.

Phase 3: Evaluation and testing of Hypothesis

This Phase was comprised the evaluation of the educationally backward students of the experimental group and control group to whom instructions were given by Blended learning approach and Traditional approach respectively, which was followed by testing of the hypothesis on the basis of obtained scores in Critical Thinking Inventory and Academic Performance Test.

Statistical Techniques

According to the requirements of the study, the researchers used Quantitative Statistical Analysis and used the Descriptive Statistics in the present study to analyze and interpretation of the data. Mean, Standard Deviation were applied as descriptive statistics and t-test was applied as inferential statistics.

Academic Achievement of Backward Students before Implementation of Blended Learning Approach and Traditional Lesson Plans

The first phase of experimentation deals with the administration of the pretest on both the groups regarding academic achievement. Before beginning the experiment, the controlled and experimental groups were equated in their academic achievement. The descriptive statistics of the groups is shown in the following table:

Table 3. Exhibiting descriptive statistics of the groups in the pretest scores for academic performance

Group	Mean	SD	t-value	Level of significance
Experimental Group	9.6	2.06	0.86	Insignificant
Control Group	9.5	1.94		

*0.01 level of Significance

The data in the above table 3 enumerates the nature of the sample. The calculated t-value is 0.86 at 0.01 level of significance. It means that there exists no significant difference in academic achievement of backward students in experimental and control group before implementation of blended learning approach. Hence groups are considered to be equal.

Critical thinking of Backward Students before Implementation of Blended Learning Approach and Traditional Lesson Plans

After the pretest done for academic achievement, the pretest was also administered for critical thinking ability on both the groups. The descriptive statistics of the groups is shown in the following table:

Table 4. Exhibiting descriptive statistics of the groups in the pretest scores for critical thinking

Group	Mean	SD	t-value	Level of significance
Experimental Group	56.62	8.51	0.23	Insignificant
Control Group	58.82	7.76		

*0.01 level of Significance

The table 4 reveals the difference of mean scores of experimental and control groups on pretest. The calculated t-value is 0.23 is less than the tabulated value at 0.01 level of significance. Therefore, it can be concluded that no significant difference is found in the pretest scores regarding critical thinking of experimental group and control group. Hence groups are considered to be equivalent.

Academic Achievement of Backward Students after Implementation of Blended Learning Approach and Traditional Lesson Plans

After implementation of the blended learning approach and traditional lesson plans on both the groups, posttest was administered. The purpose of administering the posttest was to measure academic achievement of the students in the groups, to compare their with that of pretest scores and to compare the posttest achievement between the two groups. The posttest results thus serve, as a base for testing hypothesis, i.e., there exist no significance difference in the academic achievement of backward students in experimental and control group. The obtained statistics are:

Table 5. Exhibiting descriptive statistics of the groups in the posttest scores for academic achievement

Group	Mean	SD	t-value	Level of significance
Experimental Group	17.75	3.88	3.04	Significant
Control Group	10.85	2.03		

*0.01 level of Significance

The table 5 shows that the mean values of the academic achievement of experimental and control groups in posttest are 17.75 and 10.85 respectively. The calculated t-value is 3.04 at the 0.01 level of significance that is greater than the tabulated value. Hence significance difference is found in the academic achievement of backward students in both the groups.

Critical thinking of Backward Students after Implementation of Blended Learning Approach and Traditional Lesson Plans

After implementation of the blended learning approach and traditional lesson plans on both the groups, posttest was administered. The purpose of administering the posttest was to measure critical thinking ability of the students in the groups, to compare their with that of pretest scores and to compare the posttest critical thinking ability between the two groups. The posttest results thus serve, as a base for testing hypothesis, i.e., there was no significance difference in the critical thinking ability of backward students in experimental and control group. The obtained statistics are:

Table 6. Exhibiting descriptive statistics of the groups in the posttest scores for critical thinking

Group	Mean	SD	t-value	Level of significance
Experimental Group	102.07	15.47	5.64	Significant
Control Group	59.97	7.55		

*0.01 level of significance

From the data in the above table 6, the t-value is 5.64 that is greater than the tabulated value at 0.01 level of significance. It reveals that significant difference is found in the critical thinking ability of backward students in both the groups.

Comparison of pretest and posttest scores regarding academic achievement of experimental group

Table 7. Exhibiting t-value of the experimental group regarding academic achievement

Test	N	Mean	SD	t-value	Level of Significance
Pretest	40	9.6	2.06	11.74	Significant
Posttest	40	17.75	3.88		

*0.01 level of Significance

From the data in the above table 7, the calculated t-value is 11.74 that are more than the table value at 0.01 level i.e. 2.64 with 78 degree of freedom. Therefore, significant difference is found in the academic achievement of backward students of experimental group in the pretest and posttest. There is gain found in the academic achievement of backward students. The students in the experimental group are taught by blended learning approach. The gain in the achievement can thus be attributed to the process of learning in this group. Due to it students engaged in active participation, discussion among themselves and with the teacher and also use the online material. This approach developed interest in learning among the students. This finding is supported by Usta (2007). Usta investigated the impact of blended learning and online learning environments on academic achievement and satisfaction. This study revealed that blended learning contributed more to academic achievement.

Figure 1.0. Geographical Representation of pretest mean and posttest mean values of experimental group

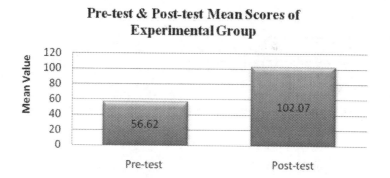

Comparison of pretest and posttest scores regarding academic achievement of control group

Table 8. Exhibiting t-value of the control group regarding academic achievement

Test	N	Mean	SD	t-value	Level of significance
Pretest	40	9.5	1.94	3.06	Significant
Posttest	40	10.85	2.03		

*0.01 level of Significance

The table 8.0 reveals the calculate t-value is 3.06 that is slightly more than the tabulated value (2.64) at 78 degree of freedom. It means that after the implementation of traditional lesson plans on the control group, minor change is found in the academic achievement of the backward students.

Comparison of pretest and posttest scores regarding critical thinking ability of experimental group

Table 9. Exhibiting t-value of the experimental group regarding critical thinking ability

Test	N	Mean	SD	t-value	Level of significance
Pretest	40	56.62	8.51	16.29	Significant
Posttest	40	102.07	15.47		

*0.01 level of Significance

The table 9 shows that the t-value is 16.29 for critical thinking ability in pretest and posttest. At the 78 degree of freedom the tabulated value is 2.64 that is less than the calculated t-value. It means there exists significant difference in critical thinking ability of backward students of experimental group in pretest and posttest. It shows that the students also gain in the critical thinking ability after implementation of blended learning approach. This blended learning approach enhances students' ability to analyze information and think critically. This finding is supported by some studies such as Bronson, 2008 and Korkmaz and Karakuzm, 2009. These studies indicate the possibility of developing critical thinking through some e-learning applications.

Besides, Burges' study suggests that Blended E-learning motivates students to be self-dependent thinkers (Burges, 2009).

Figure 2.0. Geographical representation of pretest mean and posttest mean values of experimental group

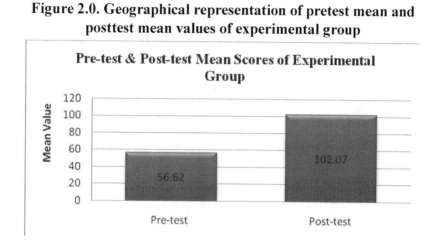

Comparison of pretest and posttest scores regarding critical thinking ability of control group

Table 10. Exhibiting t-value of the control group regarding critical thinking ability

Test	N	Mean	SD	t-value	Level of significance
Pretest	40	58.82	7.76	0.67	Insignificant
Posttest	40	59.97	7.55		

*0.01 level of Significance

Table 10 reveals the calculated t-value is 0.67 that is less than the tabulated value (2.64) at 78 degree of freedom. Therefore no significance difference is found for critical thinking ability of the backward students of control group in pretest and posttest. It shows that after implementation of traditional lesson plans on control group, there is no change found in the critical thinking ability of the backward students.

Conclusion

Educationally backward children need special attention and care for being duly helped in getting rid of their sub normality in terms of rate of learning and educational achievement. Neglecting or overlooking them may pose a serious problem for their progress and welfare besides proving a nuisance to the society. In the present study, blended learning approach is implemented on educationally backward students, and it has been found that this approach is beneficial for them. Their academic achievement and critical thinking ability is found to be changed and improved after applying this learning approach. Hence, it can be concluded that the educationally backward students can gain more and improve their achievement and critical thinking ability with the use of blended learning approach.

References

Alotaibi, K. N. R. (2013). The effect of blended learning on developing critical thinking skills.
Education Journal, 2 (4).

Elias, M. J., and Kress, J. S. (1994). Social decision-making and life skills development: A critical thinking approach to health promotion in the middle school. *Journal of School Health. 64*(2).

Horton, W. (2000). Designing Web Based Training. NY, Chichester, Weinheim, Brisbane, Singapore, Toronto: John Wiley.

Korkmaz, O. and Karakuş, U. (2009). The Impact of Blended Learning Model on Student Attitudes towards Geography Course and Their Critical Thinking Dispositions and Levels. *The Turkish Online Journal of Educational Technology (TOJET)*, 8 (4).

Lim, D. H., and Morris, M. L. (2009). Learner and Instructional Factors Influencing Learning Outcomes within a Blended Learning Environment. *Educational Technology and Society, 12* (4).

Osguthorpe, T. R., and Graham, R. C. (2003). Blended learning environments. *Quarterly Review of Distance Education, 4* (3).

Paul, R. and Elder, L. (2001). Critical Thinking: Tools for Taking Charge of Your Learning and Your Life. Upper Saddle River, NJ: Prentice Hall.

Yushau, B. (2006). The Effects of Blended E-Learning on Mathematics and Computer Attitudes in Pre-Calculus Algebra. *The Montana Mathematics Enthusiast, 3* (2).

Contributors

Dr. Augustine Obeleagu Agu. 2804, Barco, Grand Prairie, Texas 75054. E-mail: Augobele52@yahoo.com.

Dr. Ofomegbe Daniel Ekhareafo is a lecturer in the Department of Theatre Arts and mass communication, University of Benin. Edo State Nigeria. E-mail: talk2ofomegbe@gmail.com.

Professor (Mrs) B. O. J. Omatseye is a professor in the Institute of Education, University of Benin. Edo State, Nigeria. E-mail: bigbetus2001@yahoo.com.

Blessed Friedrick, Ngonso is a lecturer in the Department of Mass Communication, Samuel Adegboyega University, Ogwa, Edo State. E-mail: blessedngonso@gmail.com.

Dr. Ogunlade, B. O. is a lecturer in the Department of Curriculum Studies, Faculty of Education, Ekiti State University, Ado-Ekiti, Nigeria.ogunladebamidele1@gmail.com.

Dr. Babalola, J. O. is a lecturer in the Department of Curriculum Studies, Faculty of Education, Ekiti State University, Ado-Ekiti, Nigeria.olurotimitayo22@gmail.com.

Dr. Olabiyi, Oladiran Stephen is a lecturer in the Department of Science and Technology, Faculty of Education, University of Lagos, Akoka, Lagos State.

Prof. Denise Jonas is a professor at Robert Morris University, Moon Township, PA, 15108, USA.

Prof. Gary L. Schnellert is a professor at Robert Morris University, Moon Township, PA., 15108, USA.

Jennifer N. L. Ughelu is a doctoral student in the Department of Science and Technology, Faculty of Education, University of Lagos, Akoka, Lagos State. ughelujennifer@yahoo.com.

Dr. Sylvester Akpan is a lecturer in Department of Educational Technology and Library Science University of Uyo, Akwa Ibom State, Nigeria. Sylvester_akpan@yahoo.com.

Dr. J. A. Jimoh is a lecturer in the Department of Science and Technology Education, University of Lagos, Akoka, Lagos State. Nigeria. E-mail: jjimoh@unilag.edu.ng.

Dr. S. A. Adebayo is a lecturer in the Department of Automobile Technology, Federal College of Education (Technical), Akoka, Lagos State, Nigeria. E-mail: skdemola@gmail.com.

Dr. I. O. Oguche is a lecturer in the Department of Technical Education, Kogi State College of Education, AnKpa. Nigeria.

Okudo, Afoma Rosefelicia (Mrs.) is an assistant lecturer and a doctoral student in the Department of Arts and Social Sciences Education, Faculty of Education, University of Lagos, Nigeria.

Dr. S. A. Oladipo is an associate lecturer in the Department of Educational Administration, Faculty of Education, University of Lagos, Nigeria.

Dr. A. A. Adekunle is a lecturer in the Department of Educational Administration, Faculty of Education, University of Lagos, Nigeria.

Dr. Caleb Ademola Omuwa Gbiri is a lecturer, neurophysiotherapist, and neuroscientist in the College of Medicine, University of Lagos,

Nigeria. E-mail: calebgbiri@yahoo.com; cgbiri@unilag.edu.ng; agbiri@cmul.edu.ng.

Michele T. Cole is a professor of Nonprofit Management at Robert Morris University, Moon Township, PA. 15108, USA. E-mail: cole@rmu.edu.

Blessing F. Adeoye is a senior lecturer in the Department of Science and Technology Education, Faculty of Education, University of Lagos, Nigeria. E-mail: Docadeoye@gmail.com.

Louis B. Swartz is a professor of Legal Studies at Robert Morris University, Moon Township, Robert Morris University, Moon Township, PA., 15108, USA. E-mail: swartz@rmu.edu.

Daniel J. Shelley is a professor of Education at Robert Morris University, Moon Township, PA., 15108, USA. E-mail: shelleyd@rmu.edu.

Javid Iqbal is a lecturer in the Department of Graphics and Multimedia, University Tenaga Nasional. Department of Graphics and Multimedia, College of Information Technology, University Tenaga Nasional, 43000 Kajang, Malaysia.E-mail: manjavid@gmail.com.

Dr. Manjit Singh Sidhu is a lecturer in the Department of Graphics and Multimedia, University Tenaga Nasional. Department of Graphics and Multimedia, College of Information Technology, University Tenaga Nasional, 43000 Kajang, Malaysia.

Nathan Emanuel is a postgraduate student at the Department of Science and Technology Education, University of Lagos. Akoka-Yaba. Lagos-Nigeria. E-mail: nuelamme@yahoo.com.

T. Oluwatobiloba Olatunji is a lecturer at the Department of Virology, College of Medicine University of Ibadan Ibadan Nigeria. E-mail: Onemediastudios@gmail.com.

Dr. Adekunle Olusola Otunla is a research fellow at the Institute of Education, University of Ibadan, Ibadan, Nigeria. E-mail: otunlad@yahoo.com.

Dr. Georgy O. Obiechina is a lecturer in the Department of Human Kinetics and Health education, University of Port Harcourt. Port Harcourt, Rivers State.

Dr. Ekenedo, Golda O is a lecturer in the Department of Human Kinetics and Health Education, University of Port Harcourt. E-mail: dr.goldaoe@yahoo.com.

Kshama Pandey is a lecturer in the Department of Foundations of Education, Faculty of Education, Dayalbagh Educational Institute (Deemed University). Dayalbagh, Agra, UP India. E-mail: kshamasoham@gmail.com.

Ms Neetu Singh is a lecturer in the Department of Pedagogical Sciences, Faculty of Education.

Dayalbagh Educational Institute (Deemed University). Dayalbagh, Agra, UP India. E-mail: neetusin8@gmail.com.

Florence F. Folami is an associate professor in the Department of Nursing at Millikin University, Decatur, Illinois, United States.

Dr. (Mrs.) Georgy O. Obiechina is a lecturer in the Department of Human Kinetics and Health education, University of Port Harcourt. Port Harcourt, Rivers State.

Dr. Eimuhi, Justina Onojerena is a lecturer in the Department of Educational Foundations and Management, Faculty of Education, Ambrose Alli University. E-mail: justinaeimuhi@gmail.com.

Dr. Ikhioya, Grace Olohiomereu is a senior lecturer in the Department of Physical and Health Education, Faculty of Education, Ambrose Alli University. E-mail: graceikhioya@yahoo.com.

Mr. Nfonsang is a doctorate student in the Educational Leadership Department at the University of North Dakota. Grand Forks, ND 58202. E-mail: neba.nfonsang@my.und.edu.

Printed in the United States
By Bookmasters